Olfactory Cognition

Advances in Consciousness Research (AiCR)

Provides a forum for scholars from different scientific disciplines and fields of knowledge who study consciousness in its multifaceted aspects. Thus the Series includes (but is not limited to) the various areas of cognitive science, including cognitive psychology, brain science, philosophy and linguistics. The orientation of the series is toward developing new interdisciplinary and integrative approaches for the investigation, description and theory of consciousness, as well as the practical consequences of this research for the individual in society.
From 1999 the Series consists of two subseries that cover the most important types of contributions to consciousness studies:
Series A: Theory and Method. Contributions to the development of theory and method in the study of consciousness; Series B: Research in Progress. Experimental, descriptive and clinical research in consciousness.
This book is a contribution to Series B.

For an overview of all books published in this series, please see
http://benjamins.com/catalog/aicr

Volume 85

Olfactory Cognition. From perception and memory to environmental odours and neuroscience
Edited by Gesualdo M. Zucco, Rachel S. Herz and Benoist Schaal

Olfactory Cognition

From perception and memory to environmental odours and neuroscience

Edited by

Gesualdo M. Zucco
University of Padova

Rachel S. Herz
Brown University, Providence

Benoist Schaal
CNRS-University of Bourgogne, Dijon

John Benjamins Publishing Company
Amsterdam / Philadelphia

The paper used in this publication meets the minimum requirements of the American National Standard for Information Sciences – Permanence of Paper for Printed Library Materials, ANSI z39.48-1984.

Library of Congress Cataloging-in-Publication Data

Olfactory cognition : from perception and memory to environmental odours and neuroscience / edited by Gesualdo M. Zucco, Rachel S. Herz and Benoist Schaal.

p. cm. (Advances in Consciousness Research, ISSN 1381-589X ; v. 85)

Includes bibliographical references and index.

1. Olfactometry. 2. Olfactory receptors. 3. Smell. 4. Neurobiology. I. Zucco, Gesualdo. II. Herz, Rachel, 1963-, III. Schaal, Benoist.

QP458.O4342 2012

612.8/6 2012002091

ISBN 978 90 272 1351 8 (Hb ; alk. paper)

ISBN 978 90 272 7464 9 (Eb)

John Benjamins Publishing Co. · P.O. Box 36224 · 1020 ME Amsterdam · The Netherlands

John Benjamins North America · P.O. Box 27519 · Philadelphia PA 19118-0519 · USA

Table of contents

List of contributors IX
Foreword by Michael I. Posner XI
Preface XV
Acknowledgments XIX

Part I. Perception, psychophysics and odour environment

CHAPTER 1
Is there a measurement system for odour quality? 3
Birgitta Berglund and Anders Höglund

CHAPTER 2
There's something in the air: Effects of beliefs and expectations on response to environmental odors 23
Pamela Dalton

CHAPTER 3
Psychophysical evaluation of pain and olfaction: Many commonalities and a few significant differences 39
Richard H. Gracely

CHAPTER 4
Olfactory comfort in close relationships: You aren't the only one who does it 59
Donald H. McBurney, Sybil A. Streeter and Harald A. Euler

CHAPTER 5
Olfactory perception 73
Richard J. Stevenson

Part II. Learning and memory

CHAPTER 6
Odor memory and the special role of associative learning 95
Rachel S. Herz

CHAPTER 7
Knowing what we smell 115
Fredrik U. Jönsson and Mats J. Olsson

CHAPTER 8
Attending to olfactory short-term memory 137
Theresa L. White

Part III. Neuropsychology and olfactory dysfunctions

CHAPTER 9
Olfactory function in Parkinson's disease 155
Richard L. Doty and Hakan Tekeli

CHAPTER 10
Remembering what the nose knows 179
Robert G. Mair

CHAPTER 11
Olfactory impairment in normal aging and Alzheimer's disease 199
Steven Nordin

Part IV. Odor hedonic perception and development

CHAPTER 12
The psychophysics of olfaction in the human newborn:
Habituation and cross-adaptation 221
Lewis P. Lipsitt and Carolyn Rovee-Collier

CHAPTER 13
Emerging chemosensory preferences: Another playground for the innate-acquired dichotomy in human cognition 237
Benoist Schaal

CHAPTER 14
The acquisition of odour preferences via evaluative olfactory conditioning: Historical background and state of the art 269
Gesualdo M. Zucco

DEDICATION
Writings in remembrance of Professor Trygg Engen 295

Index 313

List of contributors

Birgitta Berglund
Department of Psychology
Stockholm University
Frescati Hagväg, 8
SE-106 91 Stockholm (Sweden)
e-mail: birgitta.berglund@ki.se

Pamela Dalton
Monell Chemical Senses Center
3500 Market Street
Philadelphia, PA 19104 (USA)
e-mail: dalton@monell.org

Richard L. Doty
Smell and Taste Center
University of Pennsylvania
School of Medicine
HUP – 3400 Spruce Street
Philadelphia, PA 19104 (USA)
e-mail: Richard.Doty@uphs.upenn.edu

Harald A. Euler
Human Wissenschaften
Institute of Psychology
University of Kassel
Hollandische Street 36–38
34127 Kassel (Germany)
e-mail: Euler@uni-kassel.de

Richard H. Gracely
Center for Neurosensory Disorders
2110 Old Dental
University of North Carolina
Chapel Hill, NC 27599-7455 (USA)
e-mail: richard_gracely@dentistry.unc.edu

Rachel S. Herz
Department of Psychiatry
and Human Behavior
Brown University
190 Thayer Street
Providence, RI 02912 (USA)
e-mail: Rachel_Herz@brown.edu

Anders Höglund
Department of Psychology
Stockholm University
Frescati Hagväg, 8
SE-106 91 Stockholm (Sweden)
e-mail: anders.hoglund@psychology.su.se

Fredrik U. Jönsson
Department of Psychology
Stockholm University
Frescati Hagväg, 8
SE-106 91 Stockholm (Sweden)
e-mail: fredrik.jonsson@psychology.su.se

Lewis P. Lipsitt
Department of Psychology
Brown University
190 Thayer Street
Providence, RI 02912 (USA)
e-mail: Lewis_Lipsitt@brown.edu

Robert G. Mair
Department of Psychology
Conant Hall, University of New Hampshire
Durham, NH 03824 (USA)
e-mail: rgm@cisunix.unh.edu

Donald H. McBurney
Department of Psychology
University of Pittsburgh
4429 Sennott Square
Pittsburgh, PA 15260 (USA)
e-mail: mcburney@pitt.edu

Steven Nordin
Department of Psychology
Umeå University
SE-901 87 Umeå (Sweden)
e-mail: Steven.Nordin@psy.umu.se

Mats J. Olsson
Karolinska Institute
Department of Clinical Neuroscience
Division for Psychology
S-171 77 Stockholm (Sweden)
e-mail: mats.j.olsson@ki.se

Michael I. Posner
Institute of Neuroscience
University of Oregon
222 Huestis Hall
Eugene, OR 97403-1254 (USA)
e-mail: mposner@uoregon.edu

Carolyn Rovee-Collier
Rutgers University
Piscataway, 08854 NJ (USA)
e-mail: crovee@earthlink.net

Benoist Schaal
Centre for Taste, Smell and Food Science
CNRS – University of Bourgogne
9E Boulevard Jeanne D'Arc
21000 Dijon (France)
e-mail: benoist.schaal@u-bourgogne.fr

Richard J. Stevenson
Department of Psychology
Macquarie University
Macquaire Square
Sydney, NSW 2109 (Australia)
e-mail: dick.stevenson@mq.edu.au

Sybil A. Streeter
Senior Research Associate
Olson-Zaltman Associates
938 Penn Avenue
Pittsburgh, PA 15222 (USA)
e-mail: sybilstreeter@gmail.com

Hakan Tekeli
Kasimpasa Asker Hastanesi
Bahryeyokusu caddesi
Beyoglu
Istanbul P.K. 34440 (Turkey)
e-mail: hakantekeli@yahoo.com

Theresa L. White
Department of Psychology
Le Moyne College
1419 Salt Springs Road
Syracuse, NY 13214 (USA)
e-mail: whitetl@lemoyne.edu

Gesualdo M. Zucco
Faculty of Medicine
Department of General Psychology
University of Padova
Via Venezia 8
35100 Padova (Italy)
e-mail: zucco@unipd.it

Foreword

An olfactory life

Michael I. Posner
University of Oregon, USA

It was a shock for me to find that my little Schnauzer Annie has a sense of smell perhaps a million times more acute than mine. This should have been obvious in her behavior each morning, particularly in the spring, when I walk her. Her nose down in the grass, she seems to commune with each smell, in the same way my eyes move to each of the editorials in my daily New York Times. She seems to have her Tom Friedman, David Brooks and Paul Krugman just as I have mine and looks or perhaps smells forward to her daily opportunity to meet them again, just as I do.

Trygg Engen's dramatic story from war time Oslo, to Lincoln Nebraska and then to become the pioneer and leader of an international consortium to study the sense of smell is the center piece of this volume in his honor. As an outsider to the Olfactory community it seems to me that Engen's seminal work and that of his many followers represented in this volume have paved the way to one of the great success stories in linking observable human and non human perception and behavior to detailed physiology and genetics. Because of this olfaction, serves as an important model system for efforts to relate physiology to perception in many other sensory, motor and cognitive systems.

Linda Buck and Richard Axel received a Nobel prize for their work on the olfactory system. They examined the system in the rat at the molecular level by first cloning receptors and then showing that all receptors expressing a given kind of protein are collected by a single glomerulus. This work showed the rat to have a multigene family that codes in excess of one hundred different odorant receptors, all related, but each unique. The discovery of how odorant receptors worked provided a molecular explanation of how the olfactory system detects odorants.

It is important to note that Buck's initial research at the University of Washington was in Psychology with visual researcher Walter Makous. Perhaps this background was important in her next goal which is to learn how signals from those receptors are organized in the brain to generate diverse odor perceptions.

She has come a long way in carrying out this goal with the bottom up approach, but examination of the top down methods of Engen will be needed to reach the perceptual goal.

Engen provides a good account of the top down approach in his 1991 book *Odor Sensation and Memory*. In the preface to the volume (pages xii–xiii) Engen says:

> one cannot predict the response to odor, verbal or behavior, from knowing the chemical … Psychological effects of odors depend on acquired association … This is not to say that the stimulus problem is unimportant. Without excitation of the olfactory receptors by chemicals there would be no odor sensation, and research on the nature of this transduction is of fundamental importance …
>
> The traditional approach to the study of the sense of smell is so-called bottom-up and is based on the assumption that odor perception is largely under the control of a prewired innate stimulus mechanism to be understood physiologically … In contrast the present approach is mainly top-down and psychological.

The bottom up approach of Buck has been successful in providing a molecular basis for the transduction of odor stimuli and has made olfaction an important model system for approaching all sensory modalities. This current volume in honor of Prof. Engen, however, strongly follows the top down psychological approach. The top down work of Engen and his students and followers raise many difficult problems for a purely stimulus driven approach. For example, Engen argues that the perception of whether an odor is pleasant or not depends not on the stimulus itself but on memory for previous experience with it. Pleasant odors are usually those with which we are familiar, while most unfamiliar odors are called unpleasant. The chapter by Dalton (this volume) stresses how beliefs and expectations also influence our reaction to odors.

It may be useful to examine how the top down effects of attention relate to sensory representation in other modalities. Attention forms an intermediate systems level of analysis, asking how top down effects may influence bottom up processing of sensory input. My own work is involved with the brain's attention system (Posner & Rothbart, 2007). In vision there is a network of neural areas involving inferior and superior parietal lobe and frontal eye fields that is specifically involved in orienting of attention toward visual events. Rather surprisingly, this same orienting network is involved in orienting to auditory and somatosensory events (Corbetta & Shulman, 2002; Driver et al., 2004), The activation of the orienting network produces increased sensory activity within the primary sensory cortex of the selected modality of input as measured by fMRI. While it seems reasonable that the chemical senses would involve the same orienting mechanisms, there are few reports of activity in parietal areas with attention to smell.

This could be a genuine difference between modalities, but it could also result from the lack of studies manipulating attention to olfactory input.

The orienting network, in adult humans is quite separate from a frontal executive network that includes the anterior cingulate and anterior insula among other frontal brain regions (Posner & Rothbart, 2007). This executive network is involved in voluntary control of output, particularly in helping to select among conflicting voluntary responses. Olfactory input appears to activate midfrontal, orbital frontal and cingulate cortical areas that are also involved in executive attention (Li et al., 2010; Plailly et al., 2008; Rolls, 2008). Thus olfaction may receive modulation via these executive frontal pathways rather than through the frontal parietal orienting network.

One advantage of knowing the anatomy of these control systems is to relate them to neuromodulators and to genes involved in their development (Posner & Rothbart, 2007). Recently resting state MRI has been used to examine attention networks in early development (Dosenbach et al., 2007; Fair et al., 2009; Gao et al., 2009). Because even young babies can be at rest, it is possible to see how these control networks develop from infancy to adulthood. Of primary interest, the orienting and executive networks tend to be closely integrated early in development and only become differentiated in later childhood as long connections develop that allow for larger scale executive control. The early development of the chemical senses may explain their direct connection to frontal areas and their failure to activate the orienting network. However, as mentioned before it is also possible that the failure to find influence of the orienting system could also arise because the paradigms that most closely reveal the orienting network in vision and audition through the use of cue to direct attention (Posner, 1980) have not been applied to odor and taste. More coordinated research comparing attention to odor and taste with audition and vision would seem useful.

The study of attention helps to merge the more strictly bottom up approach used by Buck, Axel and others and the top down methods so important in the behavioral studies of Engen and widely used in this volume. In vision, the study of endogenous cues to direct attention from stimulus input and the top down direction from instruction or memory have led to important progress in separating the voluntary from the reflexive control of attention. It is possible that similar progress or even more spectacular progress will be possible in the chemical senses. Perhaps the readers of this volume can taste (or should I say smell) another Nobel prize for someone who shows how molecular and systems neuroscience approaches can be merged to understand our perception and memories for smell.

References

Corbetta, M., & Shulman, G. L. (2002). Control of goal-directed and stimulus-driven attention in the brain. *Nature Neuroscience Reviews, 3*, 201–215.

Dalton, P. (2012). There's something in the air: Effects of beliefs and expectations on response to environmental odors. In G. M. Zucco, R. S. Herz, & B. Schaal (Eds.), *Olfactory Cognition: from Perception and Memory to Environmental Odours and Neuroscience*. Amsterdam: John Benjamins, Publisher.

Dosenbach, N. U. F., Fair, D. A., Miezin, F. M., Cohen, A. L., Wenger, K. K. R., Dosenbach, A. T., Fox, M. D., Snyder, A. Z., Vincent, J. L., Raichle, M. E., Schlaggar, B. L., & Petersen, S. E. (2007). Distinct brain networks for adaptive and stable task control in humans. *Proceedings of the National Academy of Sciences of the USA, 104*, 1073–1078.

Driver, J., Eimer, M., Macaluso, E., & van Velzen, J. (2004). Neurobiology of human spatial attention: Modulation, generation, and integration. In N. Kanwisher, & J. Duncan (Eds.), *Attention and Performance XX: Functional Brain Imaging of Visual Cognition*. Oxford (UK): OUP.

Engen, T. (1989). *Odor Sensation and Memory*. New York: Praeger Press.

Fair, D., Cohen, A. L., Power, J. D., Dosenbach, N. U. F., Church, J. A., Meizin, F. M., Schlaggar, B. L., & Petersen, S. E. (2009). Functional brain networks develop from a local to distributed organization. *Public Library of Science, 5(5)*, 1–13.

Gao, W., Zhu, H., Giovanello, K. S., Smith, J. K., Shen, D., Gilmore, J. H., & Lin, W. (2009). Evidence on the emergence of the brain's default network from 2 week-old to 2-year old healthy pediatric subjects. *Proceedings of the National Academy of Sciences of the USA, 106*, 6790–6795.

Li, W., Lopez, L., Osher, J., Howard, J. D., Parrish, T. B., & Gottfried, J. A. (2010). Right orbitofrontal cortex mediates conscious olfactory perception. *Psychological Science, 21(10)*, 1454–1463.

Plailly, J., Howard, J. D., Gitelman, D. R., & Gottfried, J. A. (2008). Attention to odor modulates thalamocortical connectivity in the human brain. *Journal of Neuroscience, 28(20)*, 5257–5267.

Posner, M. I., & Rothbart, M. K. (2007). Research on attention networks as a model for the integration of psychological science. *Annual Review of Psychology, 58*, 1–23.

Posner, M. I. (1980). Orienting of attention. The 7th Sir F. C. Bartlett Lecture. *Quarterly Journal of Experimental Psychology, 32*, 3–25.

Rolls, E. T. (2008). Function of the orbitofrontal and pregenual cingulate cortex in taste, olfaction, appetite and emotion. *Acta Physiologica Hungarica, 95(2)*, 131–164.

Preface

This book was conceived as a tribute to one of the founders of the psychological study of the sense of smell, Professor Trygg Engen. Trygg Engen was seminal in advancing our knowledge and laying the basis for further research in the psychophysical, cognitive and neuropsychological features of olfaction. With his book, *The Perception of Odors*, published in 1982, he stimulated generations of researchers in the biological and behavioural sciences and prompted a wide range of empirical investigations into the many areas of research that he developed and contributed to. A number of Trygg Engen's colleagues, including peers, students and intellectual descendents, who were inspired by his research and in different ways and in periods of their lives shared scholarly interests and activities with him, have joined in building this volume to memorialize his legacy. We are grateful to all of the contributors of this volume and thank them for their enthusiastic commitment to this project and providing new and thought provoking accounts of the state of the art in olfactory cognition. We would also like to extend our special thanks to Professor Michael Posner for writing the Foreword of this book.

The present book is divided into four academic sections that were special topics of interest to Trygg Engen. The first section reunites the fields of psychophysics and the perception of environmental odours and discusses the impact of odours on beliefs and expectations. In Chapter 1, Birgitta Berglund and Anders Höglund address a theory of odour quality measurement in which odours are conceived as homogeneous psychological states (as opposed to heterogeneous percepts in vision). In Chapter 2, Pamela Dalton surveys research on beliefs, suggestibility, and expectations as sources of inter-individual variability in olfactory cognition and in collective responses to the odour environment. Next, in Chapter 3, Richard H. Gracely discusses how measurement and psychophysical techniques in olfaction have shaped clinical and research methodology in an altogether different field – pain research. The forth chapter, by Donald H. McBurney, Sybil A. Streeter and Harald A. Euler, investigates the affective consequences of odours in people's intimate lives and uncovers important social functions of olfaction in human interpersonal behaviour. Chapter 5, by Richard J. Stevenson, then discusses the problem of odour perception against an ever-changing chemical background and proposes an Object Recognition Approach which highlights the centrality of learning and memory in olfactory perception.

The second section of this book addresses cognitive processes in olfaction, how odours are interpreted, lexicalized, associated with contexts, remembered, and able to elicit episodic memories. First, in Chapter 6, Rachel S. Herz, shows that the link between our ability to remember odours and the experience of odour-evoked memory is that they are both critically mediated by associative learning, and discusses evidence for the distinguishing and possibly unique characteristics of odour-evoked autobiographical memories. Next, in Chapter 7, Fredrik U. Jönsson and Mats J. Olsson review the nature of, and performance levels for, different basic abilities of human olfaction such as detection and discrimination, as odour source naming and identification, and as different kinds of functional judgements about odours. In Chapter 8, Theresa White, examines the literature on olfactory short-term memory, highlighting the central role of attention to this process, as well as discussing evidence supporting the uniqueness of olfactory cognition.

The third section of our volume focuses on the cerebral bases of olfactory awareness and the neuropsychological investigation of olfaction with special emphasis on olfactory dysfunctions. First, in Chapter 9, Richard L. Doty and Hakan Tekeli discuss how psychophysical methods have elucidated olfactory dysfunction for a range of neurodegenerative diseases, most notably idiopathic, familial, and pre-symptomatic Parkinson's disease and related motor disorders. In Chapter 10, Robert G. Mair summarizes his series of studies on olfactory function in Korsakoff patients, and re-examines these results in light of animal studies on the neurological basis of olfactory memory and with respect to more recent neuropsychological and functional imaging studies of olfactory perception in humans. In Chapter 11, Steven Nordin provides a broad review of early and recent findings on the neuropsychology of olfactory impairment in normal aging and Alzheimer's disease (AD) among the elderly, and what the consequences of such olfactory impairments are on quality of life, food intake and general health.

The last section of our book concerns affective and developmental processes in olfaction. In Chapter 12, Lewis Lipsitt and Rovee-Collier summarize the early work they conducted together with Trygg Engen on the psychophysical approach of olfaction in human newborns, and examine how this work laid the foundations for further research on infant odour perception. Next, in Chapter 13, Benoist Schaal surveys early and recent research on the development of olfactory preferences in human infants and examines the question of learning versus "innateness" as the basis for the development of chemosensory preferences. Finally, in Chapter 14, Gesualdo M. Zucco reviews studies on evaluative olfactory conditioning, and discusses its relevance to understanding the processes involved in the acquisition of odour preferences and the formation of human emotions.

After the academic chapters, a dedication section follows. Here all the authors briefly relate their individual stories and histories with Trygg Engen and share how he influenced and touched them both professionally and personally.

Olfaction was neglected by the sciences for many years. Starting with Trygg Engen's pioneering work, however, over the last several decades a considerable body of literature has flourished on multiple levels of enquiry ranging from the molecular processes at the periphery of the system, to the affective and cognitive processing of olfactory information, to the powers of odours to influence our everyday lives.

Our aim in producing this book is that it will help promote further research in olfactory cognition and attract new inquisitive scientists to the field, as many questions remain unanswered and there are further unexplored avenues of research to pursue. We also hope that this volume will be a useful resource for colleagues, students, and professionals who study olfaction, as well as to scientists who work in the domains of perception, cognitive neuroscience and environmental psychology more broadly.

Gesualdo M. Zucco, Rachel S. Herz, Benoist Schaal

Acknowledgments

We are grateful to Dr. Maxim Stamenov, Editor of the series Advances in Consciousness Research, and the following members of the editorial team, Hanneke Bruintjes, Els van Dongen and Patricia Leplae, for their competent and friendly assistance along the entire production process.

Trygg Engen (Oslo, 1926 – Providence, 2009)

Professor Trygg Engen seated at his desk in the Department of Psychology at Brown University in Providence, Rhode Island, where he taught for 37 years. Trygg Engen received his Ph.D. in Psychology from the University of Nebraska in 1954, and an honorary doctorate from Stockholm University in 1990. He was a pioneer in the psychological study of olfaction and inspired generations of researchers.

PART I

Perception, psychophysics and odour environment

CHAPTER 1

Is there a measurement system for odour quality?

Birgitta Berglund[1,2] and Anders Höglund[1]
[1]Stockholm University / [2]Karolinska Institute, Sweden

Although humans can discriminate a huge number of odours, the mechanisms behind the perception of odours are not well understood. The theory of the present essay and our experiments were developed with the view that odours are homogeneous perceptions (as opposed to visual objects that are heterogeneous percepts). Other typical examples of homogeneous perceptions are colours and some emotional states. Thus, through perceptual learning, which is a prerequisite for perceiving, we become aware of and may recognise odours in the environment. Typically, odours are detected at very low concentrations as compared to chemical instruments, the "electronic nose" inclusive. Four principles of investigating the human world of odour qualities are discussed: (i) classification and sorting according to odour-quality resemblance, (ii) using names and verbal attributes for odour sources, (iii) using odour qualities for a representative set of reference odorants, and (iv) measuring odour quality by similarity scaling of pairs of odorous stimuli. Although a huge number of odours can be discriminated, the quality of odour mixtures is intermediary among the odour qualities of their components. This applies to single chemical compounds or mixtures thereof and for very broadband mixtures like indoor air. Our perceptual world simply consists of a massive amount of odour qualities potentially attributable to sources in the environment. Their organisation is, however, still a mystery.

Introduction

This essay aims at improving our understanding of what odour quality is or could be. In his excellent first book on odour perception, *The Perception of Odors*, Trygg Engen (1982) showed that the theoretical and empirical literature on this subject is extensive. Then, as now, a main focus of odour-quality research is to try to better understand either the specifics of odorous stimuli or the specifics of physiological processes. The olfactory code still remains an enigma for the scientific world (e.g., Buck & Axel, 1991; Buck, 2004; Yeshurun & Sobel, 2010). Our hypothesis is that there are four worlds involved in psychophysics (rather than the two worlds of Gustav Theodor Fechner): an outer physical-chemical world, an inner human physiological-biochemical world, an inner mental-perceptual world, and an outer world of human behaviour (Galanter et al., 2010). In each of these worlds there are unique mechanisms or processing systems, and the big mystery is how these worlds communicate with each other.

A critical focus in this essay is to try to explain theoretically where odour quality may be represented (e.g. in the mental-perceptual world) and where its system(s) of measurement may be located or created (e.g., in the mental-perceptual world or physical-chemical world). In other words, what kind of outcome measures will best represent the theory of the method of measurement? For example, Amoore (1971) tried to explain odour quality by stereo-chemical measures of odorous molecules and Wright (1977) by vibration-frequency measures of odorous molecules. Thus, both chose to represent or measure odour quality in the physical-chemical world and in the case of Wright also in interaction with mechanisms of the physiological-biochemical world. A specific goal of our research is to try to measure odour quality solely inside the mental-perceptual world of odours; that is the so-called "odour space".

In analogy with the perception of colours, we define odour to be a homogeneous perception ("one coherent thing" without distinct details that is perceived, e.g. Ekman et al., 1964). Colour systems (e.g. Hunt, 1989) can be viewed as analogues to potential odour-quality systems in that colours are also homogeneous perceptions. On the contrary, many physical objects (like chair or music) are heterogeneous perceptions where the parts are perceived distinctly together with the whole. The homogeneous odour perceptions may be distinguished in the mental-perceptual world, and we refer to the odorants as the sources in the physical-chemical world. Typically, one sniff is enough for perceiving an odorant's unique odour quality, which will also be perceived more "full" or "dense" with increases in odour intensity (Engen, 1982; Zheng, 2010). Although we are not involving the physiological-biochemical world in the odour or odour quality concepts, we adhere to Hudson's (2000) terminological clarification.

The chromaticity diagram of CIE (Commission Internationale de l'Eclairage) represents human discrimination of colours (= one kind of the visual qualities) in a "colour triangle" with the approximate perceptual correlates hue and saturation representing the main coordinates. This is a psychophysical measurement system because discrimination among colour perceptions is measured as physical quantity with its physical unit of measurement. By contrast, the Natural Colour System (NCS) (Hård & Sivik, 1981) measures instead colours in terms of relative quantities (in %) of the basic colours that are perceived to be present in colour mixtures; it is a perceptual measurement system. Perceived whiteness (*w*), blackness (*b*), and chromaticness (*c*; e.g. redness and yellowness) in any colour mixture will add up to 100 %. The NCS atlas was intended to portray samples at every 10th % step.

The two measurement systems for colour (CIE and NCS) suggest that the one-to-one correspondence model typical for psychophysical relationships may be an unnecessary assumption. As a psychophysicist, Ernst Mach thoughtfully claimed that perceptions organised in one way are the study objects of physics, whereas they are the study objects of psychology if organised in another way (Boring, 1957). This way of thinking made it possible to predict the perceived odour intensity of mixtures from the perceived odour intensities of their components (e.g. Berglund et al., 1973b; Berglund, 1974; Cain & Drexler, 1974; Berglund et al., 1976; Patte & Laffort, 1979; Laing et al., 1984; Laffort, 2006), but not directly from their concentration measures in a psychophysical model. A self-evident supplementary idea is that the odour quality of mixtures reflects the component odour qualities as perceived rather than the physico-chemical structure. At least, a better understanding of the odour percepts may help reveal the nature of the psychophysical communication. Thus far, it is known that if odour intensity is kept invariant, the quality of odour mixtures up to three to five component odours (differing in perceived odour quality) will be intermediary among the odour qualities of their components (Berglund, 1974; Berglund et al., 1976; Moskowitz, 1976; Berglund & Olsson, 1993a; Livermore & Laing, 1998a; Brodin et al., 2009).

1. Theoretical considerations for odour-space measurements

Early research on odour quality was restricted to classification as the method of measurement (Engen, 1982). Herein, we define odour quality as a perceptual characteristic of an odorous "object" (or source) similar to the perception of colour of a visual object. Odour quality is a homogeneous perception because two mixed odorous substances will not be perceived simultaneously, but rather one odour perception will emerge as unitary whole (Berglund et al., 1973b; Berglund, 1974; Berglund & Olsson, 1993b). Conversely, sounds from two different musical instruments may be discerned distinctly and perceived simultaneously in a piece of music (auditory "object" or source), i.e. hearing produces a heterogeneous perception. Whereas at least two parts and their composite are discerned in a heterogeneous perception, the parts always blend in homogeneous perceptions (cf. yellow mixed with red produces orange colour only). Although a homogeneous percept results from a "synthesis", both kinds of perceptions may be analysed perceptually into their parts. For example, the homogeneous perception of the colour orange is perceived and imagined to be a mixture of red and yellow, not of lilac and yellow.

Our heterogeneous and homogeneous perceptions cannot simply be substituted by Garner's (1974) "separable" (size and colour of a cube) or "integral" (lightness and saturation of colours) attributes since both constitute attributes of the stimulus in a psychophysical theory. However, it would be easier to report on complexity of a heterogeneous percept (a 6-colour square pattern on fabric) than on a homogeneous percept (the odorous exhaust from the main stack of a pulp mill). The former parts are distinctly perceivable component colours of the pattern (the whole perception) whereas the latter are odour components imagined to exist "inside" the odorous exhaust (the whole perception). Whereas Berglund et al. (1973b) and Ekman et al. (1964) define and measure the "parts" of the "whole" (component odours) in the perceptual world, Garner (1974) defines and measures the "parts" of the "whole" (stimulus attributes) in the physical world. Berglund's and Ekman's research strategy is top-down (from perception to stimulus) whereas Garner's is bottom-up (from stimulus to perception).

Another theoretical factor in odour-quality research is the stimulus sample that is selected. On the one hand, a limited sample of single but different odour qualities as well as all their mixtures would be expected to guarantee similarities in perceived odour quality (e.g. MacRae et al., 1992). On the other hand, a large sample of different odour qualities would increase the probability of uncovering the whole odour space and its potential areas of similarity-dissimilarity (e.g. Woskow, 1968; Ekman, 1970; Berglund et al., 1973a; Schiffman, 1974; Moskowitz, 1976; Dravnieks et al., 1978; Davis, 1979; Shepard, 1991; Haddard et al., 2008). It has been shown that the odour qualities of mixtures tend to be similar to the odour qualities of their components if these are equally odour-intense and at most four plus or minus one (Livermore & Laing, 1998b; Berglund, 1974; Moskowitz & Barbe, 1977; Berglund & Olsson, 1993a). On the contrary, the odour-qualities of higher order mixtures (above five components) and among single and chemically unrelated odorous compounds show a much broader variation (Livermore & Laing, 1998a).

Most of the environmental odours which we perceive daily (e.g. indoor air or the flavour of food) involve a very larger number of odorous compounds but at relatively low concentrations (e.g. Noma et al., 1988). These kinds of broadband mixtures would represent odour qualities that are completely different from the individual odour qualities of their numerous single odorous-compounds (Berglund et al., 1982). Obviously, from the perceiver's point of view the odour space contains odour perceptions which are similar or dissimilar in odour quality. In measuring degree of similarity among odour qualities on a scale from 0 to 100%, all dissimilar pairs would get the value 0. Consequently, with many dissimilar odour qualities, a huge number of odorants (whether single compounds or mixtures) have to be researched to secure a measurement system for odour qualities (Shepard, 1991). Comparing with colours, Hunt (1989) report that 60,000 observations made on samples of coloured paper viewed in daylight resulted in a prototype NCS atlas of 1200 samples (later 16,000 notations).

A third theoretical factor is the selection of the experimental subjects. They may be naïve to the odour-quality task, have long-term experience of it, or may even be a panel deliberately trained to report according to a predetermined system of how odour qualities would be organised. According to Ericsson and Lehmann (1996), extended experience obtained by training a panel only delivers gradual improvement to the scheme learnt. Conversely, in the odour-quality research cited in this work, we would expect that the subjects have responded to what they momentarily perceive rather than to any established scheme. First, we should however check if pure database research on odorants has given any clues to the organisation of the odour space.

2. The odour space: Four major examples of database research on odorants

Several quite large databases are available which provide chemical information on odorous compounds including semantic descriptions of their odour. We here briefly summarise research on the "odour space" in which such a database constituted the psychophysical data and the research results were published in scientific peer-review journals. Chastrette et al. (1988) conducted hierarchical classification based on similarities among 74 perfumery notes characterising 2767 chemical substances. They found that there is no strict hierarchy among the 74 notes, and therefore concluded that no primary odour qualities exist among the chemical substances. Madany Mamlouk et al. (2003) used 851 stimuli of the Aldrich Flavour database and created 278 × 278 descriptor dissimilarities. The formed matrix was submitted to multidimensional scaling (MDS) and to two-dimensional self-organising mapping (SOM). The 2-Dimensional SOM provided new knowledge on the complexity and potential functionality of the database descriptors (not of human perception of the odour stimuli as claimed by the authors who used no subjects). Jaubert et al. (1995) revised their 3-dimensional odour-space model to a 2-D Field of Odours grounded in the odour qualities of 42 reference odorants available in liquid form as a kit. Together the two constitute a guide for identifying various descriptive elements of any other odour qualities. Zarzo and Stanton (2006) conducted a Principal Components Analysis (PCA) on similarities of profiles of 82 semantic descriptors for 881 perfume materials. They found their odour space to be highly dimensional and it would need 17 odour classes. Therefore, as for colours, a few primary odours cannot define the vast odour space.

Taken together, these four large database studies show that the "human odour space" is complex and highly dimensional and primary human odour qualities, if they exist at all, are large in number. Notably, this is completely opposite to Yeshurun and Sobel's (2010) recent conclusion from a meta-analysis of multidimensional similarity experiments where unidimensional odour objects only exist on a pleasantness-unpleasantness axis.

3. The odour space: Ways of measuring odour quality

Humans have a large vocabulary of colour names extending beyond the basic colours. For odour qualities, however, most of us lack words for describing adequately what we perceive. Thus far, research shows that if basic odour qualities exist (in analogue to basic colours) from which mixtures all odour qualities could be created, we are far from discovering the basic ones. The attempts to determine basic odour qualities seem to be restricted to certain kinds of odour sources adequate to application areas. These include selected chemical structures, unpleasant vs. pleasant odorants, the flavours of foods, perfumes and cosmetics, abatement of odorous air pollution outdoors, or indoor air quality (e.g. Berglund et al., 1973a; Moskowitz & Barbe, 1977; Schiffman, 1981; Berglund et al., 1982; Dravnieks, 1982; Wiltrout et al., 2003; Chrea et al., 2009).

The human world of odour qualities has been investigated in at least four principle ways:

a. The first way is by classification and sorting according to odour-quality resemblance (e.g. Harper et al., 1968; Lawless, 1989). Miller sorting means to sort odorants in as many groups needed, such that those that are similar in odour quality are in the same group and those that are dissimilar in different groups. Alternatively, the odour-quality sorting may be made according to the odour qualities of a set of reference odorants (Wright & Michels, 1964; Yoshida, 1964, 1975; Jaubert et al., 1995; Laffort, 2006).
b. A second approach is to use names of odour sources (e.g. peppermint, manure, and wood) and verbal attributes (e.g. foul, floral, and earthy) as sets of descriptors (or labels) to represent the perceived odour qualities of the odorants. Descriptor profiles are created from frequencies of presence/absence of sources or attributes or from quantitative scales of these; they may be viewed to measure qualitative complexity (Cairncross & Sjostrom, 1950; Moskowitz & Gerbers, 1974; Gregson & Mitchell, 1974; Moskowitz & Barbe, 1977;

Dravnieks, 1982; Stevens & O'Connel, 1996; Herz & von Clef, 2001; Chrea et al., 2009). Pleasant-unpleasant should be interpreted with care together with other descriptors (sources or attributes) specific to odours (Chrea et al., 2009). The sources are nouns, which are hyponymy (organised hierarchically with exclusive classes), whereas attributes are typically adjectives which are antonymy organised as "opposites" to one another (Miller, 1991). Odour descriptors in the form of sources and attributes represent their own perceptual structure of meaning (that is without smelling anything). In profiling, more specific attributes applies to a few odour qualities (sweet or sulphur), whereas others applies to all odour qualities (pleasantness and/or unpleasantness). Therefore, the antonym pleasantness and/or unpleasantness should be defined and measured separately. Miller (1991) defines four possibilities: (a) contradictory (bipolar unidimensional: from highly unpleasant through zero to highly pleasant), (b) contrary (Euclidean model: unpleasantness is one vector, pleasantness another), (c) reverse (unidimensional, 0–100% unpleasantness increases tied to 100–0% pleasantness decreases), or (d) contrasted (city-block model, either pleasant or unpleasant). High-scoring sets of descriptors of odour quality have been shown to be only a modest indicator of odorants present in two-to-four odour mixtures (Jinks & Laing, 2001). This result supports that odour mixtures are homogeneous perceptions.

c. A third approach is to use odour qualities of a representative set of reference odorants descriptors to measure the odour quality of the experimental odours in terms of the reference-odour profiles (Jaubert et al., 1995). Notably, this may be viewed as a qualitative version of Berglund's (1991) master scaling. The basic idea is to provide an invariant odour-quality context within which the subjects can perceive and scale new odour qualities.
d. A forth approach is to measure odour quality of odorous substances and their mixtures and/or emissions from materials by similarity (or dissimilarity) scaling of pairs of odorous stimuli. The data may be analysed according to two main models: (1) the content model (Ekman, 1970) – similarity matrix data submitted to a PCA for a vector solution, and cube data to a correspondence analysis (Greenacre, 1984; Noma et al., 1988) and (2) the distance model (Schiffman et al., 1981) – dissimilarity data submitted to multidimensional scaling (MDS) or individual differences scaling (INDSCAL). The similarity approach was used by Woskow (1968), Gregson (1972), Berglund et al. (1973a), Moskowitz and Gerbers (1974), Schiffman (1974, 1981), Yoshida (1976), Dravnieks et al. (1978), Davis (1979), Lawless (1989), Stevens and O'Connel (1996), and Kahn et al. (2007).

The four different approaches to measuring odour quality all have their benefits and drawbacks. A benefit of the similarity/dissimilarity approach is that it would not superimpose a verbal descriptor system (sources or attributes) onto the organisation of odour qualities in the odour space. A drawback of the similarity approach is that repeated presentations of invariant odorants in all possible pairs require much smelling even for a small sample of 15 odorants. This increases the risk for olfactory adaptation (Berglund & Engen, 1993). In the verbal profiling approach, a critical assumption is that odour qualities with similar attributes and/or source profiles would have similar perceived odour quality. This will be true only if the battery of verbal descriptors is appropriately developed, for example as items of a psychometric test (for psychometry see e.g. Lord & Novic, 1968; Jöreskog & Sörbom, 1978). Chrea et al. (2009) created a test instrument for odour quality covering a defined "odour quality space" including a balance among pleasant and unpleasant descriptors (sources or attributes). Otherwise it would not have qualified as a measuring instrument of odour quality with specified reliability and validity.

A problem specific to verbal descriptors is that not only the perceived qualities of the odours, but also the perceptions of the descriptors, will be included in the resulting scale values of similarity in odour quality. Common variance among profiles referring to perceived content of descriptors may confound the "true" perceptions of the odour qualities of the odorants. Another problem specific to verbal descriptors is that not only the investigated odour qualities but also most of the descriptors used for measuring odour qualities can be classified and scaled as either pleasant or unpleasant (true for all senses). Additionally, pleasantness or unpleasantness will co-vary with odour intensity. This will be inherent and unavoidable at least for some of the set of verbal descriptors. Since some verbal descriptors of odour quality will be inherently similar, they will also exist in an own descriptor-representational multidimensional structure. This structure of the meaning of odour descriptors would potentially exist *per se* and not require that the subjects actively smell and compare odours of the compounds during the experiment. Chrea et al. (2009) avoided part of this descriptor problem by conducting preliminary experiments to obtain a balance between pleasant and unpleasant descriptors. They also mapped not only odour perceptions but also the emotions induced by odours to develop a test instrument for measuring odour quality with specified reliability and validity.

Nosofsky (1986) points out that perceived similarity is not an invariant relation because of the influence of selective attention to attributes of complex stimuli when compared. This statement is grounded in the theoretical distinction between integrality and separability made by Garner (1974): "In perceptual classification, stimulus sets defined by integral dimensions are classified primarily in relation to similarities (Euclidean model); sets defined by separable dimensions are classified in relation to dimensional structure (city-block model)". Garner's integrality and separability refer to attributes of the stimulus that affect the resulting perception (bottom up). In our view, odour qualities from single odorants and of their mixtures and of any broadband mixture all create homogeneous perceptions (in analogue to every colour specimens of the NCS Atlas, Hunt, 1989). The "integral" stimulus attributes of two odorants (= stimuli) cannot be discerned perceptually. Based on our top-down approach, we do know that the odour quality of two-odour mixtures will always be intermediary in Euclidean space relative to the perceived odour qualities of its components (e.g. Berglund, 1974; Moskowitz, 1976; Laska & Hudson, 1993; Livermore & Laing, 1998b; Lindqvist et al., 2009).

Among the perceptual attributes of odour, which are possible to measure, perceived odour intensity is basic. However, higher intensity also makes the odour quality stronger. In what way is it then involved in the perceived odour quality? In discussion with Trygg Engen, he claimed that odour intensity was not like brightness or loudness, but rather like density or volume. The stronger the odour, the more "full" the odour quality. Perhaps odour intensity works analogously to saturation of colours? This was one reason why odour stimuli were made equal in perceived odour intensity such that multidimensional scaling would be restricted to qualitative similarities of odours (Berglund et al., 1973a; this is a top-down approach). Another reason was that Ekman et al. (1964) had developed the quantitative principle of qualitative similarity. This made it possible to measure perceived quality of homogeneous perceptions in multidimensional space (Ekman, 1970). This system of measurement of qualitative similarities as perceived was successfully tested by keeping perceived intensity invariant for emotions, odours, colours, and subjects' judgments of personality traits. Moreover, Woskow's (1968) extensive experiment on multidimensional scaling of odours (PCA analysis of similarities) supported the results of Ekman et al. (1964). Schiffman (1974) also published extensive further data analysis of Woskow's data, mapping odour quality in a distance space.

In all measurement, the quantity and its unit of measurement have to be defined (Bentley, 2005; Berglund et al., 2012). Loudness is an example of a psychophysically derived quantity which so far has been standardised internationally with the units phone and sone. Examples of other perceptual quantities that so far may be able to fulfil basic criteria for measurability are perceived exertion, feeling of happiness, odour intensity, and pain intensity. Apart from perceived exertion, there have been no serious attempts to propose and agree upon units of measurement for these quantities (Galanter et al., 2010). In our view, odour qualities as perceived are possible to measure because they are homogeneous perceptions whether they represent single odorants or mixtures of a few odorous compounds or broadband sets of chemical compounds. The number of odour-quality specimens needed for an Atlas of Odours would, however, need to surpass the 1,200 samples of the NCS Atlas of Colours. Apart from sustainable research grants, the limitations include the construction, handling, storage and form of presentation of the samples.

4. The odour space: Odour quality and air quality

Humans perceive odorous substances and their mixtures or samples of air quality as homogeneous perceptions (Ekman et al., 1964; Berglund et al., 1973b). Low-concentration air samples of indoor or outdoor air from non-industrial buildings are typically perceived as rather invariant homogeneous odours of subtle intensity (Berglund et al., 1982; Livermore & Laing, 1998b). The olfactory and trigeminal systems are involved either as warning systems or perceptual-aesthetic systems. The latter are primarily tuned towards very faint and subtle broadband low-concentration odour mixtures. Apart from faint and delicate odours, pleasant indoor air is characterised by perceptions resulting from air temperature and movement, e.g. freshness and gentle cooling (Baird et al., 1994).

4.1 Odour quality and pleasantness/unpleasantness

Inspired by the experiments of Ekman et al. (1964), Engen and McBurney (1964) and Woskow (1968) the first author conducted four experiments together with Trygg Engen in the late 1960's starting with 28 odorous chemical compounds. First, their psychophysical power functions of odor intensity in their common context were determined (Berglund et al., 1971). Concentrations of 21 of the 28 compounds were then adjusted to create equal odour intensities. Qualitative similarities were scaled for all pairs of the 21 × 21 compounds and subsequently also for the pleasantness and unpleasantness of the same odorants (Berglund et al., 1973a). Thereafter, a new group of subjects was recruited. They produced in consensus descriptors/descriptions of the odour qualities of the 21 odorants. A week later, the same subjects were called back to scale the similarities of the 21 × 21 matrix of their semantic descriptions taken in consensus to represent the 21 odorants. Each subject also scaled the pleasantness-unpleasantness of the 21 descriptors (unpublished experiment). In spite of the consensus reached on the semantic descriptions of the odour qualities, the subjects disagreed totally on what odour qualities were similar or not as represented in the descriptions. Also the pleasantness-unpleasantness scales of the semantic descriptions of the 21 odorants showed low agreement among the subjects. In discussing the results with the subjects, they pursued that they did not feel that the consensus reached matched their own perceptions very well.

Surprisingly, the subjects' similarity matrices for the 21 odorants showed very low agreement. The same was true for their pleasantness-unpleasantness scales (Berglund et al., 1973a). Still, the individual hedonic scales each correlated well with the first component of each subject's PCA on the similarity matrix. Thus, individual pleasantness and/or unpleasantness perceptions dominated the odour-quality similarities scaled by each subject, but not at all for the group data (cf. Kahn et al., 2007). Very large interindividual differences were found both in the pleasantness-unpleasantness scales and the similarity scales, though it was not known why.

For group data, it does not seem that equally pleasant or unpleasant odours are perceived to be similar in their odour qualities, nor do similarity data support Yeshurun and Sobel's (2010) idea of unidimensional odour objects. The point is that a unique pair of equally pleasant (or equally unpleasant) odours is equal for different reasons for one subject than for another subject. Correlation coefficients between similarity matrices obtained from different subjects, but with identical set of odorants, were found to be suspiciously low (cf. Berglund & Olsson, 1993b). Typically, there will be a regression towards the mean for the group data on odour similarities, which consequently is atypical for any subject. For the individual similarity matrices, pleasantness-unpleasantness will often correlate highly with the first dimension of a PCA run on the similarity matrix (or a MDS run on the dissimilarity matrix). This is due to that pleasantness-unpleasantness is an antonymy like similarity-dissimilarity. As bipolar descriptors, these two would each apply to all odours (the two also applies to all odour descriptors). Pleasantness-unpleasantness applies also to all colours, but that does not make them one-dimensional!

Pleasantness and unpleasantness are not inherent odour-specific concepts; rather they are hedonic or aesthetic evaluations freely applicable to any of our sensory perceptions (colours, music, touch, flavour). Compared to colours, which we all seem to be able to learn to name correctly, it seems strange that we have no agreed-upon set of primary odours to represent odour qualities. Rather, odour qualities are numerous and impossible to recall although potentially associated with certain life situations and easily recognised (Engen, 1991).

4.2 Indoor air quality

Odorous substances emitted from building materials to the indoor air have been found to be important to human health in non-industrial indoor environments.

Most of our time, we inhale indoor air and outdoor air, which we consider close to odourless and very faint. Research on indoor air quality would require an advanced pattern recogniser for handling the large amount of molecular information accompanying every inhalation. As an example, Berglund et al. (1982) determined concentration patterns of air samples simultaneously with the odour-strength patterns perceived by subjects. The chemical content of air samples were analysed with a gas chromatograph (GC) equipped with a flame ionisation detector (FID) and a sniff-port. The eluate from the GC was measured simultaneously with the FID and with a subject's sniff in a sniff-port. The latter provided a perceptio-olfactogramme (POG) which was compared to its GC chromatogram. The odour intensity of the eluate was measured with magnitude estimation as was also a set of five reference pyridine concentrations interspersed among the eluate presentations. The method of master scaling was used for calibrating the odour intensity scales (cf. Berglund, 1991; Berglund, in press). Collection of parallel air samples made it possible for two subjects to smell the identical stimulus samples. In total, 28 samples of outdoor air and 66 samples of indoor air were collected from two buildings. Between 40–100% of the 60–120 GC peaks of each of the 94 gas chromatograms were found to be odorous.

The chemical concentration patterns of the FID peaks were very different from the corresponding odour strength patterns of the POGs. A close to zero correlation was obtained between the two types of peak quantities because they are inherently two kinds of detectors: an instrumental detector and a human odour detector. Two separate correlation matrices were therefore formed for each of the two kinds of air-sample patterns and each matrix was submitted to a PCA. Figure 1

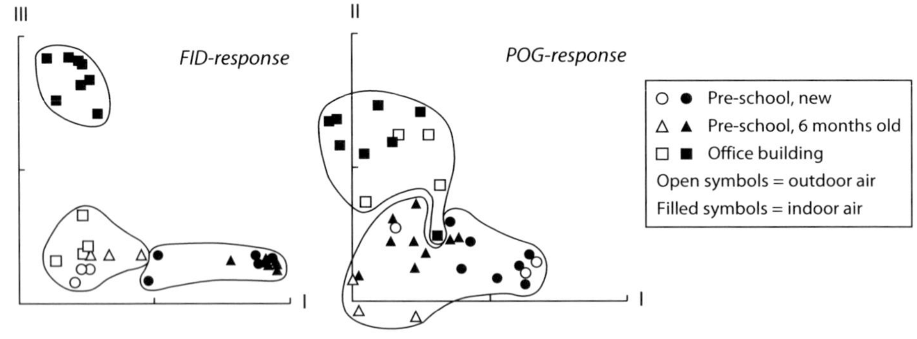

Figure 1. Results from two Principal Components Analyses. The chemical pattern of FID-responses (left) cluster in three groups: indoor air of each of two buildings and outdoor air independent of buildings. The odour-intensity pattern of the corresponding POG-responses (right) cluster in two groups: indoor air and outdoor air are together for each of two buildings. [from Berglund et al., 1982]

shows the PCA results for the FID and POG response patterns of the air samples in the left and right diagrams, respectively. The FID peak patterns of the indoor-air samples (filled symbols) identified correctly the office building (squares) or the pre-school (circles and triangles). The patterns of the corresponding two sets of outdoor air samples (open symbols) could not be recognised because they form their own cluster (left diagram). The indoor "odour print" (filled symbols) recognised correctly the partial "odour print" originating from the outdoor air (open symbols) supplied to either of the two buildings. Conversely, neither the FID nor the POG could discriminate the indoor airs of the newly built pre-school (circles) from the indoor airs of the 6-month older building (triangles). Thus, human olfaction was capable to recognise air samples by odour-strength patterns of very low concentrations. The outdoor air supplied to the buildings contributed distinctly to the odour patterns of the indoor air. This was true for the odour-strength patterns by buildings but could not be declared from the corresponding chemical concentrations patterns. The odour of the air was thus confirmed to be the best indicator of a well functioning ventilation system based on recirculation.

Air samples from various locations in the same two buildings were also presented to subjects in a hood fed by a high-flow sampling olfactometer installed in a mobile field laboratory (Berglund et al., 1986). In Figure 2, the perceived odour strength of the indoor and outdoor air samples inhaled from the hood (as single

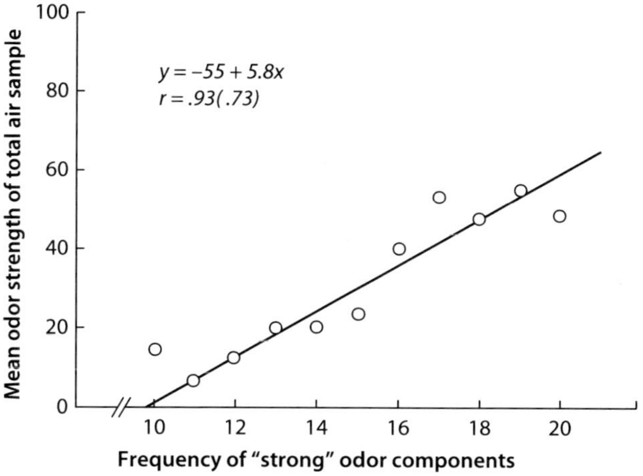

Figure 2. Perceived odour strength (arithmetic mean) of air samples from a building presented in a hood plotted against the number of "strong" odour components (among 28 strong components) in air samples identified by GC/MS. Coefficients of correlation within parenthesis refer to the relation between 88 samples before averaging. This result show that at least 10 strongly smelling VOCs are required in an air sample for it to be possible to differentiate from a clean air sample. [from Berglund et al., 1982]

sniffs) are plotted against the number of FID peaks for which the eluate in the POGs were found to be "strong" in perceived odour intensity. At least 10 (out of 28) relatively strong component odours within the air samples were needed for the indoor-air sample to have a noticeable odour above that of a clean air sample (charcoal filtered air). Above 10 strong component odours, it was shown that the odour strength of the whole air sample increased linearly with the number of "strong-smelling" components in the air sample. A strong component here means a component that in the POGs from all samples judged by any subject had received the 28 highest scale values of odour intensity. Although the component odours of indoor air are rather faint, the odour intensity of broadband mixtures increases distinctly with an increase in the number of components. It also seems that the various kinds of specific low-concentration odour qualities do not each contribute much to the overall odour quality but rather that new and more defuse and faint odour quality emerges. This phenomenon is compatible with vector summation of multiple components in broadband mixtures, i.e. when the large number of components far surpasses the number of just noticeable differences in the range of odour intensities.

5. Summary conclusions

Trygg Engen was a sharp thinker. Combined with more chaotic thinkers he generously provided structures which gave stability and endurance to chaotic insights. This essay tried to answer the question: Is there a measurement system for odour quality? We conclude:

1. Odour quality is a homogeneous perception (cf. colours) whether it originates from single odorous compounds or mixtures of a few or a vast number of chemical compounds. At equal odour intensities (not equal concentrations), odour qualities of up to four plus or minus one mixed odorants are similar to the odour qualities of their components when perceived alone. The odour quality of mixtures is intermediary of the odour qualities of their components.
2. Conclusions from large-database research (e.g. the Aldrich Flavour lists stimuli and semantic descriptors) are that the odour space must be complex and highly dimensional. If primary odour qualities exist at all, they must be large in number.
3. Odour qualities may be measured (or classified) (a) by sorting according to odour-quality resemblance (Miller sorting), (b) by assigning a multitude of semantic descriptors (and their quantities) to constitute a qualitative (or

quantitative) profile, and (c) by measuring odour-quality similarities or dissimilarities among pairs of odorants. The mapping of odour qualities into an odour space is either according to a content model (shared but homogeneous odour qualities; like colours) or a distance model (interdistances mapped in n-dimensional space). Thus far, researchers seldom use similar sets of odorants and subjects often disagree on odour-quality similaritities/dissimilarities or odour-quality classes. It is therefore common that the odour spaces mapped cannot be compared or validated.

4. It is common to measure pleasantness and/or unpleasantness of odour qualities. As descriptors these two attributes are applicable to all kinds of descriptors (nouns and adjectives) of odour qualities. Moreover, the similarities or dissimilarities among the "meanings" of a set of descriptors (e.g. as in a descriptor profile) will form a multidimensional space of their own. This semantic space will be superimposed on the odour-quality space it is intended to research. The problem with pleasantness and unpleasantness is that all verbal descriptors and all odour qualities can be classified as either pleasant or unpleasant. This does not mean that the world of odour qualities is unidimensional, only that the odour descriptors (or odour qualities) can all be scaled on one continuum. It would also be possible to scale both sets with regard to odour intensity.
5. The perception of indoor air quality is dominated by odour quality. For broadband low-concentration mixtures, like indoor air, at least 10 (out of 28) relatively strong odour components are needed for the whole indoor air sample to have a noticeable odour above that of a charcoal filtered air sample of indoor air from the same building. Above this threshold the perceived odour intensity of the whole air sample increases linearly with its number of "strong-smelling" components.
6. Although a huge number of odours can be discriminated, the quality of odour mixtures will still be intermediary between the odour qualities of their components. This seems to be true for single chemical compounds or mixtures thereof and even for very broadband mixtures like indoor air. Our perceptual world created by olfaction simply consists of a massive amount of odour qualities potentially attributable to sources in the environment.

Acknowledgments

Research presented in this work was supported by the Swedish Council for the Environment, Agricultural Sciences and Spatial Planning (FORMAS).

References

Amoore, J. I. (1971). *Molecular basis of odor*. Springfield, Ill: Charles C. Thomas.

Baird, J. C., Berglund, B., & Shams Esfandabad, H. (1994). Longitudinal assessment of sensory reactions in eyes and upper airways of staff in a sick building. *Environment International, 20*, 141–160.

Bentley, J. P. (2005). *Principles of measurement systems*. Harlow, UK: Pearson Education.

Berglund, B. (1974). Quantitative and qualitative analysis of industrial odors with human observers. *Annals of the New York Academy Sciences, 237*, 35–51.

Berglund, B. (1991). Quality assurance in environmental psychophysics. In S. J. Bolanowski, & G. A. Gescheider (Eds.), *Ratio scaling of psychological magnitudes – In honor of the memory of S. S. Stevens* (pp. 140–162). Hillsdale, NJ: Erlbaum.

Berglund, B., Berglund, U., Ekman, G., & Engen, T. (1971). Individual psychophysical functions for 28 odorants. *Perception & Psychophysics, 9*, 379–384.

Berglund, B., Berglund, U., Engen, T., & Ekman, G. (1973a). Multidimensional analysis of twenty-one odors. *Scandinavian Journal of Psychology, 14*, 131–137.

Berglund, B., Berglund, U., Johansson, I., & Lindvall, T. (1986). Research equipment for sensory air quality studies of nonindustrial environments. *Environment International, 12*, 189–194.

Berglund, B., Berglund, U., & Lindvall, T. (1976). Psychological processing of odor mixtures. *Psychological Review, 83*, 432–441.

Berglund, B., Berglund, U., Lindvall, T., & Nicander-Bredberg, H. (1982). Olfactory and chemical characterization of indoor air. Towards a psychophysical model of indoor air quality. *Environment International, 8*, 327–332.

Berglund, B., Berglund, U., Lindvall, T., & Svensson, L. T. (1973b). A quantitative principle of perceived intensity summation in odor mixtures. *Journal of Experimental Psychology, 100*, 29–38.

Berglund, B., & Engen, T. (1993). A comparison of self-adaptation and cross-adaptation to odorants presented singly and in mixtures. *Perception, 22*, 103–111.

Berglund, B., Höglund, A., & Shams Esfandabad, H. (in press). A bisensory method for odor and irritation detection of formaldehyde and pyridine. *Chemosensory Perception*.

Berglund, B., & Olsson, M. J. (1993a). Perceptual and psychophysical models for odor intensity interaction. In J. A. Garriga-Trillo, P. R. Minon, C. Garcia-Gallego, P. Lubin, J. M. Merino, & A. Villarino (Eds.), *Fechner Day '93* (pp. 35–40). Palma de Mallorca, Spain: International Society of Psychophysics.

Berglund, B., & Olsson, M. J. (1993b). Odor-intensity interaction in binary mixtures. *Journal of Experimental Psychology: Human Perception & Performance, 19*, 302–314.

Berglund, B., Rossi, G. B., & Wallard, A. (2012). Measurement across physical and behavioural sciences. In B. Berglund, G. B. Rossi, J. T. Townsend, & L. R. Pendrill (Eds.), *Measurement with persons. Theory, methods and implementation areas* (pp. 1–25). NY: Taylor & Francis.

Boring, E. G. (1957). *A history of experimental psychology*, 2nd edition. NY: Appleton-Century-Crofts.

Brodin, M., Laska, M., & Olsson, M. J. (2009). Odor interaction between bourgeonal and its antagonist undecanal. *Chemical Senses, 34*, 625–630.

Buck, L. (2004). The search for odorant receptors. *Cell, 116*, 117–119.

Buck, L., & Axel, R. (1991). A novel multigene family may encode odorant receptors: A molecular basis for odor recognition. *Cell, 65*, 175–187.

Cain, W. S., & Drexler, M. (1974). Scope and evaluation of odor counteraction and masking. *Annals of the New York Academy of Sciences, 237*, 417–439.

Cairncross, S. E., & Sjostrom, L. B. (1950). Flavor profiles–a new approach to flavor problems. *Food Technology, 4*, 3–8–11.

Chrea, C., Grandjean, D., Delplanque, S., Cayeux, I., Le Calvé, B., Aymard, L., Velazco, M. I., Sander, D., & Scherer, K. R. (2009). Mapping the semantic space for the subjective experience of emotional responses to odors. *Chemical Senses, 34*, 49–62.

Chastrette, M., Elmouaffek, A., & Sauvegrain, P. (1988). A multidimensional statistical study of similarities between 74 notes used in perfumery. *Chemical Senses, 13*, 295–305.

Davis, R. G. (1979). Olfactory perceptual space models compared by quantitative methods. *Chemical Senses & Flavour, 4*, 21–33.

Dravnieks, A. (1982). Odor quality: Semantically generated multidimensional profiles are stable. *Science, 218*, 799–801.

Dravnieks, A., Bock, F. C., Powers, J. J., Tibbetts, M., & Ford, M. (1978). Comparison of odors directly and through profiling. *Chemical Senses & Flavour, 3*, 191–225.

Ekman, G. (1970). Comparative studies on multidimensional scaling and related techniques. Reports from the Psychological Laboratories, The University of Stockholm, Suppl. 3, pp. 1–28.

Ekman, G., Engen, E., Künnapas, T., & Lindman, R. (1964). A quantitative principle of qualitative similarity. *Journal of Experimental Psychology, 68*, 530–536.

Engen, T. (1982). *The perception of odors*. NY: Academic Press.

Engen, T. (1991). *Odor sensation and memory*. NY: Praeger.

Engen, T., & McBurney, D. H. (1964). Magnitude and category scales of the pleasantness of odors. *Journal of Experimental Psychology, 68*, 435–440.

Ericsson, K. A., & Lehmann, A. C. (1996). Expert and exceptional performance: Evidence of maximal adaptation to task constraints. *Annual Review of Psychology, 47*, 273–305.

Galanter, E., Hossain, K., Berglund, B., Bourgine, P., & Morrone, M. C. (2010). *Measuring the Impossible*. Stockholm University, Sweden: Report of a MINET High-Level Expert Group under EC-FP6-NEST-MtI-Project-043297.

Garner, W. R. (1974). *The processing of information and structure*. Potomac, Md: Lawrence Erlbaum Associates.

Greenacre, M. J. (1984). *Theory and applications of correspondence analysis*. London, UK: Academic Press.

Gregson, R. A. M. (1972). Odour similarities and their multidimensional metric representation. *Multivariate Behavioural Research, 7*, 165–174.

Gregson, R. A. M., & Mitchell, M. J. (1974). Odor quality similarity scaling and odor word profile matching. *Chemical Senses & Flavour, 1*, 95–101.

Haddard, R., Lapid, H., Harel, D., & Sobel, N. (2008). Measuring smells. *Current Opinion in Neurobiology, 18*, 438–444.

Hård, A., & Sivik, L. (1981). NCS–Natural Color System: A Swedish Standard for color notation. *Color Research & Application, 6*, 129–138.

Harper, R., Bate Smith, E. C., & Land, D. G. (1968). *Odour description and odour classification*. New York: American Elsevier.

Herz, R. S., & von Clef, J. (2001). The influence of verbal labeling on the perception of odors: Evidence for olfactory illusions? *Perception, 30*, 381–391.

Hudson, R. (2000). Odor and odorant: A terminological clarification. *Chemical Senses, 25*, 693.

Hunt, R. W. G. (1989). *Measuring colour,* 2nd edition. NY: Wiley.

Jaubert, J. N., Tapiero, C., & Dore, J. C. (1995). The field of odors: Towards a universal language for odor relationships. *Perfumer & Flavorist, 20*, 157–170.

Jinks, A., & Laing, D. R. (2001). The analysis of odor mixtures by humans: Evidence for a configurational process. *Physiology & Behaviour, 72*, 51–63.

Jöreskog, K. G., & Sörbom, D. (1978). *LISREL IV: Analysis of linear structural relationships by the method of maximum likelihood.* Chicago, IL: International Educational Services.

Kahn, R. M., Luk, C.-H., Flinker, A., Aggrawal, A., Lapid, H., Haddad, R., & Sobel, N. (2007). Predicting odor pleasantness from odorant structure: Pleasantness as a reflection of the physical world. *Journal of Neuroscience, 27*, 10015–10023.

Laffort, P. (2006). Analysing and predicting synergy in sweetener blends. In W. J. Spillane (Ed.), *Optimising sweet taste in foods* (pp. 349–374). Washington: CRC Press.

Laing, D. G., Panhuber, H., Willcox, M. E., & Pittman, E. A. (1984). Quality and intensity of binary odor mixtures. *Physiology & Behavior, 33*, 309–329.

Laska, M., & Hudson, R. (1993). Discriminating parts from the whole: Determinants of odor mixture perception in squirrel monkeys, *Saimiri sciureus. Journal of Comparative Physiology A, 173*, 249–256.

Lawless, H. T. (1989). Exploration of fragrance categories and ambiguous odors using multidimensional scaling and cluster analysis. *Chemical Senses, 14*, 349–360.

Lindqvist, A., Höglund, A., & Berglund, B. (2009). The odor perception of key degradation products of polyamide 6.6. In M. A. Elliott, S. Antonijevic, S. Berthaud, P. Mulcahy, C. Martyn, B. Bargery, & H. Schmidt (Eds.), *Fechner Day 2009* (pp. 335–340). Galway: International Society for Psychophysics.

Livermore, A., & Laing, D. G. (1998a). The influence of chemical complexity and the perception of multicomponent odor mixtures. *Perception & Psychophysics, 60*, 650–661.

Livermore, A., & Laing, D. G. (1998b). The influence of odor type on the discrimination and identification of odorants in multicomponent odor mixtures. *Physiology & Behaviour, 65*, 311–320.

Lord, F. M., & Novick, M. R. (1968). *Statistical theories of mental test scores.* Reading, MA: Addison-Wesley.

MacRae, A. W., Rawcliffe, T., Howgate, P., & Geelhoed, E. N. (1992). Patterns of odour similarity among carbonyls and their mixtures. *Chemical Senses, 17*, 119–125.

Madany Mamlouk, A., Chess-Ruiter, C., Hofman, U. G., & Bower, J. M. (2003). Quantifying olfactory perception: Mapping olfactory perception space by using multidimensional scaling and self-organizing maps. *Neurocomputing, 52–54*, 591–597.

Miller, G. R. (1991). *The science of words.* NY: Scientific American Library.

Moskowitz, H. R. (1976). Multidimensional scaling of odorants and mixtures. *Lebensmittel, Wissenschaft und Technologie, 9*, 232–238. (now: *LWT–Food Science and Technology*).

Moskowitz, H. R., & Barbe, C. D. (1977). Profiling of odor components and their mixtures. *Sensory Processes, 1*, 212–226.

Moskowitz, H. R., & Gerbers, C. L. (1974). Dimensional salience of odors. *Annals of the New York Academy of Sciences, 237*, 3–16.

Noma, E., Berglund, B., Berglund, U., Johansson, I., & Baird, J. C. (1988). Joint spatial representation of chemicals and locations in a healthy and sick preschool. *Atmospheric Environment, 22,* 451–460.
Nosofsky, R. M. (1986). Attention, similarity, and the identification-categorization relationship. *Journal of Experimental Psychology: General, 115,* 39–57.
Patte, P., & Laffort, P. (1979). An alternative model of olfactory quantitative interaction in binary mixtures. *Chemical Senses & Flavour, 4,* 267–274.
Schiffman, S. S. (1974). Contributions to the psychochemical dimensions of odor: A psychophysical approach. *Annals of the New York Academy of Sciences, 237,* 164–183.
Schiffman, S. S. (1981). Characterization of odor quality utilizing multidimensional scaling techniques. In H. R. Moskowitz, & C. B. Warren (Eds.), *Odor quality and chemical structure* (pp. 1–19). Washington: American Chemical Society.
Schiffman, S. S., Reynolds, M. L., & Young, F. (1981). *Introduction to multidimensional scaling: Theory, methods and applications.* NY: Academic Press.
Shepard, R. N. (1991). Integrality versus separability of stimulus dimensions: From an early convergence of evidence to a proposed theoretical basis. In G. R. Lockhead, & J. R. Pomerantz (Eds.), *The perception of structure* (pp. 53–71). Washington: American Psychological Association.
Stevens, D. A., & O'Connell, R. J. (1996). Semantic-free scaling of odor quality. *Physiology & Behaviour, 60,* 211–215.
Wiltrout, C., Dogra, S., & Linster, C. (2003). Configurational and nonconfigurational interactions between odorants in binary mixtures. *Behavioral Neuroscience, 117,* 236–245.
Woskow, M. H. (1968). Multidimensional scaling of odors. In N. Tanyloc (Ed.), *Theories of odors and odor measurement* (pp. 147–188). Istanbul: Robert College.
Wright, R. H. (1977). Odor and molecular vibration: Neural coding of olfactory information. *Journal of Theoretical Biology, 64,* 473–502.
Wright, R. H., & Michels, K. M. (1964). Evaluation of infrared relations to odor by standards similarity method. *Annals of the New York Academy of Sciences, 116,* 535–551.
Yeshurun, Y., & Sobel, N. (2010). An odor is not worth a thousand words: From multidimensional odors to unidimensional odor objects. *Annual Review of Psychology, 61,* 219–241.
Yoshida, M. (1964). Studies of psychometric classification of odors (4 and 5). *Japanese Psychological Research, 6,* 115–124 and 145–154.
Yoshida, M. (1975). Psychometric classification of odors. *Chemical Senses & Flavour, 1,* 443–464.
Yoshida, M. (1976). Individual differences among perceived similarity of odors. *Bulletin of the Faculty of Science & Engineering, 19,* 221–232.
Zarzo, M., & Stanton, D. T. (2006). Identification of latent variables in a semantic odor profile database using principal component analysis. *Chemical Senses, 31,* 713–724.
Zheng, L. (2010). Intensity of odor and sensory irritation as a function of hexanal concentration and interpresentation intervals: An exploratory study. *Perceptual & Motor Skills, 111,* 210–228.

CHAPTER 2

There's something in the air

Effects of beliefs and expectations on response to environmental odors

Pamela Dalton
Monell Chemical Senses Center, Philadelphia, PA, USA

Although often underrated in importance, the human sense of smell has both evolutionary and contemporary significance: our ability to perceive odors provides information that allows us to evaluate and guide our responses to our environments. For example, the rich aromas wafting from a bakery encourage us to linger and savor them, whereas the smell of rotting garbage impels us to move away. No account of the role of smell in our lives would be complete, however, without an acknowledgment that people vary in their perception of and response to ambient odors in their environment. Some of this variability can be attributed to inter-individual variation in sensitivity. However, an important source of differences in response to ambient, environmental odors may also stem from cognitive factors, such as an individual's beliefs and expectations about the consequences of exposure to an odor.

This essay will provide an overview of the multiple factors-including expectations, suggestion and physiological reactions- that influence individual and community responses to environmental odors. Recognizing the interaction between the sensory and the psychological responses to odors is of significant value when trying to remediate community reactions to odors from these sources. The perception of health risks from short- or long-term exposures to environmental odors is of escalating concern to the general public and research indicates that such concerns can amplify the vigilance and attention paid to even low-level, neutral, background odors. Our research also suggests that interventions that reduce ambient pollutants, but which do not eliminate all odors may not remediate the concerns and anxieties of occupants. The historical basis for such beliefs and their consequences will provide a framework for discussing the contemporary responses to environmental odors.

1. Odors and well-being: Historical and current perspectives

Odors have important signal value. Unfamiliar or unpleasant odors can both arouse and alarm community-dwellers and building occupants. The mere presence of an odor can increase symptom reporting (Alexander & Fedoruk, 1986; Neutra et al., 1991; Roht et al., 1985; Smith et al., 1978; Stahl & Lebedun, 1996) and is often the most significant correlate of perceived health risk for individuals whose neighborhoods have been sprayed with pesticide (Ames et al., 1993; Neutra et al., 1991; McClelland et al., 1990) or near factories thought to be the source of pollution (Cavalini et al., 1991; Taylor et al., 1997).

A belief that ambient odors can influence health for better or for worse has a long history: Prior to the discovery of germ theory, unpleasant odors were deemed to be carriers of disease while good odors were viewed as potentially curative (Levine & McBurney, 1986; LeGuerer, 1994). In Paris of 1880, the shutdown of an offending waste treatment plant in Nanterre caused the company to dispose of untreated waste throughout Paris, thus resulting in the populace complaining of an asphyxiating stench that permeated the city day and night (Barnes, 2006). Due to the prevailing 19th century belief, which could be traced back to Hippocrates, that bad odors , or more generally 'bad air' could actually cause disease (the 'miasma' doctrine), it is therefore not surprising that medical and scientific authorities and lay observers attributed any and all illnesses during this period to the constant presence of the offensive odors. To be sure, neighborhoods with conditions which would lead to elevated disease rates (unsanitary water supplies or poor sanitation) were also more likely to suffer from poor air quality, thus confirming the naïve theory through a spurious association.

Although the rapid dispersion of knowledge of germ theory dispelled much of the appeal of 'miasmatism', this belief can still be manifest in concerns about becoming sick from exposures to environmental odors (Dietert & Hedge, 1996; Cain & Cometto-Muñiz, 1993). Currently, aromatherapy is the most prominent example of the belief that certain fragrances can have beneficial effects on health, mood and mental well-being (e.g., Stoddard, 1990; Van Toller & Dodd, 1988). At the same time, however, there exists a growing set of beliefs about health risks associated with exposure to airborne odors and pollutants, even those present in fragranced products (e.g., Baldwin et al., 1999; Caress & Steinemann, 2009). Interestingly, these beliefs may operate in the absence of conscious awareness. Using an implicit association task, a recent study has shown that individuals are more likely to associate the concept of odor with illness rather than health (Bulsing et al., 2009).

Before leaving this topic, it should be acknowledged that there can be direct effects of exposure to odorous volatile chemicals. There is the possibility that volatile chemicals can exert pharmacological effects via uptake into the circulatory system and dispersal to target organs throughout the body. Indeed, animal studies have shown that inhalation of an odorant can lead to detectable levels of that compound in the bloodstream (Jirovitz et al., 1992). However, in most instances, the blood concentrations that would result represent relatively trivial doses that would not be expected to cause adverse effects. A more likely mechanism for direct adverse effects is the potential for a volatile chemical to elicit irritation or chemesthetic sensations. In fact, almost all volatile chemicals produce not only an odor, but at higher concentrations, can elicit eye, nose and throat irritation, through stimulation of one or more sensory nerves in the upper airways. The resulting perception of pungency or sensory irritation is often a primary determinant in people's assessment of the quality and acceptability of indoor air (Cain, 1987; Boxer, 1990). Adverse reactions to odors are not always based on irritancy, however. An odorant (or even multiple odorants) can be present at concentrations below those known to elicit sensory irritation, yet nonetheless provoke adverse responses (Dalton et al., 1997; Wysocki et al., 1997). Moreover, perception of irritancy can depend on context: some odorants (e.g., menthol, wintergreen, eucalyptus) elicit a pungent sensation, yet produce positive responses (Dalton, 1997). Here, an individual's experience may play a key role. Studies from our laboratory have suggested that odors are reported as irritating or annoying when individuals do not recognize them and attribute erroneous effects or consequences to exposure (Dalton, 1996; Dalton et al., 1997).

2. Expectation effects on odor perception and symptoms

A far back as 1898, Slosson (1899) showed that the mere suggestion that an odor would be emitted was sufficient to elicit reports of odor perception. He told his lecture class that he wished to see how rapidly an odor would diffuse through the air and requested the students raise their hands when they smelled it. After pouring distilled water over cotton and claiming that it was a chemical with a strong, novel odor, he found that within a minute the majority of the class had raised their hands. Not surprisingly, some of those seated near the front claimed to be sickened by the odor. Although Slosson described this as an 'hallucination', it is more likely that the instructions primed the audience through expectation to pay attention to pre-existing odors in the ambient environment and misattribute them to the 'target' odor.

In a more sophisticated demonstration of this expectation effect, O'Mahoney (1978) also found that individuals would report an odor based on simple expectation. Near the end of a televised program on the chemical senses, viewers were played a tone which they were told would cause them to smell a pleasant country smell. Not only did a number of viewers call to report smelling odors ranging from hay or grass to manure, but some subjects gave accompanying reports of symptoms, such as hay fever attacks and dizziness.

Trygg Engen was an early pioneer investigating the degree to which expectation could bias odor perception. He showed that non-olfactory cues, such as color and the relative probability of an odor vs. no odor, could influence a subject's decision to say they smelled an odor even when none was presented (Engen, 1972). Colored solutions were more likely to be judged as having an odor than clear ones, even when they did not contain odor. He interpreted these findings as a criterion shift induced by the extra-modality information; however given current understanding about the degree to which olfactory perception is involved in multi-modal processing, it may also have been possible that the intensity of the solvent odor was 'enhanced' by the concurrent visual stimulation (Seo et al., 2010).

3. Top-down & bottom-up processing of odor information

Perception is in part constructive, where the representation is constructed from information garnered from the stimulus itself coupled with information present in memory. William James described this as a general law of perception in The Principles of Psychology (James, 1890), stating "... whilst part of what we perceive comes through our senses from the object before us, another part (and it may be the larger part) always comes out of our own head".

In his seminal book, *Cognitive Psychology*, Ulrich Neisser (1967) reviewed the available evidence that perception was both driven by features of the stimulus (data-driven) as well as driven by features of the observer (conceptually-driven) (Neisser, 1967). Using terms from the nascent field of computer programming, he described the flow of information processing in two directions: when a perceptual interpretation emerged only from the features of the stimulus this was called bottom-up processing; when the knowledge, expectations or past experience of the perceiver influenced perception, this was called 'top-down' processing.

Top-down processing is most useful when features of the stimulus are ambiguous and do not lend themselves to ready interpretation. Although stimulus ambiguity can be found for all manner of sensory stimuli, it is particularly true for odor – where the features that define one odor from another perceptually similar one are difficult to articulate and in some cases, even recognize. Thus, the interpretation we give to the sensory pattern that is generated by an odor is likely to be greatly influenced by the context we are in, our knowledge of the source and/or past experience. In everyday life, this holds significant implications for our response to environmental odors, particularly reactions to novel or unexpected ones.

4. Top-down mechanisms underlying adverse response to environmental odors: mental models, chemical exposures and health symptoms

Cognitive psychology has described the ways in which humans employ 'mental models' – generalized representations formed in memory to guide their reactions and expectations toward all manner of situations and environments (Leventhal et al., 1980; Clark, 1993). There are two sorts of expectancies that derive from mental models: those that are compared with incoming sensory input to organize and interpret and those that structure the output of the individual's response. Traditionally, mental model research has focused on simple physical or mathematical systems for which there exist explicit normative models, such as the naïve physics of liquids, see Norman (1980) for more examples. However, mental models of odorous chemical exposure may encompass multiple disciplines (chemistry, anatomy, physiology, psychology, toxicology) in an attempt to explain the physical principles by which chemicals are transported from their source, into the body, the biological and toxicological principles that may determine their ultimate fate (e.g., whether exhaled, detoxified or stored) and their health significance or psychological impact. These models can guide the interpretation of everyday odor experiences (Reiser et al., 1985) and provide focus to the physical signs a person monitors and ultimately perceives (Leventhal et al., 1980). Shusterman (Shusterman et al., 1991) demonstrated this in a field study where he surveyed a group of individuals living near hazardous waste sites about the frequency with which they perceived odors and symptoms. He found that symptom rates were significantly higher among those individuals who were 'worried' about the source of the odors. These results suggested that people could be cued to monitor and report symptoms that were primed by their cognitive or emotional state.

To be sure, odorous chemicals from cleaning products, solvents, gasoline and factory emissions are capable of causing adverse health effects at high airborne concentrations. However, health symptoms are frequently reported following exposures to very low level concentrations, when such effects cannot be accounted for by any known toxicological mechanism. In such cases, health symptoms, or the perception thereof, are *indirectly* related to odor exposure, most likely through activation of memory representations concerning previous exposures an individual may have experienced, or beliefs a person may have about the nature of the odor and its effects on health. One aspect of these 'mental models' are response expectancies: anticipations of automatic reactions to particular situational cues.

For example, people who drink coffee, but are unaware that it is decaffeinated, report feelings of enhanced alertness while people who are aware of its decaffeinated nature do not (Kirsch & Weixel, 1988). A sizeable body of research has shown that asthmatics who believe they are being exposed to a broncho-constrictive agent will respond with decreases in pulmonary function, while those same individuals who believe the agent is inert will not. Research in our laboratory has shown that even among healthy individuals, expectancies can play a large role in the chemosensory response to odor and irritation from volatile chemicals (Dalton, 1996; Dalton, 1999).

To explore one aspect of naïve models of odor exposures, we investigated people's understanding of acute health outcomes from exposure to odorous chemicals. We conducted two studies in which we presented subjects with various odorant/chemical exposure scenarios and asked them to try and imagine both the situation and the odor as best as they could (sometimes with visual aids) (Smeets & Dalton, 2005). They were then asked to rate how intensely (if at all) they would expect to experience a variety of different health symptoms as a consequence of the odor exposure. The list consisted of 30–40 acute symptoms, of which only some were actual effects attributable to an odor/irritant exposure and others were sham or control symptoms.

In the first study, 277 high school students (aged 15–17) provided symptom ratings to 6 exposure scenarios (e.g., a petroleum refinery, an open sewer, a chemical truck spill) using a self-report questionnaire. In the second study, 18 adult respondents (elementary school teachers) participated in personal, standardized interviews in which they were presented with a refinery exposure scenario with which they were all very familiar (the school was located within a mile of two refineries). Symptom ratings were subjected to principal components analysis with Varimax rotation to examine which symptom clusters would emerge. Although both the respondents as well as the data collection procedures were quite different, the results revealed a similar pattern for each sample, with two symptom components emerging. The first component was characterized by diffuse symptoms of psychological and physical malaise. Symptoms endorsed on this component were joint pains and swelling, feeling weak, back pains, numbness or tingling, confusion and memory problems (see Table 1). The second component was composed of focused symptoms related to sensory irritation (e.g. of nose and throat); respiratory symptoms (e.g. chest wheezing and sneezing); gastro-intestinal symptoms (e.g. nausea and vomiting); and central nervous system (CNS) symptoms (e.g. headache and dizziness).

In the student sample the eigenvalue for the first component was 17.1, explaining 34% of total variance; and 12.7 for the second component, explaining an additional 25% of total variance. In the teacher sample, the eigenvalue of the

Table 1. Component loading scores of two components obtained using Principle Component Analysis with Varimax Rotations (Kaiser Normalization) for standardized personal interviews in 18 elementary school teachers in Marcus Hook, PA, and for symptom survey in 277 high school students in Camden, NJ, and Drexel Hill, PA. Both samples evaluated health symptoms from a Refinery Exposure Scenario.

Symptom	Components Interview teachers		Components High school survey	
	1	2	1	2
Soreness muscles	0.89		0.76	0.30
Feeling tired	0.85		0.64	0.36
Trouble breathing	0.82	0.30	0.28	0.70
Lower back pain	0.82	0.29	0.71	0.34
Feeling confused	0.80	0.43	0.83	0.27
Feeling weak	0.79	0.28	0.79	0.33
Memory loss	0.78	0.44	0.84	0.23
Pain in chest	0.78	0.42	0.38	0.73
Mind going blank	0.72	0.61	0.77	0.28
Eye irritation		0.83	0.36	0.63
Coughing	0.34	0.80	0.21	0.79
Vomiting		0.78	0.40	0.65
Poor appetite	0.29	0.77	0.40	0.59
Nervousness	0.47	0.76	0.60	0.49
Blocked nose		0.74		0.75
Throat irritation	0.41	0.73	0.32	0.75
Reduced hearing	0.45	0.73	0.76	0.33
Joint pains	0.57	0.59	0.78	0.33

Note: Symptoms that were not common to both studies have not been included. These refer to symptoms used in the HS Survey that were later eliminated due to overlapping with other symptoms. Symptoms with factor loadings lower than 0.75 for any component or sample have been removed to conserve space. Component scores lower than 0.20 are not shown.

first component was 12.3, explaining 36% of total variance; and 10.0 for the second component, explaining an additional 29% of total variance. We also found evidence that symptom complaints were related to respondent characteristics. Multiple regression of symptom scores on attitudes and other respondent characteristics, revealed two types of respondents. The first was trusting of authority and the medical profession and had little desire for autonomy or to engage in activities to seek information about environmental exposure. This type of respondent reported diffuse symptoms, which were more likely to be caused by stress/anxiety than to be related to actual chemical exposure, thereby demonstrating an unelaborated mental model about chemical exposure. The second type of respondent had more science education and was active in seeking information about environmental exposure. These individuals reported focused symptoms that could be related to actual chemical exposures at some levels, thereby demonstrating a more elaborate 'response expectancy' model of chemical exposure than the first type of respondent.

Importantly, these respondent characteristics appeared to be related to the way they interpreted information about chemical risk. In a separate study, participants were given information about a specific chemical (for example, chlorine) consisting of the concentration (in ppm) at or below which no symptoms/effects would occur (the No Adverse Effect Limit: NOAEL), and the concentration at or above which serious health effects were to be expected. They were then asked to indicate their expectations of the frequency and degree of health effects across this range. We found that respondents who did not trust authority reported more odor annoyance and tended to be far more liberal in their interpretation and assignment of health effects to chemical concentration levels. In contrast, those more trusting in authority and less bothered by airborne chemical exposures tended to be far more conservative when interpreting the relationship between chemical concentration and risk.

5. Odor exposure, autonomic arousal and conditioned responses

Conditioned or learned associations can also play a role in the perceptions of odor and the health symptoms induced by them. In humans, aversive conditioning appears to occur to a broad range of odorous compounds including solvents, aldehydes, acid vapors, and phosphine gas (Tabershaw & Cooper, 1966; Schottenfeld & Cullen, 1986; Shusterman et al., 1988; Gyntelberg et al., 1986; Dager et al., 1987; Bolla-Wilson et al., 1988; Shusterman & Dager, 1991). For example, an odor that may have previously been temporally associated with a maladaptive physiological response can on subsequent presentations, elicit the response by itself. In a series of experiments by Van den Bergh et al. (Van den Bergh et al., 1995; Van den Bergh et al., 1997), psychosomatic patients with hyperventilation complaints were conditioned to respond to certain odors that had previously been presented in combination with 7.4% CO_2 enriched air, which by itself triggered changes in breathing and symptoms mimicking hyperventilation. Increased respiratory frequency and elevated levels of somatic complaints could be conditioned to negative odors (ammonia and butyric acid) but not to a pleasant odor (niaouli). Consistent with the emphasis of the role of cognitive factors in determining symptomatology, Van den Bergh and colleagues (Van den Bergh et al., 1998) found evidence suggesting that the somatic complaints produced on

CS test trials were based on the memory of US-elicited somatic symptoms rather than on the acquired physiological responses to the CS.

It is important to recognize that arousal and emotion may play an important modulating role in the learned effects discussed thus far. Odors have long been thought to function as especially potent retrieval cues for emotionally laden memories (Cann & Ross, 1989; Herz & Engen, 1996) and emotional experiences. Asthmatic individuals who may have responded adversely to one or more odorous agents in the past may, through conditioning mechanisms, generalize their aversive response to many or all odorous stimuli (Siegel & Kreutzer, 1997).

6. Factors associated with adverse response to environmental odors

6.1 Field studies

The relationship between environmental odor, annoyance and health symptoms is multi-factorial and complex. In numerous field studies over the past fifty years, investigators have sought to partition the degree to which an individual's sensory and somatic response to an odor is due to stimulus factors (frequency, duration, hedonic tone) or perceiver factors (attitudes, experience, personality), often with limited success. In addition, whether health symptoms in response to an odor (headache, nausea, loss of appetite) are directly triggered or mediated by the degree of annoyance an individual experiences has been a focus of concern, as regulators and government agencies frequently struggle with understanding the true health impact of environmental odors.

In general, an extended series of field studies by Winneke et al. (Winneke & Neuf, 1992; Steinheider & Winneke, 1993; Steinheider et al., 1998; Sucker et al., 2001) have shown that community response to environmental odor is largely determined by its' perceived intensity and hedonic tone (pleasant-unpleasant). However, both of these judgments are to some degree, subjective, and can be themselves influenced by the past experience and/or attitude of the observer. Individuals with occupational exposure to odors frequently report less annoyance or symptoms to that odor than do naïve observers (Dalton et al., 1997a). This could be due to the reduction in perceived intensity or unpleasantness following olfactory adaptation as well as to reduced concerns about the hazardous nature of the chemical itself.

Finally, the tendency to experience adversity from environmental odors may be part of a more general 'sensory hyperreactivity' factor (Winneke & Neuf, 1992): in one survey, individuals who reported being annoyed by airplane noise were also more likely to report annoyance from the smell from a nearby cellulose factory than individuals who did not report noise annoyance (43 vs. 32%) (Cederlöf et al., 1964).

6.2 Laboratory studies

In a series of laboratory studies, we have explored factors that have been shown to influence perception and response to environmental odors in field surveys. Specifically, we have examined the influence of experience, familiarity, expectations and personality. The basic procedure used in this research involves measuring people's responses to an odorant before, during, and a whole-body exposure to the odorant under controlled conditions in an exposure chamber. Numerous measures of an individual's response to the odor are obtained, including ratings of perceived odor and irritation intensity during exposure, and reports of subjective health symptoms following exposure. The principal question concerns whether a subject's reaction to odor during a 20 minute exposure is altered by their previous experience or experimentally-induced beliefs about the odorant, or is modulated by personality variables. In some conditions, volunteers have had previous occupational or community exposure to the odor in question, while others are naive. All volunteers receive either a positive, negative or neutral bias about the nature and consequences of exposure to the odorant. In some studies this information is provided prior to the exposure by the experimenter, while in other studies odorant information is conveyed by the behavior/symptoms/verbal reports of a 'sham' or 'confederate' subject (in reality, an actor whose positive or negative responses are scripted).

In all cases, we have found that the information provided to the subject greatly influences their experience during and following exposure to the chemical. Table 2 shows the ratings of symptom reports following exposure to acetone under

Table 2. Selected reported health symptoms in subjects (n = 30 for each group) after 20 min exposure to 800 ppm acetone compared. Adapted from Dalton (1997).

Symptom	Subjects exposed to odorant		
	Positive bias	Negative bias	Neutral bias
Throat irritation	4.36	8.69	8.59
Eye irritation	2.42	4.70	4.63
Nasal irritation	6.05	12.95	14.43
Lightheadedness	5.35	8.53	12.57
Headache	2.37	4.87	5.09
Nausea	1.90	2.60	5.17
Drowsiness	3.04	6.98	5.64

three different biasing conditions (Dalton et al., 1997b). As predicted, symptom reports were significantly higher in the negative bias condition than in the positive one. More surprising, however, was that the symptom reports in the neutral bias condition were as high or higher than the negative condition, suggesting that 'no' useful information about a chemical that had a sharp, pungent quality could be perceived as the most hazardous condition of all. We found similar results when the expectation about the chemical was conveyed by a "confederate" subject. In that study, the 'confederate' (an actor) relayed a series of scripted comments (positive, neutral or negative) intended to bias the true subject's perception of the quality of the odor and any symptoms or sensations the odor produced. Moreover, in both types of studies, individuals who were identified as high on the trait of 'negative affectivity' (Watson & Clark, 1984), a personality trait associated with vigilance to negative emotional states, rated symptoms more intensely.

In a subsequent study, to evaluate the degree to which both the 'bottom-up' features of the stimulus itself (hedonic quality) combined with the 'top-down' influence of the subject (bias condition), we exposed separate groups of volunteers to two odors that were normatively-rated as either 'pleasant' (wintergreen) or 'unpleasant' (butanol). We also compared their responses to a group receiving a third odor that was rated 'neutral', isobornyl acetate (balsam smell). As is shown in Table 3, responses were significantly influenced by the experimenter's characterization; however, the perceptual quality of the odorant interacted significantly with the labeling – overall, symptom responses to butanol were higher than to wintergreen or even isobornyl acetate.

In both types of studies, however, the frequency of spontaneous (Dalton, 1996) and surveyed (Dalton et al., 1997a; Dalton et al., 1997b; Dalton, 1999) symptom reports varied significantly with perceived odor intensity, suggesting that symptom perception is correlated with or triggered by the awareness of an odor.

7. Summary and conclusion

Recognizing the interaction between the sensory and the psychological responses to odors is of significant value when trying to remediate air-quality problems. The perception of health risks from short- or long-term exposures to chemical odors continues to escalate among the general public and the available research indicates that such concerns will likely amplify the vigilance and attention paid to even low-level, neutral, background odors. Our research also suggests that interventions that reduce ambient pollutants, but which do not eliminate all odors may not remediate the concerns and anxieties of occupants. The increased understanding of intuitive models will permit more effective risk communication and

Table 3. Summary of health symptom survey: Mean ratings on the Labeled Magnitude Scale (LMS) of solvent-associated and somatic symptoms for three odorants as a function of bias condition.

Symptom	Isobornyl acetate			Methyl salicylate			Butanol		
	Neutral	Positive	Negative	Neutral	Positive	Negative	Neutral	Positive	Negative
Solvent-associated symptoms									
Throat irritation	8.62	7.97	15.69	5.57	1.32	17.46	12.30	5.47	17.72
Eye irritation	5.58	6.11	8.82	2.88	1.09	6.23	4.15	1.28	5.10
Nose irritation	11.08	6.76	15.94	6.97	2.54	9.22	6.63	8.73	17.40
Dizziness	2.97	4.58	8.74	2.53	2.79	6.85	11.04	4.00	18.18
Headache	1.57	2.16	5.89	1.17	3.51	2.06	5.61	3.17	11.02
Nausea	0.41	1.58	3.61	1.55	0.70	1.89	3.92	1.59	7.90
Drowsiness	4.34	8.84	6.73	6.13	5.29	7.63	9.35	7.34	13.88
M	4.94	5.43	9.35	3.83	2.46	7.33	7.57	4.51	13.03
Somatic (control) symptoms									
Skin irritation	1.13	3.42	4.60	0.40	0.25	3.80	1.22	0.17	3.97
Bad taste	2.05	2.74	10.49	1.70	0.20	5.44	0.91	1.50	10.79
Nasal congestion	0.03	3.56	3.17	6.15	1.75	0.73	3.75	1.67	7.04
Cough	1.70	1.09	3.43	3.09	1.17	3.80	7.34	3.68	10.33
Sneeze	0.00	0.00	0.00	0.24	0.00	0.00	0.37	0.00	1.62
Stomachache	0.44	0.16	2.71	1.00	1.15	0.00	0.02	0.08	5.23
Shortness of breath	0.02	4.55	2.73	0.13	2.06	1.05	1.54	1.54	9.64
Heart palpitations	0.00	0.82	2.63	0.07	1.01	0.00	0.02	0.05	1.87
Numbness or tingling	0.00	0.21	0.49	1.64	0.23	0.27	0.95	0.00	1.33
Ear ringing	0.00	0.02	0.09	0.09	0.00	0.00	0.29	2.57	1.55
Leg cramps	0.02	0.25	1.69	0.00	0.58	0.93	0.03	0.00	1.96
Back pain	0.04	1.17	3.32	0.99	0.00	0.02	0.00	0.12	1.40
Sweating	0.59	0.00	0.65	0.24	0.20	0.00	1.22	0.08	0.78
Itching	1.48	2.27	3.23	1.08	0.25	2.43	1.15	0.37	4.15
Current irritation	2.30	3.51	8.90	1.98	1.40	6.92	3.70	1.01	9.29
M	0.65	1.58	3.21	1.25	0.68	1.69	1.50	0.86	4.73

Note: Category boundaries of the LMS are as follows: 0 = no sensation; 1.37 = barely detectable; 5.46 = weak; 15.75 = moderate; 33.57 = strong; 50.47 = very strong; and 90.45 = strongest imaginable.

amelioration of stress particularly among sensitive subpopulations impacted by odorous chemical exposures.

Odor perception can be remarkably susceptible to cognitive and contextual influence, particularly so because an odor percept divorced from source cues is often highly ambiguous. For this reason, individual factors such as expectations and beliefs regarding the nature and consequences of exposure to an odor can override veridical perception. Educational materials and effective communication regarding the relationship between odors, irritation, perceived toxicity and actual risk may be of greater value when occupants are trying to determine whether an ambiguous or unfamiliar odor in their environment poses a short or long-term hazard to their health. Moreover, continued efforts in the scientific community to conduct research that explores the relationship between odors, perceived health-risk and air-quality complaints can be of significant benefit to scientists, physicians and regulators alike.

Acknowledgments

Thanks to Justine Durmala for assistance with preparation of this manuscript. Many of the studies reported and the preparation of this manuscript were supported by a grant from the NIH-NIDCD R01 DC 03704 to the author.

References

Alexander, R. W., & Fedoruk, M. J. (1986). Epidemic psychogenic illness in a telephone operator's building. *Journal of Occupational Medicine, 28,* 42–45.

Ames, R. G., Howd, R. A., & Doherty, L. (1993). Community exposure to a paraquat drift. *Archives of Environmental Health, 48,* 47–52.

Baldwin, C. M., Bell, I. R., & O'Rourke, M. K. (1999). Odor sensitivity and respiaratory complaint profiles in a community-based sample with asthma, hay fever, and chemical odor intolerance. *Toxicology and Industrial Health, 15,* 403–409.

Barnes, D. S. (2006). Not everything that stinks kills: Odors and germs in the streets of Paris, 1880. In *The Great Stink of Paris and the Nineteenth-Century Struggle against Filth and Germs* (pp. 12–64). Baltimore, MD: Johns Hopkins University Press.

Bolla-Wilson, K., Wilson, R. J., & Bleecker, M. L. (1988). Conditioning of physical symptoms after neurotoxic exposure. *Journal of Occupational Medicine, 30,* 684–686.

Boxer, P. A. (1990). Indoor air quality: A psychosocial perspective. *Journal of Occupational Medicine, 32,* 425–428.

Bulsing, P. J., Smeets, M. A. M., & van den Hout, M. A. (2009). The implicit association between odors and illness. *Chemical Senses, 34,* 111–119.

Cain, W. S. (1987). Indoor air as a source of annoyance. In H. S. Koelega (Ed.), *Environmental Annoyance: Characterization, Measurement and Control* (pp. 189–200). Amsterdam: Elsevier.

Cain, W. S., & Cometto-Muñiz, J. E. (1993). Irritation and odor: Symptoms of indoor air pollution. In J. K. Jaakkola, R. Ilmarinen, & O. Seppänen (Eds.), *Indoor Air '93: Volume 1. Health Effects* (pp. 21–31). Helsinki: Indoor Air '93.

Cann, A., & Ross, D. A. (1989). Olfactory stimuli as context cues in human memory. *American Journal of Psychology, 102,* 91–102.

Caress, S. M. & Steinemann, A. C. (2009). Prevalence of fragrance sensitivity in the American population. *Journal of Environmental Health, 71,* 46–50.

Cavalini, P. M., Koeter-Kemmerling, L. G., & Pulles, M. P. J. (1991). Coping with odour annoyance and odour concentrations: Three field studies. *Journal of Environmental Psychology, 11,* 123–142.

Cederlöf, R., Friberg, L., Jonsson, R., Kaij, L., & Lindvall, T. (1964). Studies of annoyance connected with offensive smell from a sulphate cellulose factory. *Nord Hyg Tidsk, 45,* 39–48.

Clark, L. F. (1993). Social Cognition and Health Psychology. In R. S.Wyer & T. K. Srull (Eds.), *Handbook of Social Cognition* (Volume 2 ed., pp. 239–288). Hillsborough, NJ: Lawrence Erlbaum Associates.

Dager, S. R., Holland, J. C., Cowley, D. S., & Dunner, D. L. (1987). Panic disorder precipitated by exposure to organic solvents in the work place. *American Journal of Psychiatry, 144,* 1056–1058.

Dalton, P. (1996). Odor perception and beliefs about risk. *Chemical Senses, 21,* 447–458.

Dalton, P. (1999). Cognitive influences on health symptoms from acute chemical exposure. *Health Psychology, 18,* 579–590.

Dalton, P., Wysocki, C. J., Brody, M. J., & Lawley, H. J. (1997a). Perceived odor, irritation and health symptoms following short-term exposure to acetone. *American Journal of Industrial Medicine, 31,* 558–569.

Dalton, P., Wysocki, C. J., Brody, M. J., & Lawley, H. J. (1997b). The influence of cognitive bias on the perceived odor, irritation and health symptoms from chemical exposure. *International Archives of Occupational & Environmental Health, 69,* 407–417.

Dietert, R. R., & Hedge, A. (1996). Toxicological considerations in evaluating indoor air quality and human health: Impact of new carpet emissions. *Critical Reviews in Toxicology, 26,* 633–707.

Engen, T. (1972). The effect of expectation on judgments of odor. *Acta Psychologica, 36,* 450–458.

Gyntelberg, F., Vesterhauge, S., Fog, P., Isager, H., & Zillstorff, K. (1986). Acquired intolerance to organic solvents and results of vestibular testing. *American Journal of Industrial Medicine, 9,* 363–370.

Herz, R. S., & Engen, T. (1996). Odor memory: Review and analysis. *Psychonomic Bulletin & Review, 3,* 300–313.

James, W. (1890). *The Principles of Psychology,* 1st edition. New York: Henry Holt.

Jirovitz, J., Buchbauer, G., Jager, W., Woiiieh, A., & Nikiforov, A. (1992). Analysis of fragrance compounds in blood samples of mice by gas chromatography, mass spectrometry, GC/FTIR and GC/AES after inhalation of sandalwood oil. *Biomedical Chromatography, 6,* 133–134.

Kirsch, I., & Weixel, L. (1988). Double-blind versus deceptive administration of a placebo. *Behavioral Neuroscience, 102,* 319–323.

Le Guerer, A. (1994). *Scent.* New York: Kodansha America.

Leventhal, H., Meyer, D., & Nerenz, D. (1980). The commonsense representation of illness danger. In S. Rachman (Ed.), *Medical Psychology* (pp. 7–30). New York: Pergamon.

Levine, J. M., & McBurney, D. H. (1986). The role of olfaction in social perception and behavior. In C. P. Herman, M. P. Zanna, & E. T. Higgins (Eds.), *Physical Appearance, Stigma, and Social Behavior: The Ontario Symposium* (Vol. 3, pp. 179–217). Hillsdale, NJ: Lawrence Erlbaum.

McClelland, G. H., Schulze, W. D., & Hurd, B. (1990). The effect of risk beliefs on property values: A case study of a hazardous waste site. *Risk Analysis, 10,* 485–497.

Neisser, U. (1967). *Cognitive Psychology,* 1st edition. New York: Prentice Hall.

Neutra, R., Lipscomb, J., Satin, K., & Shusterman, D. (1991). Hypotheses to explain the higher symptom rates observed around hazardous waste sites. *Environmental Health Perspectives, 94,* 31–35.

Norman, D. A. (1980). Some Observations on Mental Models. In D. Gentner & A. L. Stevens (Eds.), *Mental Models.* Lawrence Erlbaum Associates.

O'Mahoney, M. (1978). Smell illusions and suggestion: Reports of smells contingent on tones played on radio and television. *Chemical Senses and Flavour, 3,* 183–189.

Reiser, B. J., Black, J. B., & Abelson, R. P. (1985). Knowledge structures in the organization and retrieval of autobiographical memories. *Cognitive Psychology, 17,* 89–137.

Roht, L. H., Vernon, S. W., Pier, F. W., Sullivan, P., & Ree, L. J. (1985). Community exposure to hazardous waste disposal sites: Assessing reporting bias. *American Journal of Epidemiology, 122,* 418–433.

Schottenfeld, R. S., & Cullen, M. R. (1986). Recognition of occupation-induced posttraumatic stress disorders. *Journal of Occupational Medicine, 28,* 365–369.

Seo, H. S., Arshamian, A., Schemmer, K., Scheer, I., Sander, T., Ritter, G. et al. (2010). Cross-modal integration between odors and abstract symbols. *Neuroscience Letters, 478,* 175–178.

Shusterman, D., Balmes, J., & Cone, J. (1988). Behavioral sensitization to irritants/odorants after acute overexposures. *Journal of Occupational Medicine, 30,* 565–567.

Shusterman, D., & Dager, S. R. (1991). Prevention of psychological disability after occupational respiratory exposures. *Occupational Medicine, 6,* 11–27.

Shusterman, D., Lipscomb, J., Neutra, R., & Satin, K. (1991). Symptom prevalence and odor-worry interaction near hazardous waste sites. *Environmental Health Perspectives, 94,* 25–30.

Siegel, S., & Kreutzer, R. (1997). Pavlovian conditioning and multiple chemical sensitivity. *Environmental Health Perspectives, 105,* 521–526.

Slosson, E. E. (1899). A lecture experiment in hallucinations. *Psychological Review, 6,* 407–408.

Smeets, M. A., & Dalton, P. H. (2005). Evaluating the human response to chemicals: Odor, irritation and non-sensory factors. *Environmental Toxicology and Pharmacology, 19,* 581–588.

Smith, M. J., Colligan, M. J., & Hurrell, J. J. (1978). Three incidents of industrial mass psychogenic illness. *Journal of Occupational Medicine, 20,* 399–400.

Stahl, S. M., & Lebedun, M. (1996). Mystery gas: An analysis of mass hysteria. *Journal of Health and Social Behavior, 15,* 44–50.

Steinheider, B., Both, R., & Winneke, G. (1998). Field studies on environmental odors inducing annoyance as well as gastric and general health-related symptoms. *Journal of Psychophysiology Supplement,* 64–79.

Steinheider, B., & Winneke, G. (1993). Industrial odours as environmental stressors: Exposure-annoyance associations and their modification by coping, age and perceived health. *Journal of Environmental Psychology, 13,* 353–363.

Stoddard, M. (1990). *The scented ape: The biology and culture of human odor*. Cambridge: Cambridge University Press.

Sucker, K., Both, R., & Winneke, G. (2001). Adverse effects of environmental odours: Reviewing studies on annoyance responses and symptom reporting. *Water Science and Technology, 44,* 43–51.

Tabershaw, I. R., & Cooper, W. C. (1966). Sequelae of acute organic phosphate poisoning. *Journal of Occupational Medicine, 8,* 5–20.

Taylor, S. M., Sider, D., Hampson, C., Taylor, S. J., Wilson, K., Walter, S. D. et al. (1997). Community health effects of a petroleum refinery. *Ecosystem Health, 3,* 27–43.

Van den Bergh, O., Kempynck, P. J., Van de Woestijne, K. P., Baeyens, F., & Eelen, P. (1995). Respiratory learning and somatic complaints: A conditioning approach using CO_2-enriched air inhalation. *Behaviour Research and Therapy, 33,* 517–527.

Van den Bergh, O., Stegen, K., & Van de Woestijne, K. P. (1997). Learning to have psychosomatic complaints: Conditioning of respiratory behavior and somatic complaints in psychosomatic patients. *Psychosomatic Medicine, 59,* 13–23.

Van den Bergh, O., Stegen, K., & Van de Woestijne, K. P. (1998). Memory effects on symptom reporting in a respiratory learning paradigm. *Health Psychology, 17,* 241–248.

Van Toller, S., & Dodd, G. (1988). *Perfumery: The psychology and biology of frangrances*. London: Chapman.

Watson, D., & Clark, L. A. (1984). Negative affectivity: The disposition to experience aversive emotional states. *Psychological Bulletin, 96,* 465–490.

Winneke, G., & Neuf, M. (1992). Psychological response to sensory stimulation by environmental stressors: Trait or state? *Applied Psychology: An International Review, 41,* 257–267.

CHAPTER 3

Psychophysical evaluation of pain and olfaction

Many commonalities and a few significant differences

Richard H. Gracely
University of North Carolina at Chapel Hill, USA

Psychophysical methods used to evaluate the chemical senses have been instrumental for investigations of pain and pain treatment. The evaluation of chemosensory quality has a direct parallel in the use of instruments that aid diagnosis of pain and choice of treatment by the determination of pain quality, such as pulling, throbbing, radiating and burning. The use of hedonic psychophysics in the chemical senses recognizes that tastes and smells can be described by two principal dimensions. The first is the intensity of the sensation, which is the target of most psychophysical methods. The second is the hedonic valence associated with the sensation, i.e., the degree of pleasantness or unpleasantness. These hedonic dimensions are sensitive to internal state; for example, the pleasantness of sugar solutions increases during hunger and decreases during satiety. The concept of hedonics has been successfully applied to pain, albeit with only a negative valence. The dimensions of pain intensity and unpleasantness have been shown to be independent and changes in internal state can modify unpleasantness without altering sensory intensity. In addition, the amount of unpleasantness for a given intensity has been shown to vary over painful conditions (low in childbirth, high in cancer) with consequences for treatment strategy. Methods used for olfaction have been applied to healthy control subjects using experimentally applied painful stimulation and more recently these methods have been applied directly to patients, providing information about the many unknown mechanisms of pain, and the development of more effective treatments. This chapter will describe similarities in the measurement of pain and olfaction and highlight important differences in the evaluation of these two sensory modalities.

Introduction

In natural settings, the senses of both olfaction and pain are necessary for survival of the organism. While much has been learned about these senses in physiological and behavioral experiments, necessary knowledge relies on the capabilities of humans to describe their experience. The fields of olfactory and pain psychophysics have developed independently, and both have found similar solutions to the problems of quantifying perceptual experience. This chapter will describe the principle methods of olfactory and pain assessment, noting differences and commonalities that likely apply to other senses with homeostatic functions such as taste, thermoregulation, thirst, and hunger, and apply to medically relevant conditions such as dysethesias, nausea, dizziness and tinnitus. The author is indebted to the late Trygg Engen for a personal introduction to psychophysics and for guidance from his classic descriptions of these methods (Engen, 1971a, 1971b).

1. Laboratory measures of detection threshold

Both olfaction and pain share a common feature that external stimuli can be manipulated to determine the minimal physical stimulus magnitude that evokes an odor or pain sensation. These thresholds can be determined by classical procedures such as the Methods of Limits or Method of Constant Stimuli, and by more modern procedures such as Signal Detection Theory or Sensory Decision Theory (SDT) (Engen, 1971a). The basic SDT method presents stimuli of varying intensity, including zero intensity referred to as a blank stimulus. Application of these procedures reveals fundamental differences between pain and olfaction, identifying issues that must be considered for the measurement of other sensations. For example, the need for stimulus blanks for the evaluation of detection thresholds is well documented in olfaction (Schmidt & Cain, 2010). At stimulus levels near detection, subjects make false positive errors (i.e., report a sensation evoked by a blank stimulus), referred to as "false alarms" in SDT and as Type I or alpha errors in statistical testing. Subjects also make the opposite error of a false negative response, referred to as a miss in SDT and as Type II or beta error in statistics. This may well be an error if the stimulus evoked a clear suprathreshold response, or an accurate report if the stimulus magnitude is insufficient to evoke a sensation. Without dwelling on this point, there are further parallels between statistical errors (i.e. alpha, beta) and SDT responses of false alarms and misses. Both describe a gray area in which the total uncertainty can be divided between two possible responses. In statistics, setting the alpha level higher or lower, inversely lowers or raises the beta error. In SDT, responding conservatively only when sure of a sensation decreases the frequency of false alarms and increases the frequency of misses. SDT provides a measure of the conservatism/liberalism (similar to setting the statistical alpha level) and also a measure of sensitivity, similar to the value of t in a t-test or to other tests of significance (Engen, 1971a).

The use of blanks works well for the detection of odors since varying concentration results in states of no sensation or detectable sensation. There is a problem applying these methods to pain since the two states of nothing and pain are usually, if not always, separated by an intermediate state of non-painful sensation (Gracely & Naliboff, 1996). Increasing a thermal stimulus evokes a sensation of non-painful heat before a sensation of painful burning, increasing an electrical stimulus evokes a non-painful tapping or buzzing before painful stabbing or burning. Here the blank is problematic, since the choice is not between the absence and presence of sensation, but rather the absence and presence of a certain quality of sensation. SDT cannot be applied directly to pain detection. In addition, the interpretation of SDT parameters for pain has been controversial (Chapman, 1977; Coppola & Gracely, 1983; Rollman, 1977).

Classical threshold methods can be applied to the evaluation of pain threshold. For example, the Method of Adjustment and the Method of Constant Stimuli can be applied directly (see Engen, 1971a for a description of these methods). Additionally, the Method of Limits is usually used with some modification. The Method of Limits presents alternating ascending and descending series of stimulus intensities to control for errors of habituation and anticipation (Engen, 1971a). Applications to pain avoid excessive stimulation by administering only ascending series of discrete stimulus intensities, or a continuously increasing stimulus.

The use of laboratory threshold measures is pervasive in both olfaction and pain. In olfaction, the necessary focus is on the design of stimulus delivery systems that avoid even the slightest odor cross-contamination (Schmidt & Cain, 2010). In pain, the focus is on the controlled delivery of multiple stimulus modalities such as pressure and heat that activate specific nociceptors systems. Standardization of stimulation procedures has advanced human pain and olfactory research in both the laboratory and clinic.

1.1 Clinical measures of detection threshold

The precision of pain stimulators and olfactometers is a simple engineering problem of matching the measurement resolution to the size of the problem. In the case of laboratory investigations of the neural processing of pain and smell, the needed resolution is supplied by complex and expensive equipment. In the clinic, common clinical deficits are often large and evaluated quickly by more simple

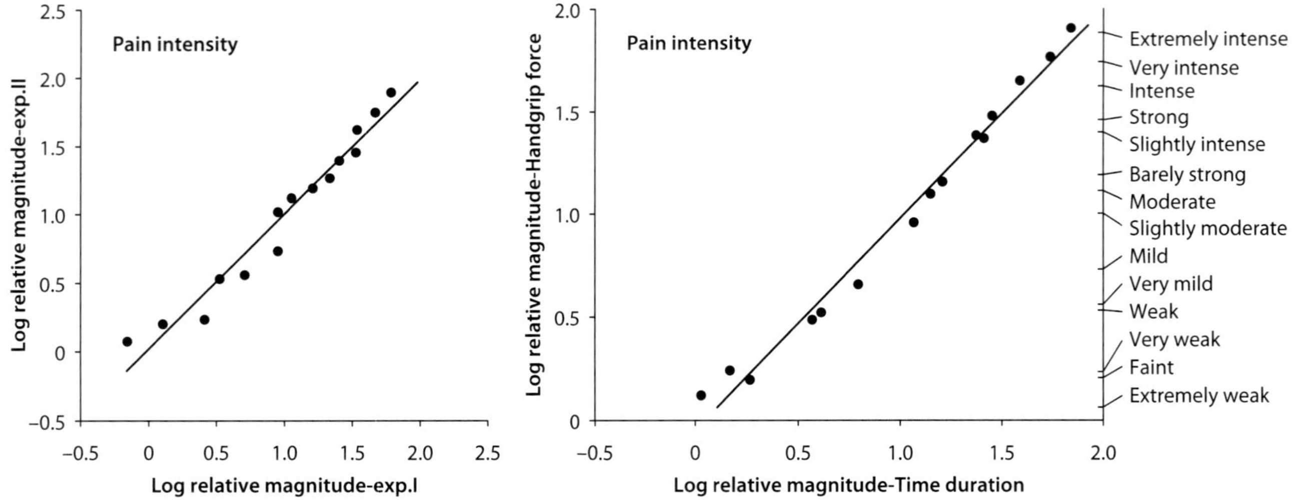

Figure 1. Quantified verbal descriptors of the intensity of pain sensations. The left panel shows scales of common verbal descriptors from two groups of subjects. The group scale from Experiment II is plotted against the group scale from Experiment I in log-log coordinates. The fitted line shows the agreement ($R^2 = 0.96$) for words common in the two experiments that used different methods and subjects. In each experiment, subjects cross-modality matched two different response measures (magnitude estimation and handgrip force in Exp I, time duration and handgrip force in Exp II) to the descriptors presented as stimuli and to the length of projected lines. The line length functions were used to convert each mean response to each word to a common unit of line length. The right panel shows the agreement ($R^2 = 0.98$) between these theoretically equal scales for Experiment II.

methods. A host of clinical olfactory tests are available; Eibenstein and colleagues (Eibenstein, et al., 2005) describe 15 tests of olfaction varying in ease of administration and sensitivity. A broad array of tests are also used to assess pain sensitivity to modalities such as pin prick, heat and dull pressure, ranging from brief tests in neurological exams to extensive laboratory procedures (Gracely, Price, Roberts, & Bennett, 1996).

Many of these brief, simple clinical methods are sufficient for detecting gross alterations in sensory sensitivity. Ideally, a hierarchy of methods can be standardized to match the needed sensitivity to the clinical or experimental demand. In both pain and olfaction, investigators have developed their own methods and until recently there has been little agreement on a standard. Recent efforts in pain assessment have approached an international standard through a consortium of investigators (Magerl et al., 2010; Rolke et al., 2006). This effort is concerned with the proper behavior of examiners to avoid biases and to confirm the veracity of subject responses. This concern is shared by olfactory research, including false reporting by subjects (Linschoten & Harvey, 2004). Malingering is especially a concern in medico-legal pain determinations, and methods have been proposed for detection of false reports (Chapman & Brena, 1990; Leavitt, 1987; Main & Waddell, 1998; Wallis & Bogduk, 1996).

2. Measures of suprathreshold sensation

Engen's classic chapters in Kling and Riggs *Textbook of Psychology* (Engen, 1971a, 1971b) were devoted to threshold and suprathreshold assessment, respectively. Olfactory and pain threshold are still evaluated in many studies, although investigators in both fields point out that these measures only assess the very bottom of the perceptual range (Bartoshuk, Duffy, Hayes, Moskowitz, & Snyder, 2006; Gracely, 2005). These measures indicate where the psychophysical function describing sensory intensity versus stimulus intensity begins, but not necessarily the shape of this function as stimulus intensity is increased. Threshold sensitivity does not necessarily predict suprathreshold sensitivity, and a manipulation that alters sensitivity may have different effects at low and high levels of sensory intensity. The psychophysical function describing the intensity of sensations evoked over a range of stimulus intensity is evaluated by a number of methods described at length by Engen (Engen, 1971b).

One of the most prominent methods described and used by Engen (Engen, 1971b) is S. S. Stevens' method of ratio scaling (Stevens, 1975). The instructions are gratifyingly simple; for example, *make a response to the first stimulus and then proportional responses to subsequent stimuli. If the next stimulus is twice as great as the first, provide a response twice as great.* The result is a function that is usually curvilinear but linear when plotted in logarithmic coordinates. The slope of the logarithmic function describes the exponent of a power function when plotted in arithmetic units. A simple explanation of this exponent is based on the fact that each stimulus and response modality (many modalities, such as numbers, line length or brightness can serve as either) has a characteristic range that varies from small (electric shock) to large (brightness, 11 log units) and that can be described by a characteristic exponent that is inversely related to range, such as 3.5 for electric shock and 0.33 for brightness of a small target (Stevens, 1975). Every scaling procedure involves two of these modalities, one for the stimulus and one for the response, and the resulting power function exponent reflects the combination of these two exponents (Abbey, 1962; Stevens & Marks, 1965). The use of numbers has been arbitrarily defined as an exponent of unity, and the resulting method termed 'magnitude estimation.' This family of methods has been described generally as 'direct scaling' and specifically as 'cross-modality matching' or 'magnitude matching,' in which magnitude estimation is the special case in which the response dimension is the use of numbers. Other common response dimensions include line length, handgrip force, time duration, visual brightness and auditory loudness (Stevens, 1975).

Trygg Engen's view of direct scaling and the power function is summarized succinctly, "Empirically, the power function is probably as well established as is possible for any quantitative relations involving the whole man, and better than any other in psychology" (Engen, 1971b, p. 84). Engen was among those who applied the ratio scaling method to olfaction (Engen, 1971b; Engen & McBurney, 1964), and influenced Bernie Tursky's (Tursky, 1976) and the author's (Gracely, McGrath et al., 1978a) applications of ratio-scaled descriptors to pain assessment. The concept of ratio scaling is firmly rooted in experimental evaluations of both disciplines. The methods have been applied almost directly with one important difference. Psychophysical scaling focuses on the growth of sensation with increasing physical stimulation. The height of the function and the intercept are largely ignored and, in the case of direct ratio scaling, are purposefully meaningless (Engen, 1971b). In contrast, clinical and experimental pain studies need to know the actual absolute magnitude of pain, and this need is shared by investigations of olfaction. A classic approach to absolute pain assessment uses common language to describe meaningful levels of pain or odor such as "weak, mild, moderate, intense." Verbal category scales fulfill this property, with analysis methods that assume specific units or determine the units by internal metrics such as Thurstone's Method of Successive Categories (Thurstone, 1959) or external metrics determined from scaling the descriptors in a separate experiment (Gracely, McGrath, & Dubner, 1978b). These methods rely on the stability of semantic meanings within a language to provide concrete anchors for perceptual judgment. Pain pharmacology has long depended on simple category scales of pain, (none, slight, moderate and severe) or pain relief (none, some, lots, complete) (Gracely, 1979). These scales provided semantic anchors, but the apparent lack of resolution provided by only four categories led to the development and use of visual analog scales (VAS), usually a 10 cm horizontal line labeled at the ends with descriptors such as 'no pain' and 'worst pain imaginable' or 'no sensation' and 'most intense sensation imaginable.'

An alternative approach to increased resolution combines ratio scaling with the anchoring with multiple verbal categories. Verbal descriptors such as "mild", "moderate" and "intense" are quantified by ratio methods and then presented in a category format, or mixed analog and category format, to provide semantic anchoring with each response choice. It is interesting to note that such scales, termed Generalized Labeled Magnitude Scales (gLMS) in the chemical sense literature (Bartoshuk et al., 2004; Green et al., 1996) have been developed independently for pain (Gracely, McGrath, & Dubner, 1978a; Gracely et al., 1978b), the chemical senses (Green, Shaffer, & Gilmore, 1993) and physical effort (Borg, 1982). These methods provide semantic anchors that guide responses, with each response converted to a meaningful magnitude of sensory intensity or other dimension (such as pleasantness or unpleasantness described below). An important issue is whether these descriptor values, especially those describing a maximal sensation, vary in a systematic way between identified groups (Bartoshuk et al., 2006). Such group differences could be an important problem in static group comparisons (e.g., olfactory sensitivity between men and woman, or between young and old subjects), but may be less of an issue in assessing the effect of an intervention such as the effects of a novel analgesic.

3. Nonverbal suprathreshold measurement: Cross modality matching

An alternative method of measurement for the intensity of suprathreshold sensations of pain and odors dispenses with descriptors. Participants match the intensity of sensations evoked by one modality to a response modality that the participant is free to vary. For example, subjects could match the intensity of odors by adjusting the loudness of a tone or the brightness of a light. As noted in the section on ratio scaling, even magnitude estimation can be considered a form of cross-modality matching with the response being 'number use'. In applications of cross-modality matching, or the more recent term, "magnitude matching," the response is usually a sensory continuum such as brightness or loudness. One advantage of cross-modality matching over magnitude estimation is that previous responses cannot be remembered, and biases of number choice inherent in magnitude estimation (e.g., 10, 15 20 instead of 12, 17, 23) are eliminated. Another bias is concerned with remembering the stimulus. If the spacing between stimulus intensities is too large, subjects can actually identify the stimulus such as "that is the third one from the bottom that I call 25". This identification of the stimulus defeats the use of repeated delivery of the same stimulus to reduce random scaling error. Engen referred to this problem as "artifactual reliability," and stressed the need to use sufficiently small stimulus spacing to avoid identifying specific stimuli (personal communication). Under such conditions of sufficiently small stimulus spacing and a cross-modality response that cannot be remembered, these methods result in an independent sensory judgment on each trial (Bartoshuk et al., 2004; Duncan, Feine, Bushnell, & Boyer, 1988; Gracely, 1977; Marks, 1991).

4. Subjective judgment, behavior and physiology

The nervous system is often dichotomized into afferent (input) and efferent (output) neural activity. Most research in neurology focuses on efferent neural responses, by assessing measures of movement and muscle activity. There usually is little or no research into afferent or sensory systems, in which the main dependent variable is individuals' descriptions of their own experience. This may be due in part to an inherent distrust of subjective reports and the perceived rigor provided by objective measures of movement and motor activity.

Those who choose to study afferent systems, including olfaction and pain, cannot escape the use of subjective measures ranging from surveys and clinical questionnaires to psychophysical analyses of sensations evoked by controlled stimulation. These approaches vary considerably in sophistication and vulnerability to biases. Those uncomfortable with subjective measures have at least three options. The first option is to rely only on observable behavior. A number of behavioral observation methods have been developed with good psychometric properties (Keefe, Wilkins, & Cook, 1984). One important issue is the interpretation of mismatches between subjective pain and pain behavior. An individual who carries on all physical activities while complaining bitterly about pain is usually suspect. On the other hand, a stoic personality who uses the distraction of work to manage persistent pain might be mistakenly classified by behavioral methods that correlate subjective pain with behavior as not being in pain.

In addition to behavior, both pain and olfaction have sought a second option using objective measures to assess sensory magnitude by physiological measures such as cortical evoked potentials (Allison & Goff, 1967; Carmon, Mor, & Goldberg, 1976; Dowman, Rissacher, & Schuckers, 2008; Thesen & Murphy,

2002) and more recently functional brain imaging (Jones, Brown, Friston, Qi, & Frackowiak, 1991; Schweinhardt & Bushnell, 2010; Sobel et al., 1998; Zatorre, Jones-Gotman, Evans, & Meyer, 1992). These research tools have been applied to the investigation of mechanisms of pain and olfaction by using averaged group data. While an increasing number of studies of individual differences promise clinical utility for measures of a single person, the general consensus is that the technology has not yet matured to the point of clinical diagnosis or choice of treatment for individual cases.

The third option uses psychophysical procedures to assess rating behavior rather than subjective judgment. For example, distrust of subjective judgments of olfaction has led to a preference for assessment using discrimination performance (Wise, Olsson, & Cain, 2000). In the case of pain, SDT has been applied to suprathreshold scaling by taking each pair of adjacent stimulus intensities and using the lesser as the blank to assess the proportion of category responses among the two stimuli, with each category serving as a different response criterion. In this manner, the discrimination of suprathreshold sensations could be linked to the presentation of a blank by an equation that considers the discrimination of all adjacent pairs of stimuli between the blank and the stimulus intensity of interest (Chapman, Gehrig, & Wilson, 1975; Clark & Yang, 1974). As noted above, this and conventional threshold use of discrimination measures, such as SDT, has led to lively debates about pain mechanisms and the difference between pain sensitivity and pain discrimination (Chapman, 1977; Coppola & Gracely, 1983; Rollman, 1977).

Wise et al. (Wise, Olsson, & Cain, 2000) have proposed that psychophysical discrimination performance may provide a more bias-free measure of olfactory sensitivity. However, like pain, odor discrimination may be associated with increased or decreased sensitivity, or be independent of sensitivity. In addition to sensitivity, discrimination is influenced by both physiological and cognitive variability (Coppola & Gracely, 1983). Similar to a statistical *t*-test, discrimination is based on a ratio of difference divided by variability. Variability in discrimination tasks is due both to variability in the sensory system and cognitive variability in the choice of a rating. Recent evidence suggests that olfactory discrimination may reflect a cognitive performance factor unrelated to sensory intensity. Executive function and semantic memory were associated with discrimination performance and also odor identification, but had no effect on odor threshold (Hedner, Larsson, Arnold, Zucco, & Hummel, 2010).

The question remains, however, whether subjective responses, behavior or physiology, could serve as the gold standard for pain and olfactory assessment. In the case of clinical conditions of spontaneous pain or malodors, subjective report must be the primary measure. No one would accept waking up in the middle of the night in extreme discomfort and being denied medication because of a reading on a physiological meter that did not support the perceived sensation. Thus in clinical settings, subjective measures remain the standard, and the validity of other measures is established by comparisons with subjective reports.

5. Hedonic psychophysics: Evaluating the pleasantness/unpleasantness of evoked sensations

Many sensations are intimately involved in the regulation of biological processes. These can be characterized by common dimensions of sensory intensity and a feeling state that can be either unpleasant or pleasant. Hunger, thirst, and pain are unpleasant feeling states arising from homeostatic imbalance. The unpleasant sensations consequently motivate behaviors to reduce the unpleasant feeling and restore homeostatic balance. Thermal comfort is an interesting bivalent dimension in which departure in either direction (cool, warm) can be unpleasant and the return to neutrality by the opposing sensation is pleasant. Odors can provide bivalent feelings of either pleasantness or unpleasantness, such as the pleasantness of flowers and the stench of garbage. Multidimensional scales of odors usually find pleasantness/unpleasantness to be a dominant dimension, while pain is characterized by unpleasantness, an aversive quality that motivates avoidance and escape. The measurement of hedonics in olfaction is straightforward, odors can be pleasant or unpleasant to varying degrees. Measurement of pain hedonics is more difficult, especially since the hedonic is only (or usually) unpleasant, and this unpleasantness increases monotonically with increasing sensory intensity. When subjects are asked to rate pain unpleasantness or intensity, they may not discriminate between the two, and instead respond to the more salient or meaningful dimension (e.g., intensity for experimental pain scaling, unpleasantness for chronic pain). There is evidence that the use of language, besides anchoring judgments to absolute levels, facilitates the discrimination of pain intensity and unpleasantness (Gracely, 1977). When a descriptor scale is limited to words of a specific dimension, subjects appear to "tune in" to that dimension. For example, in contrast to number use or cross modality matching, descriptor scales have demonstrated different psychophysical functions for pain intensity and unpleasantness, and differential effects of administration for the minor tranquilizer, diazepam, or the opioid, fentanyl (Gracely, Dubner, & McGrath, 1979; Gracely et al., 1978b).

Food becomes more appealing when we are hungry, while even favorite foods become unappealing when we have overeaten (Rolls, Van Duijvenvoorde, & Rolls, 1984). This shows that ratings of sensory pleasantness or unpleasantness may be influenced by internal state. Trygg Engen's students, George Mower and Robert Mair, illustrated these effects in a study in which extreme conditions of hunger (fully sated after a glucose drink, fasting for 22 hrs) shifted the pleasantness/unpleasantness ratings of sugar solutions without altering ratings of taste intensity (Mower, Mair, & Engen, 1977). Additional studies have identified subgroups that show different relationships between hedonic ratings and concentrations of sucrose solutions, and changes in this relationship due to body weight or ingestion of a caloric preload (Thompson, Moskowitz, & Campbell, 1976). Mower (Mower, 1976) found a similar effect for intensity and hedonic ratings of thermal sensations. Manipulations of internal body temperature (hypothermic, neutral, hyperthermic) shifted pleasantness/unpleasantness ratings of immersion of a hand into a water bath with temperatures varying from 21–45°C, while sensory ratings of coolness or warmness were unchanged. This independence of intensity and unpleasantness ratings validates these dimensions and demonstrates the need to measure both in studies of pain and olfaction and in studies of interventions designed to modify either.

5.1 The concept of affective gain or relative unpleasantness

The unpleasantness of an olfactory or painful sensation is likely associated with the intensity of the sensation; increasing sensory intensity increases unpleasantness. In addition, it is possible that the amount of unpleasantness associated with a particular sensory intensity can vary, either between individuals, type of pain or smell, or in response to an intervention. Unpleasantness can be conceptualized as an amplifier with sensory intensity as the input and unpleasantness as the output (Gracely, 1992a). Unpleasantness can be altered by changing the sensory input into the amplifier (reducing sensory intensity) or by changing the volume control, the 'affective gain' of the amplifier (reducing the amount of unpleasantness associated with a specific intensity) (Gracely, 1992a). Thus, sensory intensity and unpleasantness can be expected to be highly associated since the output of the amplifier will be a function of the input. Investigators in the pain field have recognized this association and further used it as evidence that unpleasantness is redundant and only a single dimension of pain magnitude is needed (Gracely, 1992b). However, this view ignores the important variable of affective gain – the position of the volume control. While sensory intensity and unpleasantness are likely associated, sensory intensity and the real variable of interest, the affective gain, are likely independent. This gain can be expressed as a ratio of unpleasantness to sensory intensity. This ratio has been described as "relative unpleasantness." In pain, relative unpleasantness varies with type of pain; minimal for experimental pain, very high for clinical cancer pain, and less for the clinical pain of childbirth (Price, Harkins, & Baker, 1987; Rainville, Feine, Bushnell, & Duncan, 1992). Manipulations such as hypnosis, affective learning, information about the stimulus, and administration of minor tranquilizers have selectively reduced rated pain unpleasantness (Gracely et al., 1978b; Malone, Kurtz, & Strube, 1989; Price, Barrell, & Gracely, 1980; Rainville, Duncan, Price, Carrier, & Bushnell, 1997; Wunsch, Philippot, & Plaghki, 2003) as well as the rated unpleasantness of odors (Loggia, Mogil, & Bushnell, 2008) The unpleasantness of odors is reduced by exposure (Cain & Johnson, 1978), and odor intensity and unpleasantness reveal other divergences such as the lateralization of odor intensity sensitivity (greater in the right nostril), but not hedonic sensitivity (Thuerauf et al., 2008). Reduction of unpleasantness for a given sensory intensity may constitute one of few available options for intractable pain syndromes, such as pain from CNS lesions that are treated poorly by conventional methods.

The field of pain has distinguished three types of negative affect (Gracely, 1992a). The term "unpleasantness" was borrowed from hedonic psychophysics to describe the immediate disagreeable nature of pain, similar to the unpleasantness of a disagreeable smell. These negative feeling states prompt behaviors of avoidance and escape and are present in both animals and humans.

A second type of unpleasantness is implied by *affect*. This type is an emotional state that can be described in terms of "distress", "hopelessness" and "despair". This emotional state can be considered to result from thoughts about both present pain and a future without pain relief. Part of this state is pain specific and a part generalizes to most diseases, resulting in a third level that results from being a patient, health concerns, interaction with the health care system, and dismal future prospects (Gracely, 1992a).

6. Measurement of sensory quality

Many scales treat "pain" as a single qualitative dimension, however, there is considerable recognition of multiple pain qualities and instruments designed to assess these qualities (Melzack, 1975, 1987; Melzack & Torgerson, 1971; Torgerson, Bendebba, & Mason, 1998). The McGill Pain Questionnaire (MPQ) is one of the most widely used instruments for assessment of pain quality (Melzack, 1975). This questionnaire was based on original work by Dallenbach (Dallenbach, 1939) and Melzack and Torgerson (Melzack & Torgerson, 1971) which resulted in scaled descriptors of 16 different pain qualities divided into three categories of "sensory" words describing mechanical, thermal, spatial and temporal qualities of sensation; "affective" words, describing tension, fear and autonomic qualities;

and "evaluative" words, describing the general intensity of the pain experience. Later, four "miscellaneous" categories were added to include descriptors which did not fit in any of the three general categories. The questionnaire approach represents a simple technology of specified dimensions with quantified descriptor scales within each dimension. Each descriptor is assigned to only one dimension. Subsequent methods have used multidimensional scaling to define the semantic space for pain descriptors (Clark et al., 2001). An example is the Ideal Type Model by the late Warren Torgerson and colleagues, which placed descriptors in multidimensional space defined by 17–19 Ideal Types (Torgerson et al., 1998). In contrast to the MPQ, this approach allowed descriptors to be a mix of more than one dimension. In addition, while some dimensions were described by a close association with a single descriptor such as "sharp", "tugging", and "pounding", many of the dimensions were not described by any single descriptor, representing an "Ideal Type" semantically without a vocabulary to describe them. Furthermore, this method made distinctions within MPQ dimensions, for example, recognizing the rotational character of "drilling" and "boring" and placing them in separate class from pricking and stabbing, while the MPQ includes all of these sensations in a punctate pressure class.

Efforts to assess the extensive domain of olfactory quality have followed a similar course. Much like the MPQ or dimensions of color, early methods attempted to define a basic set of olfactory primary qualities defined by a small alphabet of basic odors (Amoore, 1974; Guillot, 1948; Haddad, Lapid, Harel, & Sobel, 2008; Henning, 1916). Problems with these simplified scales led to the application of multidimensional scaling methods much like the use of the Ideal Type measure for the assessment of pain quality. Two early studies of olfactory quality discrimination generated eight or three dimensions using parametric analysis methods, respectively (Woskow, 1964, 1968; Wright & Michels, 1964). Re-analyses using a nonparametric approach found somewhat similar two-dimensional solutions for both studies (Schiffman & Erickson, 1972). Interpretation of these dimensions suggest a continuum of changing quality and also problems when subjects are asked to provide verbal descriptions in addition to quantitative judgments of similarity (Moskowitz & Gerbers, 1974; Schiffman, 1974; Stevens & Greenbaum, 1966; Stevens & Oconnell, 1996). However, later studies using descriptors and more extensive statistical approaches found robust results using verbal descriptor methods (Dranieks, 1982; Mamlouk, Chee-Ruiter, Hofmann, & Bower, 2003).

Current progress in evaluating odor quality is described in several chapters in this volume. The methods used for olfaction have also been applied to the evaluation of pain quality. In the clinic, altered sensory qualities may provide important diagnostic information. Stimulus modalities can be linked to physiologically distinct afferent systems and the results of clinical tests may inform diagnosis and choice of treatment. However, as emphasized below, the clinical issue in pain is not the response to stimuli, but rather the treatment of spontaneous, chronic pain. Verbal description is necessary, constituting the main dependent variable in a wide variety of clinical research and treatment scenarios. While many of these studies are concerned with the intensity and unpleasantness of spontaneous pain, scales that assess the multiple dimensions of clinical pain are critical for accurate assessment and treatment planning. The many types of clinical pain cannot be directly compared by numerical similarity judgments, but rather must be assessed by multi-dimensional instruments. These instruments are necessarily constructed using judgments of similarity and dissimilarity of instrument items.

While the focus on olfaction is on the experimental determination of function, there are a number of olfactory clinical conditions that parallel pain conditions. Anosmia and hyposmia, the loss or diminished ability to smell, puts persons at risk in terms of detection of smoke in a fire, or detection of a natural gas leak. Consumption of spoiled food is also a concern, as is the general issue of providing pleasure normally associated with food. The parallel in pain is congenital insensitivity to pain, with its attendant morbidity. The presence of parosmias or troposmias, altered smell of an odor, are similar to abnormal physical sensations (paresthesias) and sensations with an unpleasant character (dysesthesias). The major clinical problem, suffering associated from persistent pain is paralleled by unpleasant phantosmias, perceptions of smell without an olfactory stimulus (Ship & Chavez, 2001). In cases of spontaneous pain and smell, multidimensional scales provide information for diagnosis and treatment choice that cannot be obtained by any other means.

7. Measuring changes in sensory function

The manipulations of core temperature, hunger, and pain sensitivity introduce a new goal of measurement, the assessment over time to evaluate the effect of aging, the influence of other clinical conditions, or the effects of deliberate interventions that alter the intensity or hedonic aspect of smell or pain. Such interventions can be experimental, such as the use of hypnosis to alter selective dimensions of pain, or the interaction of two odors. In the case of pain, these interventions are often clinical, employing strict clinical trial methodology to assess the analgesic action of a test compound or other intervention in comparison to a placebo and/or standard intervention. In these situations issues about the absolute anchoring of intensity between groups are less of a concern than the ability of a scale to respond to a change in sensitivity. Factors such as remembering the previous category response could facilitate biases of either showing no effect due to a belief that a placebo was administered, or showing a dramatic effect if the experience of side-effects leads to the belief that an analgesic has been administered (Max, Schafer, Culnane, Dubner, & Gracely, 1988). However, the use of descriptors may facilitate the discrimination of unpleasantness from sensory intensity, distinguishing between effects on affective gain (the sensations have the same intensity but are less unpleasant) and on sensory intensity (the sensations are less intense and correspondingly less unpleasant) that may be mediated by different mechanisms.

8. Conclusion

Pain and olfaction share many common features. The intensity of sensations evoked by controlled stimulation is measured in the laboratory and in the clinic by classical psychophysical measures of threshold, and by advanced techniques based on Sensory Decision Theory. Threshold sensitivity may not predict suprathreshold sensitivity, and similar methods of suprathreshold scaling have been applied to both. These scales have evolved from category scales to ratio scales and to methods that quantify category values with ratio scale properties or that use language-free magnitude matching between a continuous stimulus (e.g., temperature of painful heat or odor concentration) and a response continuum such as line length or visual brightness. Investigators wary of subjective judgment have turned to behavioral performance measures, including discrimination performance, and physiological indices such as cortical potentials and the recent explosion in brain imaging methods. Both pain and olfaction are described by sensory magnitude and a hedonic feeling that motivates behavior. Methods have been developed to assess these separate dimensions, which are differentially altered by a wide range of interventions and changes in internal state. In addition, pain and olfaction vary greatly in sensory quality, and multidimensional scaling approaches are applied to evaluate the underlying structure of qualitative differences.

The commonalities in pain and olfaction, however, are matched by major differences. Detection of odors can be assessed by Sensory Decision Theory approaches that use blank stimulation. In contrast, the states of no sensation and pain are almost always separated by a band of non-painful stimulation, which changes detection tasks to identification of painful quality, and complicates the necessary use of stimulus blanks. The hedonic aspect of a smell can be either pleasant or unpleasant and may grow nonlinearly with increasing intensity. By contrast, the hedonic aspect of pain is only unpleasant, and grows monotonically with intensity. The distinction between intensity and hedonics is also more difficult for pain. Measurement of pain quality is naturally divided into dimensions such as thermal, mechanical, spatial extent, and movement, and scaling theory has evolved from descriptors sorted into classes to descriptors considered as mixtures of primary types. Verbal description remains the primary measure of pain, with measures of behavior providing secondary adjuncts to this measure. By contrast, olfactory studies are equivocal about the use of semantic measures in the assessment of magnitude and quality.

In summary, investigations of olfaction and pain have developed similar methods to describe the intensity quality of evoked and spontaneous sensations and the feeling states associated with these sensations. The major difference is in feeling state, where odors can be pleasant or unpleasant, while pain is only unpleasant. In the laboratory, increasing the intensity of a noxious stimulus usually results in a transition from no odor to odor in olfaction, but in pain from no sensation to a non-painful sensation then a painful sensation. In the clinic, abnormalities in evoked pain and the presence of spontaneous pain are the major reasons individuals seek treatment, and many conditions still have no effective treatment. Similar problems of evoked abnormalities and spontaneous unpleasant sensations are found with olfaction. Continued investigations in both disciplines will advance our understanding of mechanisms mediating pain and olfaction and advance the ability to diagnose and treat human suffering.

Acknowledgements

The author is indebted to Trygg Engen, PhD thesis advisor and friend, for his guidance in science and much more.

References

Abbey, D. S. (1962). Cross-modality matching of numerosity and pitch. *Canadian Journal of Psycholology, 16*, 283–290.

Allison, T., & Goff, W. R. (1967). Human cerebral evoked responses to odorous stimuli. *Electroencephalograohy and Clinical Neurophysiology, 23*, 78.

Amoore, J. E. (1974). Evidence fo the chemcial olaftory code in man. *Annals of the New York Acadamy of Sciences, 237*, 137–143.

Bartoshuk, L. M., Duffy, V. B., Green, B. G., Hoffman, H. J., Ko, C. W., Lucchina, L. A., Marks, L. E., Snyder, D. J., & Weiffenbach, J. M. (2004). Valid across-group comparisons with labeled scales: the gLMS versus magnitude matching. *Physiology and Behavior, 82*, 109–114.

Bartoshuk, L. M., Duffy, V. B., Hayes, J. E., Moskowitz, H. R., & Snyder, D. J. (2006). Psychophysics of sweet and fat perception in obesity: Problems, solutions and new perspectives. *Philosophical transactions of the Royal Society of London. Series B, Biological sciences, 361*, 1137–1148.

Borg, G. (1982). A category scale with ratio properties for intermodal and interindividual comparisons. In H. G. Geissler & P. Petzold (Eds.), *Psychophysical Judgment and the Process of Perception*. Berlin: VEB Deutscher Verlag der Wissenschaften.

Cain, W. S., & Johnson, F., Jr. (1978). Lability of odor pleasantness: Influence of mere exposure. *Perception, 7*, 459–465.

Carmon, A., Mor, J., & Goldberg, J. (1976). Evoked cerebral responses to noxious thermal stimuli in humans. *Experimental Brain Research, 25*, 103–107.

Chapman, C. R. (1977). Sensory decision theory methods in pain research: a reply to Rollman. *Pain, 3*, 295–305.

Chapman, C. R., Gehrig, J. D., & Wilson, M. E. (1975). Acupuncture compared with 33 per cent nitrous oxide for dental analgesia: A sensory decision theory evaluation. *Anesthesiology, 42*, 532–537.

Chapman, S. L., & Brena, S. F. (1990). Patterns of conscious failure to provide accurate self-report data in patients with low back pain. *Clinical Journal of Pain, 6*, 178–190.

Clark, W. C., Janal, M. N., Hoben, E. K., & Carroll, J. D. (2001). How separate are the sensory, emotional, and motivational dimensions of pain? A multidimensional scaling analysis. *Somatosensory and Motor Research*, 18, 31–39.

Clark, W. C., & Yang, J. C. (1974). Acupunctural analgesia? Evaluation by signal detection theory. *Science, 184*, 1096–1098.

Coppola, R., & Gracely, R. H. (1983). Where is the noise in SDT pain assessment? *Pain, 17*, 257–266.

Dallenbach, K. M. (1939). Pain: History and present status. *American Journal of Psychology, 52*, 331–347.

Dowman, R., Rissacher, D., & Schuckers, S. (2008). EEG indices of tonic pain-related activity in the somatosensory cortices. *Clinical Neurophysiology, 119*, 1201–1212.

Dranieks, A. (1982). Odor quality: Semantically generated multi-dimensional profiles are stable. *Science, 2218*, 799–801.

Duncan, G. H., Feine, J. S., Bushnell, M. C., & Boyer, M. (1988). Use of magnitude matching for measuring group differences in pain perception. In R. Dubner, G. R. Gebhart, & M. R. Bond (Eds.), *Proceedings of the Vth World Congress on Pain*. Amsterdam: Elsevier.

Eibenstein, A., Fioretti, A. B., Lena, C., Rosati, N., Amabile, G., & Fusetti, M. (2005). Modern psychophysical tests to assess olfactory function. *Neurolological Science, 26*, 147–155.

Engen, T. (1971a). Psychophysics I. Discrimination and Detection. In J. Kling, W. & L. A. Riggs (Eds.), *Experimental Psychology*, 3rd edition (pp. 11–46). New York: Holt.

Engen, T. (1971b). Psychophysics II. Scaling Methods. In J. Kling, W. & L. A. Riggs (Eds.), *Experimental Psychology*, 3rd edition (pp. 47–86). New York: Holt.

Engen, T., & McBurney, D. H. (1964). Magnitude and Category Scales of the Pleasantness of Odors. *Journal of Experimental Psychology, 68*, 435–440.

Gracely, R. H. (1977). *Pain Psychophysics*. Providence: PhD, Brown University.

Gracely, R. H. (1979). Psychophysical assessment of human pain. In J. J. Bonica (Ed.), *Advances in Pain Research and Therapy* (Vol. 3). New York: Raven Press.

Gracely, R. H. (1992a). Affectivce dimensions of pain: How many and how measured? *American Pain Society Journal, 71*, 243–247.

Gracely, R. H. (1992b). Evaluation of multidimensional pain scales. *Pain, 48*, 297–300.

Gracely, R. H. (2005). Studies of Experimental Human Pain. In S. McMahon & M. Koltzenburg (Eds.), *Wall and Melzack's Textbook of Pain*, 5th edition (pp. 267–289). Philadelphia, PA: Elsevier, Health Sciences Division, Churchill Livingstone.

Gracely, R. H., Dubner, R., & McGrath, P. A. (1979). Narcotic analgesia: Fentanyl reduces the intensity but not the unpleasantness of painful tooth pulp sensations. *Science, 203*, 1261–1263.

Gracely, R. H., McGrath, F., & Dubner, R. (1978a). Ratio scales of sensory and affective verbal pain descriptors. *Pain, 5*, 5–18.

Gracely, R. H., McGrath, P., & Dubner, R. (1978b). Validity and sensitivity of ratio scales of sensory and affective verbal pain descriptors: Manipulation of affect by diazepam. *Pain, 5*, 19–29.

Gracely, R. H., & Naliboff, B. (1996). Measurement of pain sensation. In L. Kruger (Ed.), *Handbook of Perception and Cognition: Somatosensory Systems* (pp. 243–313). New York: Raven Press.

Gracely, R. H., Price, D. D., Roberts, W. J., & Bennett, G. J. (1996). *Quantitative sensory testing in patients with CRPS-I & II*. Seattle: IASP Press.

Green, B. G., Dalton, P., Cowart, B., Shaffer, G., Rankin, K., & Higgins, J. (1996). Evaluating the 'Labeled Magnitude Scale' for measuring sensations of taste and smell. *Chemical Senses, 21*, 321–334.

Green, B. G., Shaffer, D. J., & Gilmore, M. M. (1993). A semantically-labeled magnitude scale of oral sensation with apparent ratio properties. *Chemical Senses, 18*, 683–702.

Guillot, M. (1948). Anosmies partielles et odeurs fondamentales. *Comptes rendus de l'Académie des sciences, 226*, 1307–1309.

Haddad, R., Lapid, H., Harel, D., & Sobel, N. (2008). Measuring smells. *Current Opinion in Neurobiology, 18*, 438–.

Hedner, M., Larsson, M., Arnold, N., Zucco, G. M., & Hummel, T. (2010). Cognitive factors in odor detection, odor discrimination, and odor identification tasks. *Journal of Clinical and Experimental Neuropsychology, 32*, 1062–1067.

Henning, H. (1916). *Der Geruch*. Leipzig: Barth.

Jones, A. K., Brown, W. D., Friston, K. J., Qi, L. Y., & Frackowiak, R. S. (1991). Cortical and subcortical localization of response to pain in man using positron emission tomography. *Proceedings. Bioliological Sciences, 244*, 39–44.

Keefe, F. J., Wilkins, R. H., & Cook, W. A. (1984). Direct observation of pain behavior in low back pain patients during physical examination. *Pain, 20*, 59–68.

Leavitt, F. (1987). Detection of simulation among persons instructed to exaggerate symptoms of low back pain. *Journal of Occupational Medicine, 29*, 229–233.

Linschoten, M. R., & Harvey, L. O., Jr. (2004). Detecting malingerers by means of response-sequence analysis. *Perception and Psychophysics, 66*, 1190–1201.

Loggia, M. L., Mogil, J. S., & Bushnell, M. C. (2008). Experimentally induced mood changes preferentially affect pain unpleasantness. *Journal of Pain, 9*, 784–791.

Magerl, W., Krumova, E. K., Baron, R., Tolle, T., Treede, R. D., & Maier, C. (2010). Reference data for quantitative sensory testing (QST): Refined stratification for age and a novel method for statistical comparison of group data. *Pain, 151*, 598–605.

Main, C. J., & Waddell, G. (1998). Behavioral responses to examination. A reappraisal of the interpretation of "nonorganic signs". *Spine (Phila Pa 1976), 23*, 2367–2371.

Malone, M. D., Kurtz, R. M., & Strube, M. J. (1989). The effects of hypnotic suggestion on pain report. *American Journal of Clinical Hypnosis, 31*, 221–230.

Mamlouk, A. M., Chee-Ruiter, C., Hofmann, U. G., & Bower, J. M. (2003). Quantifying olfactory perception: Mapping olfactory perception space by using multidimensional scaling and self-organizing maps. *Neurocomputing, 52–54*, 591–597.

Marks, L. E. (1991). Reliability of magnitude matching. *Perception and Psychophysics, 49*, 31–37.

Max, M. B., Schafer, S. C., Culnane, M., Dubner, R., & Gracely, R. H. (1988). Association of pain relief with drug side effects in postherpetic neuralgia: A single-dose study of clonidine, codeine, ibuprofen, and placebo. *Clinical Pharmacological Therapeutics, 43*, 363–371.

Melzack, R. (1975). The McGill Pain Questionnaire: Major properties and scoring methods. *Pain, 1*, 277–299.

Melzack, R. (1987). The short-form McGill Pain Questionnaire. *Pain, 30*(2), 191–197.

Melzack, R., & Torgerson, W. S. (1971). On the language of pain. *Anesthesiology, 34*(1), 50–59.

Moskowitz, H. R., & Gerbers, C. L. (1974). Dimensional salience of odors. *New York Acadamy of Sciences, 237*, 1–16.

Mower, G. D. (1976). Perceived intensity of peripheral thermal stimuli is independent of internal body temperature. *Journal of Comparative and Physiological Psychology, 90*, 1152–1155.

Mower, G. D., Mair, R., & Engen, T. (1977). Influence of internal factors on the perception of olfactory and gustatory stimuli. In M. Hare & J. Mailer (Eds.), *The Chemical Senses and Nutrition*. New York: Academic Press.

Price, D. D., Barrell, J. J., & Gracely, R. H. (1980). A psychophysical analysis of experimential factors that selectively influence the affective dimension of pain. *Pain, 8*, 137–149.

Price, D. D., Harkins, S. W., & Baker, C. (1987). Sensory-affective relationships among different types of clinical and experimental pain. *Pain, 28*, 297–307.

Rainville, P., Duncan, G. H., Price, D. D., Carrier, B., & Bushnell, M. C. (1997). Pain affect encoded in human anterior cingulate but not somatosensory cortex. *Science, 277*, 968–971.

Rainville, P., Feine, J. S., Bushnell, M. C., & Duncan, G. H. (1992). A psychophysical comparison of sensory and affective responses to four modalities of experimental pain. *Somatosensory & Motor Research, 9*, 265–277.

Rolke, R., Baron, R., Maier, C., Tolle, T. R., Treede, R. D., Beyer, A., & Wasserka, B. (2006). Quantitative sensory testing in the German Research Network on Neuropathic Pain (DFNS): Standardized protocol and reference values. *Pain, 123*, 231–243.

Rollman, G. B. (1977). Signal detection theory measurement of pain: A review and critique. *Pain, 3*, 187–211.

Rolls, B. J., Van Duijvenvoorde, P. M., & Rolls, E. T. (1984). Pleasantness changes and food intake in a varied four-course meal. *Appetite, 5*, 337–348.

Schiffman, S. S. (1974). Contributions to the physiochemical dimensions of odor: A psychophysical approach. *Annals of the New York Academy of Sciences, 237*, 164–183.

Schiffman, S. S., & Erickson, R. P. (1972). *Organization of olfactory stimuli and neurons.* Paper presented at the VII Symposium Meriterraneen sur L'Odorat, Cannes, France.

Schmidt, R., & Cain, W. S. (2010). Making scents: dynamic olfactometry for threshold measurement. *Chemical Senses, 35*, 109–120.

Schweinhardt, P., & Bushnell, M. C. (2010). Pain imaging in health and disease – how far have we come? *Journal of Clinical Investigation, 120*, 3788–3797.

Ship, J. A., & Chavez, E. M. (2001). Special senses: Disorders of taste and smell. In S. J. Silverman, L. R. Eversole, & E. L. Truelove (Eds.), *Essentials of Oral Medicine* (pp. 277–288). Hamilton, Ontario: BC Decker.

Sobel, N., Prabhakaran, V., Desmond, J. E., Glover, G. H., Goode, R. L., Sullivan, E. V., & Gabrieli, J. D. (1998). Sniffing and smelling: separate subsystems in the human olfactory cortex. *Nature, 392*, 282–286.

Stevens, J. C., & Marks, L. E. (1965). Cross-modality matching of brightness and loudness. *Proceedings of the National Academy of Sciences USA, 54*, 407–411.

Stevens, S. S. (1975). *Psychophysics – Introduction to Its Perceptual, Neural and Social Prospects*. New York: John Wiley & sons, Inc.

Stevens, S. S., & Greenbaum, H. B. (1966). Regression effect in psychophysical judgment. *Perception and Psychophysics, 1*, 439–446.

Stevens, S. S., & Oconnell, R. J. (1996). Semantic-free scaling of odor quality. *Physiology and Behavior, 60*, 211–215.

Thesen, T., & Murphy, C. (2002). Reliability analysis of event-related brain potentials to olfactory stimuli. *Psychophysiology, 39*, 733–738.

Thompson, D. A., Moskowitz, H. R., & Campbell, R. G. (1976). Effects of body weight and food intake on pleasantness ratings for a sweet stimulus. *Journal of Applied Physiology, 41*, 77–83.

Thuerauf, N., Gossler, A., Lunkenheimer, J., Lunkenheimer, B., Maihofner, C., Bleich, S., & Reulbach, U. (2008). Olfactory lateralization: Odor intensity but not the hedonic estimation is lateralized. *Neuroscience Letters, 438*, 228–232.

Thurstone, L. I. (1959). *The Measurement of Values*. Chicago: University of Chicago Pres.

Torgerson, W. S., Bendebba, M., & Mason, K. J. (1998). Varieties of Pain. In R. Dubner, G. R. Gebhart, & M. R. Bond (Eds.), *Proceedings of the Vth World Congress on Pain* (pp. 368–374). Amsterdam: Elsevier.

Tursky, B. (1976). The development of a pain perception profile: A psychophysical approach. In M. Wesinberg & B. Tursky (Eds.), *Pain: New Perspectives in Therapy and Research* (pp. 171–194). New York: Plenum Press.

Wallis, B. J., & Bogduk, N. (1996). Faking a profile: can naive subjects simulate whiplash responses? *Pain, 66*, 223–227.

Wise, P. M., Olsson, M. J., & Cain, W. S. (2000). Quantification of odor quality. *Chemical Senses, 25*, 429–443.

Woskow, M. H. (1964). *Multidimensional scaling of odors*. PhD, University of California at Los Angeles, Los Angeles, CA.

Woskow, M. H. (1968). Multidimensional scaling of odors. In N. Tanyolac (Ed.), *Theories of Odor and Odor Measurement* (pp. 147–191). Istanbul: Robert College.

Wright, R. H., & Michels, K. M. (1964). Evaluation of far infrared relations to odors by a standard similarity method. *Annals of the New York Acadamy of Sciences, 116*, 535–551.

Wunsch, A., Philippot, P., & Plaghki, L. (2003). Affective associative learning modifies the sensory perception of nociceptive stimuli without participant's awareness. *Pain, 102*, 27–38.

Zatorre, R. J., Jones-Gotman, M., Evans, A. C., & Meyer, E. (1992). Functional localization and lateralization of human olfactory cortex. *Nature, 360*, 339–340.

CHAPTER 4

Olfactory comfort in close relationships

You aren't the only one who does it

Donald H. McBurney,[1] Sybil A. Streeter[1,3] and Harald A. Euler[2]
[1]University of Pittsburgh / [2] University of Kassel / [3]Olson Zaltman, Inc.

Smelling the unwashed clothing of a loved one when separated is at the same time little known but widely practiced (McBurney et al., 2006). A majority of persons surveyed reported that they did it to remember an absent romantic partner and because it smells good; additionally, it made them feel happy and comfortable. Shoup et al. (2008) found that such olfactory comfort behavior extended to relatives as a function of degree of relatedness. In both American and German subjects, olfactory comfort correlated with adult attachment style: positively with Secure attachment, and negatively with Dismissive attachment. Number of personal photos displayed in a student's dorm room correlated with olfactory comfort behavior and showed the same relationships with attachment. Streeter (2008) experimentally demonstrated the role of olfactory comfort by having subjects smell various T-shirts after experiencing a stressor. Smelling a partner's T-shirt increased comfort, and reduced anxiety and negative affect. The smell of an unknown individual also reduced anxiety and negative affect, but did not influence comfort. Those with Secure attachment showed greater reduction in anxiety than those with other attachment styles. Euler (unpublished) found that smelling of clothing correlated positively with the love styles (Lee, 1973) mania, agape, and eros. In this contribution we review these findings and suggest that they indicate a system of olfactory comfort that relates to Harlow's contact comfort (e.g, Harlow, Harlow, & Suomi, 1971) as well as Bowlby's (1969) theory of attachment as extended to adult romantic attachment (Hazan & Shaver, 1987).

Introduction

The role of olfaction in social relationships is well known, although relatively little studied (Levine & McBurney, 1986). Less well known, however, is the importance of smell in close (romantic) relationships. Herz and Inzlicht (2002), for example, found that a potential lover's odor was a very important consideration in mate choice, either 2nd or 4th out of 10 items. But there remains a class of olfactory-related behaviors that is at the same time extremely common and little known in the scientific literature, namely smelling a loved one's clothing during separation. The first mention of this behavior to our knowledge was by Krafft-Ebing (1931), who noted that

> Among the natives of the Philippine Islands … when it becomes necessary for an engaged pair to separate, they exchange articles of apparel, by means of which each becomes assured of faithfulness. These objects are carefully preserved, covered with kisses and smelled. (p. 32)

The fact that Krafft-Ebing's book was entitled *Psychophathia Sexualis*, however, suggests that the behavior is not quite normal. In addition, the fact that Krafft-Ebing attributes the behavior to only one culture suggests that it is limited geographically. When one of us asked a clinical psychologist how common he thought it was for people to smell each other's clothing during separation, he was visibly repulsed by the idea, and said he supposed possibly one in a hundred might do it. Prior to our study (McBurney, Shoup, & Streeter, 2006), which showed the behavior to be engaged in by the majority of people we asked, there were only scattered reports in the media (e.g., Eisman, 2004) and the Kraft-Ebing anecdote in the literature, to our knowledge.

This behavior, then, is one of those behaviors that falls outside the common, unremarkable, public, ritualized activities of everyday life. But it is well known that there are other behaviors that are commonly done in private, such as picking one's nose, passing gas, and scratching oneself in places called, appropriately, private parts. We find it interesting that olfactory comfort behavior does not even get the public mention that these other behaviors do. Only occasionally in a movie do we see a character smell an absent partner's clothing.

The work of Harlow on attachment in infant monkeys is among the best known in psychology, and has given us the concept of contact comfort (Harlow, Harlow, & Suomi, 1971). We note that Harlow demonstrated that contact comfort is a prerequisite for attachment bonds to develop. Because of the relationship between olfaction and attachment that we discuss below, our working hypothesis is that his argument can be extended beyond the concept of "comfort" as simply a tactile sense to an olfactory one too.

We suggest that olfaction plays a crucial role in social behavior that has been overlooked in humans partly because the behaviors are so private, and partly because olfaction has not received as much attention as other senses. But social olfaction has been well studied in other animals (e.g.,Sanchez-Andrade & Kendrick, 2009). The rise of evolutionary psychology in recent years (e.g., Gaulin

& McBurney, 2004) provides both the framework and motivation to examine this virtually unexplored area.

1. Smelling the clothing of loved ones

McBurney, Shoup, and Streeter (2006) asked American college students who were or had ever been in a committed relationship with a member of the opposite sex a number of questions about smelling their partner's clothing during separation. Seventy two percent of women and 26% of men had slept in or with an item of unwashed clothing of a partner when separated. Euler, Shoup, Streeter, and McBurney (2005) found with German students that 66% of women and 29% of men had done the same. The American students were asked their reasons for smelling the clothing. Women most often said that they did it to remember him or feel close to him, followed by being comforted, and the less specific reason that it smells good. The men were less articulate than the women. Their most common response was that it smells good, and they gave no response twice as often as the women did. But their more specific reasons were the same as the women's, although they gave them about half as often (see Table 1).

The women were also more specific about the resulting emotion from smelling their partner's clothing. They felt happy, relaxed or safe when they smelled the clothing. Males most commonly gave no response, although those who did reported similar emotions as the women. Interestingly, both men and women listed feeling safe at the same frequency, although it was the third most common articulated reason for both (see Table 2).

Shoup, Streeter, and McBurney (2008) replicated the previous study, and extended the questions to others besides lovers. They found that although smelling

Table 1. Open-ended responses to the question, "If you have ever done any of these (i.e. wear partner's clothing to sleep; intentionally smell partner's clothing; go into partner's closet to intentionally smell clothing), please describe your reasons for doing so."

Reason given	Female	Male
To remember him/her, feel close	53%	28%
Smells good	22%	38%
Comforts me	22%	13%
No response/no reason	8%	19%
Feels good	4%	0%
Other	29%	28%

Note. The percentages do not total 100% because of multiple responses (McBurney et al., 2006, reproduced by permission).

Table 2. Open-ended responses to the question, "If you have ever done any of these (i.e. wear partner's clothing to sleep; intentionally smell partner's clothing; go into partner's closet to intentionally smell clothing), please describe how it made you feel."

Resulting emotion	Female	Male
Happy/good	43%	25%
Comfortable/content/relaxed	43%	16%
None/no response	16%	35%
Secure/safe	10%	10%
Close	8%	6%
Other	20%	31%

Note. The percentages do not total 100% because of multiple responses (McBurney et al., 2006, reproduced by permission).

Table 3. Frequency of response to each question by sex and degree of relatedness, from Shoup et al. (2008), reproduced by permission.

Relatedness	Smelled clothing		Slept with clothing		Given/taken clothing	
	Male	Female	Male	Female	Male	Female
First degree	9.0%	25.0%	0%	4.3%	0%	14.3%
Second degree	6.0%	13.0%	0%	4.3%	0%	0.0%
Third degree	0.0%	1.9%	0%	2.1%	0%	0.0%

of clothing was much less common among relatives, it was not uncommon. In accordance with evolutionary theory (Gaulin & McBurney, 2004) the degree of relationship was a strong predictor of smelling clothing: It was most common with first degree relatives (parents and siblings) than second degree relatives (grandparents, aunts and uncles) followed by third degree relatives (cousins and great grandparents (see Table 3). (Because of their ages, these subjects did not have enough children to include as first degree relatives.)

2. The relationship between olfactory comfort and attachment

Bowlby (1969) studied the attachment of infants to their primary caregivers. His groundbreaking work led to a large body of research. Ainsworth and others (e.g., Ainsworth, Blehar, Waters, & Wall, 1978) defined three fundamental styles of infant attachment: secure, anxious/ambivalent, and avoidant. It was not long before the concept of attachment was extended to adults. Hazan and Shaver (1987) were the first to consider romantic love as an attachment process, and make an important distinction between the infant system and adult process. The logic underlying the argument is relatively simple: The same qualities that Bowlby describes as hallmarks of the infant system are preserved and co-opted in adults to promote pair bonds. Attachment theorists have explored the individual differences between these types in the service of many goals: parenting skills, psychopathology, adjustment, and relationship quality (e.g., Schmitt et al., 2003; Simpson, Rholes, & Nelligan, 1992; Watt, McWilliams, & Campbell, 2005; Weems, Berman, Silverman, & Rodriguez, 2002).

Bowlby's theory of attachment is widely considered a mid-level evolutionary theory, like the theory of parent-offspring conflict, or the theory of reciprocal altruism. These mid-level theories all relate to the theory of natural selection above it, but make more specific predictions related to the particular question of interest (Buss, 1995). Accordingly adult attachment should be similarly grounded by early experiences and sensitive to the local environment. Consequently, we may not expect that individual differences in adult attachment should be necessarily stable. When disparities in mate value exist, different strategies should be explored for potential in a manner similar to the Belsky, Steinberg, and Draper (1991) model, whereby father-absence leads to early puberty as a response to uncertain environmental conditions.

While there is considerable variability in the way humans approach pair-bonding or romantic relationships, the mechanisms proposed by attachment theory suggest that early relationships (e.g., one with a warm, responsive caregiver) are a factor that facilitate stability in attachment styles. Further, the evolutionary viewpoint suggests that environmental factors are incorporated into new relationships (Fraley & Spieker, 2003). Thus, the infant attachment system is parsimoniously co-opted by evolution for the maintenance of pair-bonds and continues to operate in essentially the same context (Kirkpatrick, 1998; Zeifman & Hazan, 1997).

We examined the relations between olfactory comfort behaviors and attachment cross-culturally (McBurney, Euler, Streeter, & Shoup, 2005). The same American and German subjects who were asked about smelling clothing of lovers and family members were also given the Relationship Questionnaire, which is a measure of Adult Attachment (Bartholomew & Horowitz, 1991). It asks individuals to select which of the statements listed below best describe them:

> Secure attachment: "It is easy for me to become emotionally close to others. I am comfortable depending on them and having them depend on me. I don't worry about being alone or having others not accept me."
>
> Fearful attachment: "I am uncomfortable getting close to others. I want emotionally close relationships, but I find it difficult to trust others completely, or to depend on them. I worry that I will be hurt if I allow myself to become too close to others."

Table 4. Correlations between intentional smelling a partner's clothing and styles of Adult Attachment in a cross-cultural sample.

Secure	Fearful	Preoccupied	Dismissing
$r = .08$	$r = .02$	$r = .10$	$r = -.18$
($p = .06$)	n.s.	($p = .02$)	($p = .00$)
(n = 545)	(n = 543)	(n = 543)	(n = 544)

> Preoccupied attachment: "I want to be completely emotionally intimate with others, but I often find that others are reluctant to get as close as I would like. I am uncomfortable being without close relationships, but I sometimes worry that others don't value me as much as I value them."
>
> Dismissive attachment: "I am comfortable without close relationships. It is very important for me to feel independent and self-sufficient, and I prefer not to depend on others or have others depend on me."

There is some disagreement regarding whether these types should be considered as categorical styles or dimensional measures (e.g., Bartholomew & Shaver, 1998). We followed the more common practice of using continuous, seven-point rating scales in the present study to permit more quantitative analysis.

Because cultural differences are not relevant to the present discussion and for the sake of clarity, we collapsed across nationality and combined the two samples for correlational analysis. These correlations are presented in Table 4.

The frequency of intentionally smelling a loved one's clothing correlated significantly and positively with higher scores on Preoccupied attachment and approached significance for Secure attachment. The same behavior correlated significantly but negatively with higher scores on Dismissing attachment. Women smelled clothing much more often than men in both samples. These correlations suggest that olfactory comfort behaviors are an expression of attachment style.

3. Another attachment-related behavior: Displayed photos

It occurred to us that number of photos students put up in their dorm room might be another measure of attachment style. We asked both American and German subjects to report number of photos they had on display of people they knew, leaving out celebrities and the like. Because number of photos was markedly skewed, we logarithmically transformed the data. We will, however, report the antilog mean log (retransformed) numbers. There was no difference between German and American subjects in the number of photos, although there was a significant interaction between nationality and sex, such that the sex difference was larger with American subjects. Consistent with the sex difference in olfactory comfort behavior, females had many more photos than males: $M = 12.2$, vs. 2.4; $t(268) = 11.45$, $p < .001$. Supporting our intuition, we found a strong correlation between number of photos and intentional smelling, $r = .25$ ($p < .001$) for German data, and $r = .39$ ($p < .001$) for the US data.

Even more interesting is the fact that number of photos paralleled smelling in the pattern of relationships with attachment style. Among the US subjects, secure attachment correlated with photos, $r = .19$ ($p < .001$) and also with intentional smelling, $r = .18$ ($p < .003$). The pattern with the German subjects was more complex: For secure attachment, only number of photos correlated significantly: $r = .17$ ($p < .005$). Both photos and smelling correlated negatively with dismissive attachment: $r = -.15$ ($p < .015$) and $r = -.18$ ($p < .003$), respectively. Preoccupied attachment correlated with intentional smelling, $r = .16$ ($p < .01$), but not with photos.

It is possible that with the German subjects there is less emphasis on body odor, such that olfactory comfort behaviors are only expressed among less stable/secure individuals who are seeking attachment (i.e. preoccupied attachment style).

4. An experimental test of olfactory attachment

Streeter (2008) explored olfactory attachment at a different level of analysis: Do olfactory comfort behaviors serve a functional purpose? Why do we take comfort in the scent of a significant other, and could individual differences in adult attachment be a moderating factor?

4.1 Methods

Three established instruments to measure anxiety, affect, and attachment style were used. The first was Spielberger's (Spielberger, Gorsuch, & Lushenc, 1970) State-Trait Anxiety Inventory (STAI). The State Anxiety (S-Anxiety) subscale is designed to measure transient changes in anxiety, as opposed to the Trait Anxiety subscale (T-Anxiety), which measures persistent and relatively stable individual differences in tendencies toward anxiety-proneness. Streeter modified the S-Anxiety subscale to include affective words that were most common from our earlier open-ended questionnaires. The second instrument was the Positive and Negative Affect Schedule (PANAS) (Watson, Clark, & Tellegen, 1988). This scale is designed to measure transient changes in positive affect (PANAS-P) and negative affect (PANAS-N) orthogonally. The third instrument was the Relationship Questionnaire (RQ) discussed above.

4.2 Participants

One hundred eighteen university students (94 females and 23 males) participated in the study. The only criterion for inclusion was being in possession of an item of clothing that retains the smell of a romantic partner. Their mean age was 19.64 (±4.66), and their mean relationship length 1.52 years (±16.09) (median = 1 year; mode = 6 months). One participant was removed from the dataset as an outlier on several demographic characteristics. Participants were asked to bring an item of clothing that retained the odor of their romantic partner.

4.3 Design

The experiment was a between-subjects design with 3 randomized conditions: (1) Experimental condition, Partner's Odor: a clothing item with the smell of their partner; (2) Control condition, Neutral Odor: cotton t-shirt item laundered in unscented detergent; (3) Placebo condition, Other Odor: a clothing item with the odor of an unknown volunteer. This item of clothing was never more than three days old and was different for every participant.

4.4 Procedure

Baseline PANAS and STAI were measured at the beginning of the session. Because the hypotheses are based upon reduction of negative affect and anxiety, and because Bowlby (1979) claims that attachment is most strongly activated in times of distress it is under these conditions that attachment processes should be most evident (Simpson, Rholes, & Nelligan, 1992). In order to increase the likelihood of seeing changes in negative affect caused by the experimental manipulation, Streeter asked participants first to perform a challenging mental task known to be a mild stressor (Feeney & Kirkpatrick, 1996): They were to count backwards by, e.g., 17 from a randomly chosen 4-digit number for 5 minutes. (The interval was sometimes changed to 25, 7, or 13 based on the subject's facility with the task, in order to keep it appropriately difficult.) PANAS and STAI were administered immediately following the stressor task (post-stress, but pre-odor exposure).

Each participant was blindfolded, and instructed to smell the contents of an opaque odorless bucket that contained the item from one of the experimental conditions described above.

PANAS and STAI were then administered again (post-odor exposure). The final questionnaire given was the RQ to measure adult attachment.

4.5 Results

Participants exposed to their partners' odor had a significant ($p < .001$) increase in reported feelings of Comfort ($M = 3.05$, $sd = .74$) relative to both the control (Neutral odor; $M = 2.34$, $sd = .90$) and placebo (Other odor; $M = 2.60$, $sd = .89$) conditions as illustrated in Figure 1.

Exposure to a partner's scent significantly reduced Anxiety ($M = -13.49$, $sd =$ .9.26) only when compared to a neutral odor ($M = -7.63$, $sd = 11.60$) but not when compared to other persons ($M = -12.82$, $sd = 8.94$). Figure 2 illustrates this effect.

The pattern for Negative Affect was similar to that of Anxiety. Figure 3 shows that exposure to a partner's scent significantly improved Anxiety ($M = 10.70$, $sd =$ 4.02) only when compared to a neutral odor ($M = 12.19$, $sd = 5.26$) but not when compared to the odor of other persons ($M = 10.64$, $sd = 3.82$). Because the figure is based on percentiles rather than the mean, and the data are skewed, the figure minimizes the size of the effect.

Multiple regression analysis was performed to determine the percent of variance accounted for on the various dependent variables. For Comfort, the regression equation accounted for 25% of the variance; for Anxiety the equation accounted for 36% of the variance, and for Negative Affect 39% of the variance was accounted for by the regression equation.

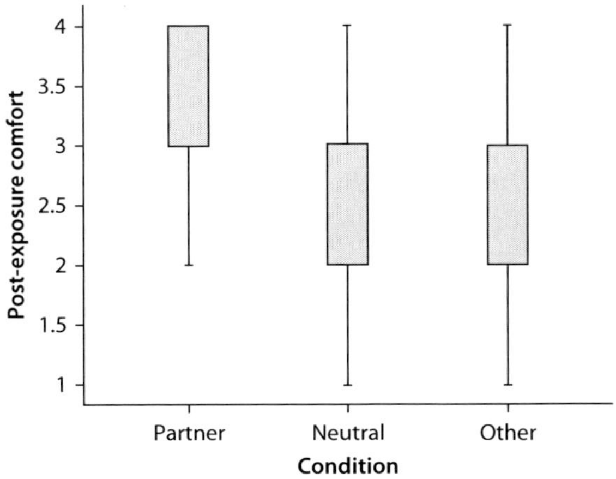

Figure 1. The effect of a partner's scent on Comfort. Participants exposed to their partner's odor showed increased comfort compared to another's odor or a neutral odor. From Streeter (2008). The boxes indicate the 25th and 75th percentiles and the whiskers show the 10th and 90th percentiles.

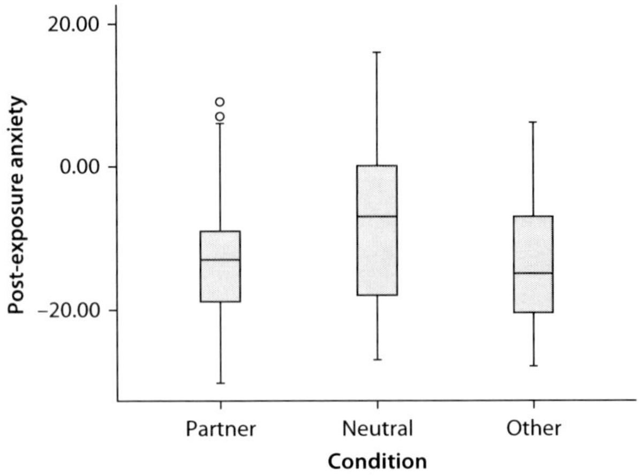

Figure 2. The effect of a partner's scent on Anxiety. Participants exposed to their partner's odor or to another's odor showed less anxiety compared to a neutral odor. From Streeter (2008). The boxes indicate the 25th, 50th, and 75th percentiles and the whiskers show the 10th and 90th percentiles. Circles indicate ratings outside the 10th or 90th percentiles.

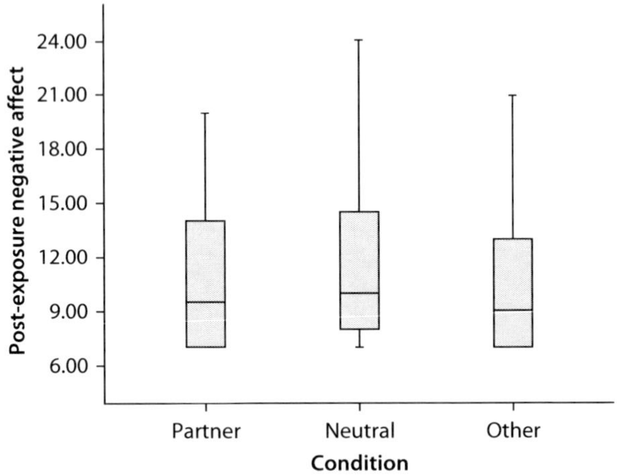

Figure 3. The effect of a partner's scent on Negative Affect. Participants exposed to their partner's odor or another's odor showed less negative affect compared to a neutral odor. From Streeter (2008). The boxes indicate the 25th, 50th, and 75th percentiles and the whiskers show the 10th and 90th percentiles.

4.6 Summary of experimental findings

The scent of a romantic partner improved feelings of comfort compared to the scent of another person or a neutral scent. This experimental finding validates our conceptualization of "olfactory comfort." The scent of one's partner also improved anxiety and negative affect when compared to a neutral odor, but its ability to reduce these aversive states was not different from the scent of unknown persons. This implies that there may be a specific quality to human odors, possibly an olfactory cue to the presence of another person, is sufficient to improve negative emotional states.

Overall, the smell of a romantic partner increased feelings of comfort compared to either a neutral odor or that of another unknown person. This result validates our previous findings (McBurney, Shoup, & Streeter, 2006) from questionnaires confirming that people do derive comfort from the smell of a loved one. Even though participants are not always able to accurately report the reasons for engaging in a particular behavior, in this case they appear to be accurately describing the qualia associated with the scent of a romantic partner. This also validates our conceptualization and coining of the term "olfactory comfort." The results presented in this paper suggest that this behavior is not only common, but it is also functional. In this manipulation, Comfort was experimentally decreased and a clear main effect of the scent of the romantic partner emerged: A partner's scent increased comfort when compared to both placebo (unknown persons) and control (neutral).

5. Love styles and olfactory comfort

Another theoretical approach to the phenomenon of olfactory comfort is given by the concept of love styles, which Lee (1973) developed on the basis of extensive interviews. Hendrick and Hendrick (1986) constructed a questionnaire, the results of which showed suitable internal reliability and reasonable independence for each of the following love styles: Eros (passionate love); Ludus (game playing love); Storge (friendship love); Pragma (logical, or "shopping list" love); Mania (possessive, dependent love); Agape (selfless love). We thought that this theory might be provide some triangulation on the theoretical significance of olfactory comfort. Euler (unpublished) administered a German version of the Love Styles questionnaire (Bierhoff & Klein, 1991) to German students and correlated the results with those of the following question: "How often (when you were separated) did you smell the clothes of another person to be closer to him or her?" The strongest correlation was with Mania: $r = .41$ ($p < .001$, $N = 69$) for males, and $r = .21$

($p < .001$, N = 227) for females. Significant correlations were also found for Agape (r = .16, $p < .005$, N = 296) and Eros (r = .14, $p < .016$, N = 296) for all data. The data separated by sex were similar but the significance was marginal except for females on Agape (r = .19, $p < .003$, N = 227). Note that the number of males surveyed was much smaller than that of females. The magnitude of the correlations, especially for Mania, suggests that further exploration of the relationship between love styles and olfactory comfort would be fruitful. Further, the Mania love style may be similar to the Preoccupied attachment style, and Agape may relate to the Secure attachment style.

6. Summary and conclusion

The data presented here demonstrate that smelling the body odor of a romantic partner or close relative is not only surprisingly common, but that it is functional in reducing anxiety and increasing comfort. It is our hope that these findings will help these behaviors come out of the closet and take their place in research and clinical practice.

References

Ainsworth, M. D. S., Blehar, M. C., Waters, E., & Wall, S. (1978). *Patterns of attachment: A psychological study of the strange situation.* Hillsdale, NJ: Erlbaum.

Bartholomew, K., & Horowitz, L. M. (1991). Attachment styles among young adults: A test of a four-category model. *Journal of Personality and Social Psychology, 61(2),* 226–244.

Bartholomew, K., & Shaver, P. R. (1998). Methods of assessing adult attachment: Do they converge? In J. A. Simpson, & W. S. Rholes (Eds.), *Attachment theory and close relationships* (pp. 25–45). New York: Guilford Press.

Bierhoff, H. W., & Klein, R. (1991). Dimensionen der Liebe: Enwicklung einer deutschsprachigen. Skala zur erfassung von Liebesstilen. /Dimensions of love: Development of a German scale for measuring love styles. *Zeitschrift für Differentielle und Diagnostische Psychologie. 12,* 53–71.

Belsky, J., Steinberg, L., & Draper, P. (1991). Childhood experience, interpersonal development, and reproductive strategy: An evolutionary theory of socialization. *Child Development, 62,* 647–670.

Bowlby, J. (1969). *Attachment and loss: Attachment* Vol. 1, New York: Basic.

Bowlby, J. (1979). *The making and breaking of affectional bonds.* London: Tavistock.

Buss, D. M. (1995). Evolutionary psychology: A new paradigm for psychological science. *Psychological Inquiry, 6,* 1–30.

Eisman, K. (2004). Secrets every woman keeps from her man. Men's Health. Retrieved May 7, 2004, from http://www.menshealth.com/cda/article/0,2823,s1-2-72-0-976,00.html.

Euler, H., Shoup, M., Streeter, S. A., & McBurney, D. H. (2005). Olfactory comfort: Replication with partners and extension to family members, *Annual meeting of the Human Behavior and Evolution Society*. Austin, TX.

Feeney, B. C., & Kirkpatrick, L. A. (1996). Effects of adult attachment and presence of romantic partners on physiological responses to stress. *Journal of Personality and Social Psychology, 70(2)*, 255–270.

Fraley, R. C., & Spieker, S. J. (2003). What are the differences between dimensional and categorical models of individual differences in attachment? Reply to Cassidy (2003), Cummings (2003), Sroufe (2003) and Waters and Beauchaine (2003). *Developmental Psychology, 39(3)*, 423–429.

Gaulin, S. J. C., & McBurney, D. H. (2004). *Evolutionary Psychology*, 2nd edition. Upper Saddle River, NJ: Prentice Hall.

Harlow, H. F., Harlow, M. K., & Suomi, S. J. (1971). From thought to therapy: Lessons from a primate laboratory. *American Scientist, 59*, 538–549.

Hazan, C., & Shaver, P. (1987). Romantic love conceptualized as an attachment process. *Journal of Personality and Social Psychology, 52(3)*, 511–524.

Hendrick, C., & Hendrick, S. (1986). A theory and method of love. *Journal of Personality and Social Psychology, 50*, 392–402.

Herz, R. S., & Inzlicht, M. (2002). Sex differences in response to physical and social factors involved in human mate selection: The importance of smell for women. *Evolution and Human Behavior, 23*, 359–364.

Kirkpatrick, L. A. (1998). Evolution, pair-bonding, and reproductive strategies: A reconceptualization of adult attachment. In J. A. Simpson, & W. S. Rholes (Eds.), *Attachment theory and close relationships* (pp. 353–393). New York: Guilford.

Krafft-Ebing, R. F. von (1931). *Psychopathia Sexualis*. Brooklyn, NY: Physicians and Surgeons Book Co.

Lee, J. A. (1973). *The colors of love: An exploration of the ways of loving*. Englewood Cliffs, NJ: Prentice-Hall.

Levine, J. M., & McBurney, D. H. (1986). The role of olfaction in social perception and behavior. In C. P. Herman, M. P. Zanna, & E. T. Higgins (Eds.), *The Ontario Symposium on Personality and Social Psychology* Vol. 3 (pp. 179–217). Hillsdale, NJ: Lawrence Erlbaum Associates, Inc.

McBurney, D. H., Euler, H. A., Streeter, S. A., & Shoup, M. (2005). Olfactory comfort and attachment styles. *Poster presented at the Annual meeting of the Human Behavior and Evolution Society*. Austin, TX, June 1–5, 2005.

McBurney, D. H., Shoup, M. L., & Streeter, S. A. (2006). Olfactory comfort: Smelling a partner's clothing during periods of separation. *Journal of Applied Social Psychology, 36*, 2325–2335.

Schmitt, D. P., Alcalay, L., Allensworth, M., Allik, J., Ault, L., Austers, I., et al. (2003). Are men universally more dismissing than women? Gender differences in romantic attachment across 62 cultural regions. *Personal Relationships 10*, 307–331.

Sanchez-Andrade, G., & Kendrick, K. M. (2009). The main olfactory system and social learning in mammals. *Behavioral Brain Research, 200*, 323–336.

Shoup, M. L., Streeter, S. A., & McBurney, D. H. (2008). Olfactory attachment: The prevalence and function of smell within relationships. *Journal of Applied Social Psychology, 38*, 2954–2963.

Simpson, J. A. (1990). Influence of attachment styles on romantic relationships. *Journal of Personality and Social Psychology, 59(5),* 971–980.

Simpson, J. A., Rholes, W. S., & Nelligan, J. S. (1992). Support seeking and support giving within couples in an anxiety-provoking situation: The role of attachment styles. *Journal of Personality and Social Psychology, 62,* 434–446.

Streeter, S. A. (2008). *Odor, adult attachment and emotions in romantic relationships.* Unpublished PhD Dissertation, University of Pittsburgh. (Streeter2008.ETD.pdf)

Spielberger, C. D., Gorsuch, R. L., & Lushene, R. X. (1970). STAI Manual for the State-Trait Anxiety Inventory. Palo Alto, CA: Consulting Psychologists Press.

Watt, M. C., McWilliams, L. A., & Campbell, A. G. (2005). Relations between anxiety sensitivity and attachment style dimensions. *Journal of Psychopathology and Behavioral Assessment,* 27, 191–200.

Watson, D., Clark, L. A., & Tellegen, A. D. (1988). Development and validation of brief measures of positive and negative affect: The PANAS scales. *Journal of Personality and Social Psychology, 54,* 1063–1070.

Weems, C. F., Berman, S. L., Silverman, W. K., & Rodriguez, E. T. (2002). The relation between anxiety sensitivity and attachment style in adolescence and early adulthood. *Journal of Psychopathology and Behavioral Assessment, 24,* 159–168.

Zeifman, D., & Hazan, C. (1997). A process model of adult attachment formation. In S. Duck, & W. Ickes (Eds.), *Handbook of personal relationships: Theory, research and interventions,* 2nd edition (pp. 179–195). Hoboken, NJ: John Wiley & Sons, Inc.

CHAPTER 5

Olfactory perception

Richard J. Stevenson
Macquarie University, Australia

From a functional perspective odor perception involves the detection and recognition of complex chemical blends against an ever-changing chemical background. This chapter examines one account of how this might occur using an object recognition approach (ORA). Drawing first on sensory physiology, it is apparent that basic parameters of olfactory processing are consistent with an ORA. There is the capacity to support figure/ground separation based upon rapid cortical adaptation and via content addressable memory, a capacity to recognize even degraded signals. The second part of the chapter examines whether psychological evidence is consistent with an ORA. Based upon data from behavioral genetics, psychophysical studies, child development, cross-cultural research and neuropsychology, these different fields all suggest the centrality of learning and memory in olfactory perception, and support various predictions derived from an ORA. The final part of the chapter examines the limitations and implications of an ORA, for multimodal flavor perception, for the centrality of affect in olfaction and the nature of top-down processing.

Introduction

This chapter considers the problem central to Trygg Engen's most influential book (*The Perception of Odors*, 1982), namely how the human brain extracts meaning from the myriad of volatile chemicals that bind to the olfactory receptors in the nose. In addressing how we get from an odorant to an odor it is profitable to start by considering both the functions that the olfactory system serves in humans and the nature of the stimuli involved in these functions. The reason for adopting this ecological perspective lies in identifying the particular problems that the brain has to solve – the process of perception – in order to achieve these functional goals. With these problems in mind, the chapter then focuses on one proposed solution – an object recognition approach to olfactory perception. This particular approach, suggested in various forms by several authors over the last 30 years (e.g. Bower, 1991; Hopfield, 1999; Lynch, 1986; Malnic et al., 1999; Plailly et al.,

Table 1. Volatile chemical present in representative food stuffs (from Maarse, 1991).

Food stuff	Number of volatiles identified
Cooked rice	124
Cooked beef	880
Cooked leek	100
Potato chips	150
Grapes	225
Cognac	71
Roasted coffee	655

2005; Stevenson & Boakes, 2003; Wilson & Stevenson, 2006), treats sets of largely invariant volatile chemicals as discrete entities – odor 'objects' – which are then recognized by the brain. The main part of the chapter focuses on this memory based recognition theory of odor perception and the evidence that supports it. The final part of the chapter then considers some problems associated with this object recognition approach and its implications.

Although there is likely to be disagreement about the particulars, it is possible to define with some certainty the broad functions of the human olfactory system (Stevenson, 2010). Olfaction is involved in several aspects of ingestive behavior, it serves as an acquired signal for danger (e.g. gas, smoke, rotting flesh, feces) and has a role in social communication. What then are the chemical signals that relate to these different functions? From the functions identified above, most is probably known about the chemical composition of food – flavor. Here, almost without exception, flavor composition is characterised by highly complex blends of volatiles (see Table 1 for examples). More broadly, for many of the danger signals these too are likely to be complex blends, and similarly for social communication cues (e.g., Gallagher et al., 2008). The olfactory system then needs to be able to recognize complex chemical blends rather than single pure chemicals to satisfy each of the functions identified above. In addition, the level of analysis that would seem to be of most functional value is that aimed at the combination not at its individual components (which may be present in other mixtures), because it is the combination that reflects the environmental entity that the olfactory system needs to recognize to execute a particular functional goal.

Recognizing complex chemical blends appears then to be the principal means of achieving each of the functional goals identified above. However, this primary task is made considerably harder by the fact that the environment is generally full of other volatile agents, some of which may be competing signals and some of which may be chemical noise (e.g. Schroder & Hilker, 2008). Thus not only must the olfactory system recognize complex chemical blends, it must do so against a shifting olfactory background. This task is made even harder by an additional consideration, namely that a variety of environmental influences will act to degrade the invariance of a particular chemical mixture (e.g. climatic conditions). For these reasons the olfactory system may often have to recognize degraded chemical sets. In sum, the problem of olfactory perception is how to recognize *broadly* invariant chemical blends against an ever-shifting chemical background. The next section considers how this might be achieved.

1. Perception

The object recognition approach advanced here has four components, which act together to detect biologically salient combinations of volatile chemicals against a chemical background (Wilson & Stevenson, 2003; Stevenson & Wilson, 2007). First, it utilizes rapid cortical adaptation to background odors, so that new olfactory objects can be detected – figure/ground separation. Second, it can learn and store new odor objects, namely particular patterns of stimulation, which correspond to an environmental entity (a particular combination of volatiles). Third, it recognizes odor objects via a pattern matching system, which draws upon the latter memory store. Fourth, it generates a representation – an odor object – which is discrete and largely irreducible (the figure). While this system can arguably recognize complex chemical blends against a chemical background, it is entirely hypothetical unless it is consistent with findings from both olfactory physiology and perception.

1.1 Sensory physiology

Volatile chemicals bind to G-protien receptors located on neurons on the olfactory mucosa (Buck & Axel, 1991). These receptors are broadly tuned, and respond to a range of different volatiles (Malnic et al., 1999). Each olfactory receptor neuron expresses one type of receptor, there being 350–400 different types in humans (Malnic et al., 2004). Broadly, each of the neurons expressing a particular receptor type converges onto one or two glomeruli located in the olfactory bulb (Ressler, Sullivan, & Buck, 1993). Thus the glomerular layer provides a two-dimensional map reflecting the degree to which volatile chemicals have depolarized each of the different receptor types. There is a further temporal dimension to this coding scheme, with the pattern of activation across the glomerular layer evolving over time (Schaefer & Margrie, 2006). An important implication of this summary is that direct correspondence between particular activity of one receptor and one smell (i.e. a labeled line) is highly unlikely, because of the multitude of olfactory receptor types, their broad tuning, and the complex nature of most odorants. A further implication is that differences in expression of individual receptors types are unlikely to make a major difference to a person's ability to recognize a typical environmental odorant, which is consistent with recent human behavioral genetic data (Knaapila et al., 2008).

The output of the initial phase of sensory processing then feeds into the piriform cortex (PC). At least three major processing events occur here. First, in animals and humans, the PC rapidly adapts to consistent presentation of the same odorant (Sobel et al., 2000) and this occurs even in the face of largely consistent peripheral input (e.g., Best & Wilson, 2004). Stevenson and Wilson (2007) have suggested that this rapid cortical adaptation allows for recognition of new olfactory objects against the current chemical background (see Kadohisa & Wilson, 2006a, for evidence) and this process of figure/ground separation may be further enhanced by sniffing (Verhagen, Wesson, Netoff, White, & Wachowiak, 2007) – at least in rats.

The second process in the PC concerns the loss of information relating to the structure and composition of the stimulating odorants. In other words, the PC may be responsible for the unitary nature of olfactory perception. Human neuroimaging work (Gottfried, Winston, & Dolan, 2006) indicates that the anterior PC appears to retain information regarding structural features. In contrast, qualitative similarities among structurally different chemical odorants are mediated by the posterior PC, and conscious access appears to be restricted to this more holistic level of processing. Further support for this view comes from another neuroimaging study (Howard et al., 2009), which suggests that it is activity within the posterior PC, which most closely correlated with participants' ratings of odor quality.

The third process the PC supports is a capacity to learn new input patterns and to match inputs to stored patterns. For example, in animals, neuronal responses to a ternary mixture of odorants becomes progressively more distinct with exposure relative to the response of these neurons to the individual components, suggesting that the piriform cortex is learning to respond selectively to the mixture (Kadohisa & Wilson, 2006b). More generally, several authors have suggested that the PC is an ideal medium to support a content addressable memory system (Haberly, 2001; Hasselmo & Barkai, 1995). Content addressable memory offers two highly useful features: (1) it can identify whether a particular input corresponds with a stored input (i.e. pattern recognition); and (2) it is sensitive to similarity, such that inputs similar to a stored pattern can be recognized. This latter capacity would allow the recognition of degraded inputs – one of the problems that olfactory system has to overcome.

The capacity of the PC to learn, to pattern match, to recognize degraded input, along with combinatorial encoding, loss of chemical feature based information and figure/ground separation, are all processes that appear central to perceiving an odor. Human neuroimaging work also suggests that the PC is involved in learning odor combinations (Li, Luxenberg, Parrish, & Gottfried, 2006) and although there are no direct studies as yet indicating a role in pattern recognition, it is possible that we utilize a more extensive network of structures, including the orbitofrontal cortex and hippocampus. However, the most important conclusion here is that an object recognition approach is entirely consistent with olfactory sensory physiology in animals and humans – indeed it suggests such an approach.

2. Olfactory perception

This section examines the psychological evidence for an object recognition approach, by reviewing evidence pertinent to each major claim.

2.1 Object-level processing and figure/ground separation

As discussed earlier, most odorants are characterized by their complexity. The brain appears to identify the combination and one aspect of this combinatorial coding is that information is lost about the components. There are several reasons for this. First, as described above, the object is the most appropriate level of information processing so there appears little reason to retain information about the components. Second, with degraded signals, the combination is likely to be a more useful in recognition than a reliance on individual features. Third, with rapid cortical adaptation to background odors, detection of a mixture is more likely even if one or more components of the background odor overlap with components in the mixture – thus the mixture is again the more reliable entity with which to detect an event.

An apparent source of confusion in the literature about the analytical capacity of the olfactory system arises from a failure to distinguish between feature-based information that *directly* corresponds with chemical structure or components (which has a physical basis), against similarities to other odorants (which has a psychological basis). In the gustatory system sucrose, for example, reliably generates a sweet taste, and it is possible to detect a sweet taste even in the presence of other tastants (Laing, Link, Jinks, & Hutchinson, 2002). In this case the quality sweetness directly corresponds to properties of the stimulus that lead to the activation of the sweetness receptor, and information from this channel is largely preserved even when other channels of gustatory information are present. With complex odors – in other words *typical* environmental stimuli – this is not the case. Rather when people report odor qualities these more likely reflect similarities to other complex odors and these similarities are a product of mind, not directly of the stimulus as with tastes. Thus when an odor is described as having multiple qualities, it does not automatically follow that these correspond to its component *chemical* parts.

The human psychophysical literature suggests that we have only rudimentary access to the parts that compose an odor – and this conclusion is based upon experiments, which arguably test this supposition under the most favorable circumstances. Under conditions where single pure odorants are used, and participants are trained to identify these odorants alone, when combined into binary, ternary, quaternary etc mixtures, participants are unable to discern the individual components in mixtures composed of three or more odors (e.g., Laing & Francis, 1989). This limit appears to be fairly universal, and applies equally under conditions of lower cognitive load, such as attempting to identify whether one target odor is present in a mixture (e.g., Laing & Glenmarec, 1987), and for combinations of odors identified as good or poor blenders (e.g., Livermore & Laing, 1998a). Even more tellingly, this restriction also applies equally to odor experts and novices (Livermore & Laing, 1996). These findings suggest that we have minimal access to the chemical feature based information that characterizes the early stages of olfactory processing.

The evidence pertinent to figure/ground separation in humans is less well developed. One finding that is suggestive comes from a further experiment by Livermore and Laing (1998b) on odor mixtures. In this study, rather than using single pure chemicals, they instead used 'real' familiar odorants – in other words each odorant was a highly complex chemical mixture. Revealingly, participants' ability to determine which odorants were present, when they were mixed together was identical to that observed with single pure chemicals. That is individual odors could be reliably detected above chance in binary and ternary mixtures, but more complex mixtures could not be reliably analyzed into their parts. This study suggests, first, that a 'real' familiar odor composed of many chemicals is treated as a single entity in just the same way as a single pure chemical. Second, it suggests that these 'real' familiar odors can be as well discriminated from the olfactory background as single pure chemicals, at least when that background consists of no more than one or two simultaneously present odors (in other words a less than ideal test).

A more general feature of olfactory perception, which is consistent with an ability to engage in figure/ground separation, is the finding that olfaction is susceptible to rapid adaptation (Dalton, 2000). Relatedly, this rapid adaptation has both a central and peripheral component in humans (Dalton, 2000), and that at least in animals the central component may be of prime importance in achieving figure/ground separation (e.g. Kadohisa & Wilson, 2006a). So while the limited psychophysical data in humans supporting figure/ground separation is consistent, there is as yet no definitive evidence. In contrast, there is robust evidence favoring object-level processing, with only limited access to component-based information.

2.2 Learning and pattern matching

If perceiving an odor involves being able to recognize it via the type of content addressable memory system that may be instantiated in the PC, then novel odors should be processed differently to familiar odors. Before turning to discuss this, it is important to note that novelty may reflect a new category of odor or a new member of a familiar category and it is not yet clear how important this distinction might be. For the studies reviewed here, most novel odors would be of the latter kind. Two predictions then can be made regarding the processing of novel odors. First, as novel odors would not have been encountered before, they should be harder to discriminate from other novel odors relative to discriminations made with familiar odors. This is because when the activation pattern arising from the glomerular layer relevant to that novel odor is processed within content addressable memory, although it may activate many similarities, it will not be a particularly good match to any stored pattern. As a result the odor will be perceived as redolent of many other odors and thus, would be hard to discriminate from another novel odor (this too being redolent of many, and particular to none). So not only would we expect discrimination to be poorer for novel odors, we would also expect that novel odors would be judged to have more similarities (i.e. be redolent of many other odors) relative to familiar odors.

Consistent with the first prediction, Rabin (1988) reported that participants exposed to a set of unfamiliar odors were better able to discriminate between them on a same/different task, than a set of participants exposed to an unrelated set of unfamiliar odors. This type of study is potentially confounded by the paired comparison approach, in which a small interval separates both to-be-discriminated odorants. Rather than a failure of perception, perhaps it is a failure to hold a stable representation in memory that leads to poorer discriminative performance. This does not, however, appear to be the case. In a second experiment, Rabin (1988), presented participants with a target odor, which could either be familiar or unfamiliar, followed by an odor mixture, which might or might not contain that target. On trials where a target was present in the mixture, the distractor odor in this mixture might be familiar or unfamiliar. The findings from this study revealed that the best discriminative performance occurred with a familiar target and a familiar distractor and that the worst discriminative performance occurred under conditions in which both target and distractor were unfamiliar. In this case both to-be-discriminated odors were simultaneously present and performance was impaired when either odor, and especially so when both odors, were unfamiliar. Thus both successive and simultaneous discrimination procedures indicate that familiarity with an odorant is a significant factor in successful telling the two odors apart. This may occur because it reflects the degree to which each odor is represented in content addressable memory.

A further type of study by Mingo and Stevenson (2007) examined whether there were qualitative differences between familiar and unfamiliar odors that corresponded with the discriminative differences described in the studies above. As expected, Mingo and Stevenson (2007) observed that unfamiliar odors were indeed redolent of a great number of other odors than familiar ones were. In a further experiment, they then examined whether exposure would act to reduce redolence. This was observed, in that exposed unfamiliar odors (i.e. no longer so unfamiliar now) were judged to be redolent of fewer other odors than unexposed unfamiliar odors.

Together, these findings suggest that without prior exposure, a novel odor is judged to be redolent of many different smells and is harder to discriminate from other odors, relative to the familiar smells. This would imply that the input pattern arising from the glomerular layer, when a novel odor is smelled, comes to be learnt and then forms part of the store of previously encountered odors in content addressable memory. More direct evidence of this capacity to learn and of a capacity to pattern match, has been obtained in a rather different sort of experiment. In the first sort, participants are passively exposed to two odor mixture either AX and BY or AY and BX (with each letter indicating a different odorant), with odors A and B being typically more familiar than X and Y (e.g. Stevenson, Case, & Boakes, 2005). Over several studies, employing a range of different odors, exposure to AX and BY, for example, leads to a range of related effects. A becomes more X-like, and X more A-like, and B becomes more Y-like and Y more B-like, when these four odors are smelled alone and rated for their respective odor qualities (Stevenson, 2001a & b). As well as these shifts in perceived quality, odors A and X, and B and Y, are judged to smell more alike following conditioning than A and Y, and B and X, even though all of these odors have been experienced with equal frequency (Stevenson et al., 2005). Finally, A and X, and B and Y are harder to discriminate than A and Y, and B and X, suggesting that shifts in quality, similarity and discriminability are related, and reflect different facets of the same underlying process (Stevenson, 2001c).

If learning were required to support odor perception, then a further prediction would be that any group who has had less opportunity for exposure to a range of odors (in general or in particular) should be poorer at discrimination than a group that has had such exposure. Three such 'natural experimental' populations have been studied. The first is children, who should progressively acquire a broader and broader range of odor experiences during development. It would be expected that younger children – all else being equal – would be poorer at discriminating between sets of odors than older children and adults, as the younger ones would likely have encountered a smaller range of odors during their lifetime, and evidence consistent with this prediction has been obtained (Cain et al., 1995; Stevenson, Mahmut, & Sundqvist, 2007; Stevenson, Sundqvist, & Mahmut, 2007).

A further type of 'natural experiment' occurs when participants from different cultures are compared. Yet again, and particularly for culturally specific odorants, we might expect discriminative benefits for such odors within a culture, relative to another culture, and vice versa. While the evidence base here is less well developed than for children it is still consistent. For example, Ueno (1993) found that when Japanese and Nepalese participants were asked to organize a set of odors by their relative similarity to each other fish-like odors were clustered into a discrete group by Japanese participants, but no such discrete grouping was observed for Nepalese participants. Presumably fish-like odors are less frequently encountered in Nepal.

The final type of 'natural experiment' concerns groups that have developed some particular form of olfactory perceptual expertise. In this regard, wine and beer tasters are of some interest. Needless to say these stimuli are multimodal, but the olfactory component is a significant contributor to what discriminates one wine or beer from another. Consistent with data above on culture specific odors and, developmental differences in children, three conceptually similar observations have emerged from the beer and wine expertise literature. First, novice wine drinkers perform more poorly on a wine discrimination task than experts (e.g. Solomon, 1990). Second, regular wine drinkers are as effective at discriminating differences between wines as experts, when verbal based cues are controlled for (e.g. Melcher & Schooler, 1996). Third, exposure enhances discriminative performance (Peron & Allen, 1988).

These 'natural experiments' seem to confirm what the laboratory based studies suggest, that passive exposure – presumably leading to learning of the particular glomerular input pattern associated with that odor – results in an improved capacity to discriminate between that learned odor and others. This perspective places a premium on memory. If then the content addressable memory store were damaged, resulting in the progressive or sudden loss of its cache of odor

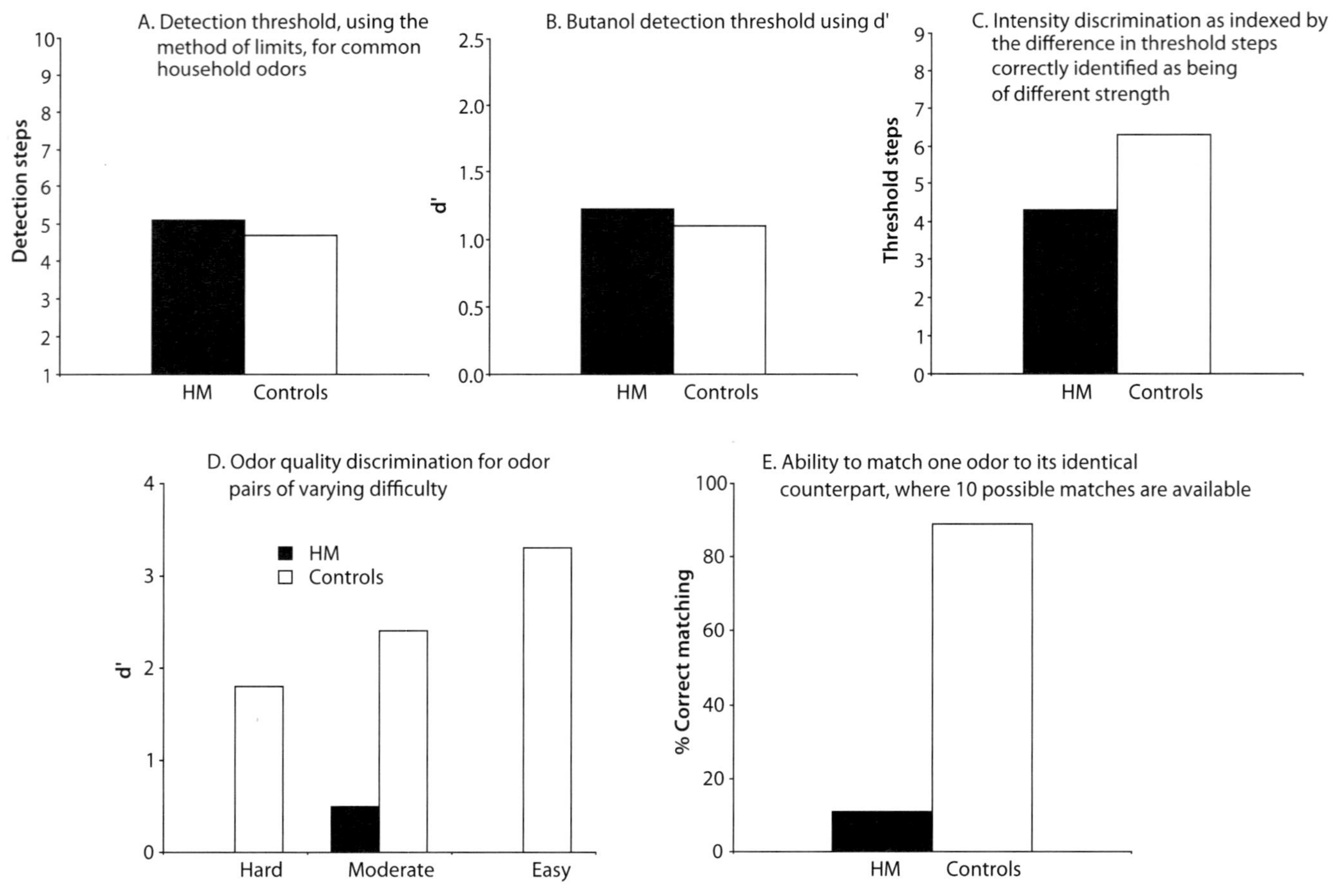

Figure 1. HM and control participants' performance on tests of odor detection (panel A and B), intensity discrimination (panel C), quality discrimination (panel D) and matching (panel E). Data adapted from Eichenbaum, Morton, Potter, and Corkin (1983).

input memories, this should result in significantly impaired discriminative ability. Moreover, if this odor memory store were completely destroyed, a person should still be able to tell that an odor is present or absent, but would have no ability to discriminate that odor from others or to appreciate its similarities (i.e. qualities) to other odors. In fact as detailed below, a considerable body of neuropsychological data points towards just such a pattern of impairment following damage to the PC.

The best example of impaired object perception with intact sensitivity is HM, who received a bilateral temporal lobotomy for intractable epilepsy, which eliminated nearly all of both PCs. A detailed study of his olfactory abilities was undertaken by Eichenbaum, Morton, Potter, and Corkin (1983). They found that HM's ability to detect an odor (i.e. sensitivity) was intact and as good as controls (see Figure 1, panels A & B). Similarly, HM's ability to discriminate differences in intensity between odorants did not differ from that of controls (see Figure 1, panel C). However, HM was completely impaired on tasks that depended upon perceiving odor quality. On a same/different discrimination task, using pairs of odors graded for difficulty, HM performed at chance level, and significantly below that of controls (see Figure 1, panel D). On a matching task, in which he had to match the smell of one odor to the *same* odor, where 10 different potential matches were available, he again performed at chance (see Figure 1, panel E). This last task also included a naming component, but he was unable to name these common odors, even though he could identify these items quite readily via visual and tactile means.

Similar deficits, albeit not as absolute as HM's, have also been observed in patients who have experienced unilateral temporal lobectomies. In these cases, the extent of the dissociation between detection and discrimination may be less apparent because the participant still has one largely intact temporal lobe, and thus associated olfactory structure with which to process incoming olfactory information (e.g. Eskenazi, Cain, Novelly, & Friend, 1983). A further patient population of interest are those with Alzheimers disease. This condition has long been known to affect olfactory processing, both centrally and peripherally (e.g. Luzzi et al., 2007). Notwithstanding the potential for impairment at multiple levels a number of studies suggest that discrimination deficits can occur with intact detection, and that identification deficits (i.e. naming an odor) can occur with intact discriminative ability. Luzzi et al. (2007) found that Alzheimer's patients were significantly impaired on a discrimination and naming task, while Fronto-temporal dementia and Semantic dementia patients had normal discriminative ability, but highly impaired naming ability – relative to age-matched controls. Djordjevic, Jones-Gotman, De Sousa, and Chertkow (2008), found that impaired discriminative ability in an early Alzheimer's group (one akin to the Luzzi et al., 2007, study) was independent of detection threshold impairments that were also observed in this same group. Other studies, which have only focused on identification in Alzheimer's patients have observed that such deficits are associated with abnormalities in the PC (Kareken et al., 2001), and as intact discriminative capacity is required for successful identification (i.e. the latter being higher in the processing hierarchy), this suggests one neural locus for this impairment. Finally, a parallel, although not as profound dissociation between detection and discrimination, can also be observed during normal ageing (e.g. Cain, de Wijk, Nordin, & Nordin, 2008). In sum, these neuropsychological findings suggest that impaired discriminative capacity (i.e. perception of odor quality – object perception) can occur independently from impaired detection capacity (i.e. odor present/absent) and impaired identification. Selective impairments in olfactory discrimination seem to represent an inability to experience odor objects. This arguably results from damage to the store of previous olfactory experiences and the pattern matching hardware – content addressable memory – located in the PC.

The evidence presented in this section draws from several diverse areas of the olfactory literature. Nonetheless, it seems to point to a particular form of perception based upon object recognition. This model takes as its input a spatial and temporal code, which emerges from the glomerular layer. This code contains structural information, but this structural information is largely lost by convergence in the posterior PC. Repeated presentation of an input pattern results in rapid adaptation, so that the system remains optimized for detecting novel olfactory events against this adapted background. The posterior PC also appears to operate as content addressable memory store, able to learn new inputs and with its output representing the odor percept – a discrete (i.e. discriminable) representation of one broadly invariant olfactory event currently occurring in the environment. This percept lacks features in the sense that much information relating to its chemical structure and component volatiles is lost, but the object can be readily compared to other odor objects stored in memory, which produces both its unique quality and its similarities. In the final section of this chapter some of the unresolved features of this model – and olfactory perception – are examined, along with their broader implications.

3. Discussion

While it seems difficult to dispute some sort of recognition model broadly along the lines considered here, there are several aspects of this approach that are unresolved. In a recent article, Yeshurun and Sobel (2010) suggest that the focus on an olfactory object as being a representation of some aspect of the environment (e.g. predator, mate, food etc.), may be incorrect and that an odor object would be better conceived as something that when perceived stirs a particular emotion – an idea foreshadowed by Trygg Engen who noted that "odor is to emotion what vision is to cognition" (*The Perception of Odors*, p. 155). In this alternate perspective the affective nature of the odorant *is* the object. This is an intriguing line of argument and one that is appealing because the affective nature of olfactory experience has frequently been remarked upon (e.g. Engen, 1982; Herz & Schooler, 2002). One major plus of this suggestion is that an emotional response is simply a vehicle for conveying information, and one that may bypass the need for (in humans) verbal identification of the odor source. Nonetheless, the odor source still has to be discriminated from the multitude of potentially meaningful environmental odors and so the fundamental problems of perception remain. What Yeshurun and Sobel's (2010) new perspective adds is that the *meaning* of a particular odor may principally be reflected in the affective state that it produces in the perceiver. It is conceivable then that olfaction generates both a quality-based and an affectively-based object.

With a focus on affect as a key product of the perceptual process, there would be little need to formally identify the odorant. This might be as well, because many studies have observed that odor identification is particularly poor and this may result from misperception (Jonsson, Tchekjova, Lonner, & Olsson, 2005), interference with language processing (Lorig, 1999), and unstable access to semantic information during identification (Cain, de Wijk, Lulejian, Schiet, & See, 1998). Whether hedonic reactions to odorants are more stable (setting aside variation produced by current motivational needs) remains to be seen, but what research there is suggests that affective reaction, as with odor quality, can be affected by various 'top-down' means. A major question is whether such top-down influences on odor perception and affect reflect changes in the percept (object) or changes in the way that percept is described and reported. For example, in Herz and Von Clef's (2001) studies on the influence of verbal descriptors on affective judgments (e.g. more positive affective reactions when a more positive odor label is employed) participants here did seem to think that the same odors with different labels were *not* the same. Whether this was case in Morrot, Brochet, and Dubourdieu's (2001) study, where coloring white wine red led to olfactory judgments consistent with red wines, is not known.

If labels and color cues do affect the very nature of the olfactory percept, how then do these influences work – do they, for example, lead to the recovery of different olfactory memories (e.g. red wines or bad smells)? This possibility seems problematic, because it is hard to voluntarily access an odor memory (e.g. forming a mental image of an odor or 'rehearsing' an odor to keep it in short term memory), but perhaps involuntary access is different. Not many studies have attempted to

Table 2. Mean discriminative errors under various color conditions (from Stevenson & Oaten, 2008).

Experiment	Uncolored	Inappropriate color	Appropriate color
No articulatory Suppression	0.9	1.5	1.0
With articulatory Suppression	1.3	1.1	1.1

address this sort of question directly, but Stevenson and Oaten (2008) attempted to do so by comparing the effects on discriminative performance of presenting odors in appropriate colors (e.g. lime vs mint in green colored fluid), inappropriate colors (e.g. lime vs mint in red colored fluid) or in colorless fluids. Participants made significantly more discriminative errors in the inappropriate colored fluids, suggesting a 'top-down' influence on perception (see Table 2). However, when the experiment was repeated using an articulatory suppression task to minimize the use of a verbal strategy, discriminative performance did not differ between these three conditions, and discriminative performance in the inappropriate color condition significantly improved (see Table 2). This would seem to suggest that rather than color affecting the percept, it affected participants' capacity to identify the odor and this then interfered with discrimination. Eliminating this influence, improved participants' capacity to tell the odors apart. This would imply for odor quality at least, that top-down processing might not always influence the perception of the object. So while there may be examples of where perception is influenced, this experiment illustrates that top-down processing can work via non-perceptual channels.

A further issue is the sometimes multimodal nature of olfactory perception. The object based approach presented so far is specific to the olfactory modality, but evidence has emerged that pure olfactory stimuli can give rise to perceptions that seem to mimic those generated by other sensory systems (e.g. Stevenson & Tomiczek, 2007). This has been most extensively documented for tastes, but may also apply for certain somatosensory qualities as well (e.g. Sundqvist, Stevenson, & Bishop, 2006). Two major questions arise about these type of observations. The first concerns whether the reports of smells with gustatory or somatosensory qualities reflect perceptual content or are more indicative of participants recollection of with what such odors occurred with before (note the essential similarity of this question to the one discussed above). In this case the evidence seems to point to a perceptual phenomenon, at least for tastes, as somatosenory qualities have not been studied in any depth (see Stevenson, 2009, for review). The second question, and one which does not have an answer as yet, is how the object based perceptual system described in this chapter meshes with a capacity to generate multimodal percepts. Some have suggested that the PC may be capable of storing multimodal memories (Haberly, 2001), while other have suggested more extensive networks encompassing the hippocampal system and the orbitofrontal cortex (Verhagen & Engelen, 2006).

Odor-induced tastes appear to arise from the dual nature of olfactory perception. Odors can be sensed orthonasally and retronasally (Rozin, 1982). Retronasal sensing is essential for flavor perception. Flavors seem to be learnt (e.g. Stevenson, Boakes, & Prescott, 1998) and accessed in a fashion that parallels that of odors discussed earlier in this chapter. Orthonasal perception of an odor that once formed part of flavor seems to result in a percept that is essentially similar to that flavor (e.g. sniff strawberry odor and perceive strawberry-sweet). It has been suggested that these two modes of odorant delivery – orthonasal and retronasal – may engage different neural systems, and so represent two discrete but overlapping systems of odor perception (Rozin, 1982; Small et al., 2005). While neuroimaging evidence is consistent with this suggestion (Small et al., 2005), there is at present little evidence at a psychological level for this distinction (Stevenson, 2009).

As has been noted by many authors, olfactory perception has some apparently unique features relative to the other senses (e.g. Herz & Engen, 1996; Smythies, 1997; Stevenson, 2009b; Zucco, 2003). Multimodal perception is one such example, and unstable access to semantic memory is another. A further difference, and one directly emerging from the perceptual model considered in this chapter is the serial unitary nature of olfaction. In the visual system, more information is phenomenally available (e.g. rich visual scene) than can be accessed (e.g. particular parts of that scene; see Block, 2005). However, for olfaction, phenomenal experience is unitary and here accessible content seems to equal phenomenal content (Stevenson, 2009). This difference, which has important implications for the generality of contemporary models of consciousness (see Block, 2005), along with the multimodal nature of olfactory experience and unstable access to semantic memory – and other differences too – point to the way that the functional requirements of a sensory system and the nature of the stimuli it detects, appear to dictate the way in which that system represents information and thus the nature of our conscious experience. Thus to end, where the chapter began, it appears essential to try and understand the functional requirements and the ecological constraints that apply to a particular perceptual system in order to determine how it might work.

Acknowledgements

The author thanks the Australian Research Council for their support and Don Wilson and Bob Boakes for their insights into olfactory perception.

References

Best, A. R., & Wilson, D. A. (2004). Coordinate synaptic mechanisms contributing to olfactory cortical adaptation. *Journal of Neuroscience, 24*, 652–660.

Block, N. (2005). Two neural correlates of consciousness. *Trends in Cognitive Sciences, 9,* 46–52.

Bower, J. M. (1991). Piriform cortex and olfactory object recognition. In J. Davis, & H. Eichenbaum (Eds.), *Olfaction as a model system for computational neuroscience* (pp. 265–285). Cambridge, MA: MIT Press.

Buck, L. B., & Axel, R. (1991). A novel multigene family may encode odorant receptors: A molecular basis for odor recognition. *Cell, 85*, 175–187.

Cain, W. S., Stevens, J. C., Nickou, C. M., Giles, A., Johnston, I., & Garcia-Medina, M. R. (1995). Life-span development of odor identification, learning, and olfactory sensitivity. *Perception, 24*, 1457–1472.

Cain, W. S., de Wijk, R., Lulejian, C., Schiet, F., & See, L.-C. (1998). Odor identification: Perceptual and semantic dimensions. *Chemical Senses, 23*, 309–326.

Cain, W. S., de Wijk, R., Nordin, S., & Nordin, M. (2008). Independence of odor quality and absolute sensitivity in a study of aging. *Chemosensory Perception, 1*, 24–33.

Dalton, P. (2000). Psychophysical and behavioral characteristics of olfactory adaptation. *Chemical Senses, 25*, 487–492.

Djordjevic, J., Jones-Gotman, M., De Sousa, K., & Chertkow, H. (2008). Olfaction in patients with mild cognitive impairments and Alzheimer's disease. *Neurobiology of Aging, 29,* 693–706.

Eichenbaum, H., Morton, T. H., Potter, H., & Corkin, S. (1983). Selective olfactory deficits in case H.M. *Brain, 106*, 459–472.

Engen, T. (1982). *The Perception of Odors.* NY: Academic Press.

Eskenazi, B., Cain, W. S., Novelly, R. A., & Friend, K. B. (1983). Olfactory functioning in temporal lobectomy patients. *Neuropsychologia, 21*, 365–374.

Gallagher, M., Wysocki, J., Leyden, J., Spielman, A., Sun, X., & Preti, G. (2008). Analyses of volatile organic compounds from human skin. *British Journal of Dermatology, 159*, 780–791.

Gottfried, J. A., Winston, J. S., & Dolan, R. J. (2006). Dissociable codes of odor quality and odorant structure in human piriform cortex. *Neuron, 49*, 467–479.

Haberly, L. B. (2001). Parallel-distributed processing in olfactory cortex: New insights from morphological and physiological analysis of neuronal circuitry. *Chemical Senses, 26,* 551–577.

Hasselmo, M. E., & Barkai, E. (1995). Cholinergic modulation of activity-dependent synaptic plasticity in the piriform cortex and associative memory function in a network biophysical simulation. *Journal of Neuroscience, 15*, 6592–6604.

Herz, R. S., & Engen, T. (1996). Odor memory: Review and analysis. *Psychonomic Bulletin and Review, 3*, 300–313.

Herz, R. S., & Schooler, J. W. (2002). A naturalistic study of autobiographical memories evoked by olfactory and visual cues: Testing the Proustian hypothesis. *American Journal of Psychology, 115*, 21–32.

Herz, R. S., & von Clef, J. (2001). The influence of verbal labelling on the perception of odors: Evidence for olfactory illusions? *Perception, 30*, 381–391.

Hopfield, J. J. (1999). Odor space and olfactory processing: collective algorithms and neural implementation. *Proceedings of the National Academy of Sciences USA, 96*, 12506–12511.

Howard, J. D., Plailly, J., Grueschow, M., Haynes, J.-D., & Gottfried, J. A. (2009). Odor quality coding and categorization in human posterior piriform cortex. *Nature Neuroscience, 12*, 932–940.

Jonsson, F. U., Tchekhova, A., Lonner, P., & Olsson, M. J. (2005). A metamemory perspective on odor naming and identification. *Chemical Senses, 30*, 353–365.

Kadohisa, M., & Wilson, D. A. (2006a). Olfactory cortical adaptation facilitates detection of odors against background. *Journal of Neurophysiology, 95*, 1888–1896.

Kadohisa, M., & Wilson, D. A. (2006b). Separate encoding of identity and similarity of complex familiar odors in piriform cortex. *Proceedings of the National Academy of Sciences, 103*, 15206–15211.

Kareken, D. A., Doty, R. L., Moberg, P. J., Mosnik, D., Chen, S. H., Farlow, M. R., & Hutchins, G. D. (2001). Olfactory-evoked regional cerebral blood flow in Alzheimer's disease. *Neuropsychology, 15*, 18–29.

Knaapila, A., Tuorila, H., Silventoin, K., Wright, M. J., Kyvik, K., Keskitalo, K., Hansen, J., Kaprio, J., & Perola, M. (2008). Environmental effects exceed genetic effects on perceived intensity and pleasantness of several odors: A three-population twin study. *Behavioral Genetics, 38*, 484–492.

Laing, D. G., Link, C., Jinks, A. L., & Hutchinson, I. (2002). The limited capacity of humans to identify the components of taste mixtures and taste-odour mixtures. *Perception, 31*, 617–635.

Laing, D. G., & Francis, G. W. (1989). The capacity of humans to identify odors in mixtures. *Physiology and Behavior, 46*, 809–814.

Laing, D. G., & Glemarec, A. (1987). Selective attention and the perceptual analysis of odor mixtures. *Physiology and Behavior, 52*, 1047–1053.

Li, W., Luxenberg, E., Parrish, T., & Gottfried, J. A. (2006). Learning to smell the roses: Experience-dependent neural plasticity in human piriform and orbitofrontal cortices. *Neuron, 52*, 1097–1108.

Livermore, A., & Laing, D. G. (1996). Influence of training and experience on the perception of multicomponent odor mixtures. *Journal of Experimental Psychology: Human Perception and Performance, 22*, 267–277.

Livermore, A., & Laing, D. G. (1998a). The influence of odor type on the discrimination and identification of odorants in multicomponent odor mixtures. *Physiology and Behavior, 65*, 311–320.

Livermore, A., & Laing, D. G. (1998b). The influence of chemical complexity on the perception of multicomponent odor mixtures. *Perception and Psychophysics, 60*, 650–661.

Lorig, T. S. (1999). On the similarity of odor and language perception. *Neuroscience and Biobehavioral Reviews 23*, 391–398.

Luzzi, S., Snowden, J., Neary, D., Coccia, M., Provinciali, L., & Lambon-Ralph, M. (2007). Distinct patterns of olfactory impairment in Alzheimer's disease, semantic dementia, frontotemporal dementia, and corticobasal degeneration. *Neuropsychologia, 45,* 1823–1831.
Lynch, G. (1986). *Synapses, circuits and the beginnings of memory.* Cambridge, MA: MIT Press.
Maarse, H. (1991). *Volatile compounds in foods and beverages.* NY: Marcel Dekker.
Malnic, B., Hirono, J., Sato. T., & Buck, L. B. (1999). Combinatorial receptor codes for odors. *Cell, 96,* 713–723.
Malnic, B., Godfrey, P. A., & Buck, L. B. (2004). The human olfactory receptor gene family. *Proceedings of the National Academy of Sciences, 101,* 2584–2589.
Melcher, J. M., & Schooler, J. W. (1996). The misrememberance of wines past: verbal and perceptual expertise differentially mediate verbal overshadowing of taste memory. *Journal of Memory and Language, 35,* 231–245.
Mingo, S., & Stevenson, R. J. (2007). Phenomenological differences between familiar and unfamiliar odours. *Perception, 36,* 931–947, 2007.
Morrot, G., Brochet, F., & Dubourdieu, D. (2001). The color of odors. *Brain and Language, 79,* 309–320.
Peron, R. M., & Allen, G. L. (1988). Attempts to train novices for beer flavor discrimination: A matter of taste. *Journal of General Psychology, 115,* 403–418.
Plailly, J., Bensafi, M., Pachot-Clouard, M., Delon-Martin, C., Kareken, D. A., Rouby, C., Segebarth, C., & Royet, J.-P. (2005). Involvement of right piriform cortex in olfactory familiarity judgments. *NeuroImage, 25,* 1032–1041.
Rabin, M. D. (1988). Experience facilitates olfactory quality discrimination. *Perception and Psychophysics, 44,* 532–540.
Ressler, K. J., Sullivan, S. L., & Buck, L. B. (1993). A zonal organization of odorant receptor gene expression in olfactory epithelium. *Cell, 73,* 597–609.
Rozin, P. (1982). "Taste-smell confusions" and the duality of the olfactory sense. *Perception and Psychophysics, 31,* 397–401.
Schaefer, A. T., & Margrie, T. W. (2006). Spatiotemporal representations in the olfactory system. *Trends in Neurosciences, 30,* 92–100.
Schroder, R., & Hilker, M. (2008). The relevance of background odor in resource location by insects: A behavioral approach. *BioScience, 58,* 308–316.
Small, D. M., Gerber, J. C., Mak, Y. E., & Hummel, T. (2005). Differential neural responses evoked by orthonasal versus retronasal odorant perception in humans. *Neuron, 47,* 593–605.
Smythies, J. (1997). The functional neuroanatomy of awareness: With a focus on the role of various anatomical systems in the control of intermodal attention. *Consciousness and Cognition, 6,* 455–481.
Sobel, N., Prabhakaran, V., Zhao, Z., Desmond, J. E., Glover, G. H., Sullivan, E. V., & Gabrieli, J. D. E. (2000). Time course of odorant-induced activation in the human primary olfactory cortex. *Journal of Neurophysiology, 83,* 537–551.
Solomon, G. E. A. (1990). Psychology of novice and expert wine talk. *American Journal of Psychology, 103,* 495–517.
Stevenson, R. J. (2001a). Associative learning and odor quality perception: How sniffing an odor mixture can alter the smell of its parts. *Learning and Motivation, 32,* 154–177.
Stevenson, R. J. (2001b). The acquisition of odour qualities. *The Quarterly Journal of Experimental Psychology, 54A,* 561–577.

Stevenson, R. J. (2001c). Perceptual learning with odors: Implications for psychological accounts of odor quality perception. *Psychonomic Bulletin and Review, 8*, 708–712.

Stevenson, R. J. (2009). *The psychology of flavour*. Oxford: Oxford University Press.

Stevenson, R. J. (2009b) Phenomenal and access consciousness in olfaction. *Consciousness and Cognition, 18*, 1004–1017.

Stevenson, R. J. (2010). An initial evaluation of the functions of human olfaction. *Chemical Senses, 35*, 3–20.

Stevenson, R. J., & Boakes, R. A. (2003). A mnemonic theory of odor perception. *Psychological Review, 110*, 340–364.

Stevenson, R. J., Boakes, R. A., & Prescott, J. (1998). Changes in odor sweetness resulting from implicit learning of a simultaneous odor-sweetness association: An example of learned synesthesia. *Learning and Motivation, 29*, 113–132.

Stevenson, R. J., Case, T. I., & Boakes, R. A. (2005). Implicit and explicit tests of odor memory reveal different outcomes following interference. *Learning and Motivation, 36*, 353–373.

Stevenson, R. J., Mahmut, M., & Sundqvist, N. (2007). Age-related changes in odor discrimination. *Developmental Psychology, 43*, 253–260.

Stevenson, R. J., & Oaten, M. (2008). The effect of appropriate and inappropriate stimulus color on odor discrimination. *Perception and Psychophysics, 70*, 640–646.

Stevenson, R. J., Sundqvist, N., & Mahmut, M. (2007). Age-related changes in discrimination of unfamiliar odors. *Perception and Psychophysics, 69*, 185–192.

Stevenson, R. J., & Tomiczek, C. M. (2007). Olfactory induced synesthesias: A review and model. *Psychological Bulletin, 133*, 294–309.

Stevenson, R. J., & Wilson, D. A. (2007). Olfactory perception: An object recognition approach. *Perception*, 36, 1821–1833.

Sundqvist, N. C., Stevenson, R. J., & Bishop, I. R. J. (2006). Can odours acquire fat-like properties? *Appetite, 47*, 91–99.

Thomann, P. A., Dos Santos, V., Sidl, U., Toro, P., Essig, M., & Schroder, J. (2009). MRI-derived atrophy of the olfactory bulb and tract in mild cognitive impairment and Alzheimer's disease. *Journal of Alzheimer's Disease, 17*, 213–221.

Ueno, Y. (1993). Cross-cultural study of odor perception in Sherpa and Japanese people. *Chemical Senses, 18*, 352–353.

Verhagen, J. V., Wesson, D. W., Netoff, T. I., White, J. A., & Wachowiak, M. (2007). Sniffing controls an adaptive filter of sensory input to the olfactory bulb. *Nature Neuroscience, 10*, 631–639.

Verhagen, J. V., & Engelen, L. (2006). The neurocognitive bases of human multimodal food perception: sensory integration. *Neuroscience and Biobehavioral Reviews, 30*, 613–650.

Wilson, D., & Stevenson, R. J. (2003). The fundamental role of memory in olfactory perception. *Trends in Neuroscience, 26*, 243–247.

Wilson, D., & Stevenson, R. J. (2006). *Learning to smell*. Baltimore: John Hopkins University Press.

Yeshurun, Y., & Sobel, N. (2010). An odor is not worth a thousand words: From multidimensional odors to unidimensional odor objects. *Annual Review of Psychology, 61*, 219–241.

Zucco, G. M. (2003). Anomalies in cognition: Olfactory memory. *European Psychologist, 8*, 77–86.

PART II

Learning and memory

CHAPTER 6

Odor memory and the special role of associative learning

Rachel S. Herz
Brown University, Providence, USA

The field of odor-memory comprises both memory for odors and the ability for odors to evoke autobiographical recollections. Trygg Engen pioneered the psychological investigation of memory for odors. His observation that odor recognition memory was based on associative learning was pivotal to igniting the study of odor-evoked memory and to developing an understanding of the fundamental mechanisms of olfactory perception and cognition. In this chapter, I review Engen's contributions to the field of odor memory and research in odor-evoked memory from its origins to the present. Engen's work on odor memory led him to posit that it may be a unique memory system, and this chapter discusses what has led to the conclusion that odor memory is distinct from other memory systems. In particular, I highlight that the special features of odor-associative learning, from proactive interference to the unique predisposition for change on the basis of experience, are the basis for odor memory's distinguishing characteristics.

Introduction

Among his many achievements, Trygg Engen pioneered the field of odor memory. The broad framework of "odor memory", however, comprises two distinct cognitive-perceptual processes. One is the ability to recognize and/or identify that one has smelled a particular odor before. This is our memory for odors. In contrast to our ability to recognize that an odor has been previously encountered, is the experience of having an odor evoke the memory of a past episodic event. This experience is often colloquially referred to as the "Proust phenomenon" for the famed literary anecdote where Marcel Proust recounts dipping a madeleine biscuit into linden tea and being transported to a long forgotten event from his childhood (Proust, 1928). Scientifically, these occurrences are designated as "odor-evoked memories".

The empirical odor memory issues that Engen was most interested in were questions concerning memory for odors. Specifically, how particular odors are remembered and recognized after immediate exposure (short-term memory), as well as after long delays (long-term memory); how semantic and verbal labels are involved in odor recognition; how odors are forgotten; and whether or not we can "recall" odors. What follows is a brief review of Engen's contribution to these areas of study. I will then explain how this research paved the way for investigating odor-evoked memory and review the major findings to date.

1. Memory for odors

Engen's odor recognition research was modeled after the classical Ebbinghaus memory paradigm where the ability to distinguish "old" (previously encountered items) from "new" (novel items) is compared (Ebbinghaus, 1913). In several experiments, Engen and colleagues (Engen & Ross, 1973; Engen, Kuisma, & Eimas, 1973) found that there was essentially no difference between short-term and long-term memory for odors. For example, in tests of the ability to recognize a previously presented odor after time intervals that varied from three seconds to one year, there was almost no difference in performance accuracy; after three seconds, memory performance was at 70% and after one year it had dropped to 65%, with only a borderline statistical difference between the two response rates. Regarding these findings, Engen concluded: "Time seems to play no role in odor memory" (*The Perception of Odors*, 1982, p. 107).

Importantly, however, Engen showed that the number of "new" odors and how similar the "new" odors were to the "old" odor influenced the rate of correct recognition. Engen et al. (1973) found that requiring participants to inspect and remember five odors rather than one odor significantly decreased the number of hits and increased the false alarm rate. Robert Mair and his colleagues along with Engen further revealed that odor recognition accuracy was better if, for example, the "old" odor *garlic* was contrasted with a highly discrepant "new" odor such as *whisky*, rather than a similar odor such as *onion* (Mair, Capra, McEntee, & Engen, 1980).

Verbal coding and the influence of semantic labels in olfactory perception and cognition has been addressed in detail elsewhere (see de Araujo, Rolls, Velazco, Margot, & Cayeux, 2005; Herz, 2005; Lorig, 1999). Engen's particular interest lay in how words influenced the ability to recognize odors and how accessible verbal labels were in odor perception. In one of his experiments investigating verbal accessibility, participants were asked to list ten odors and ten colors as quickly as possible. Two interesting findings emerged. First it took participants four times longer to come up with the odor list than the word list (48 sec v. 12 sec). Second, the variability within the odor list was far greater than in the color list. For example "orange" color was named by 90% of participants, while the top odor named by participants, "perfume", was only given by 35% of participants (Engen & Eaton, 1975). In another experiment Engen and Ross (1973) found that providing canonical names (e.g., raspberry, vinegar) for various odors led to better later recognition for these odors than providing descriptive odor names (e.g., bubble gum, salad dressing). They determined that the reason for this discrepancy was because the descriptive names were based on personal referents and since individual odor experiences are variable there is low consistency in the applicability of such labels. Engen summarized these findings as indicating that the role of language in odor perception was "both idiosyncratic and impoverished" (*The Perception of Odors*, 1982, p. 106). This observation has now been substantiated by numerous experiments and other theorists (see Herz, 2005 for a review).

Another major contribution to olfactory cognition was Engen's research on the nature of odor forgetting. With his then graduate student, Harry Lawless, a paired associate (PA) learning paradigm was implemented where a series of odors were paired with specific pictures, and then two days later repaired with other pictures (Lawless & Engen, 1977). The results of this famous study showed that the first association made to an odor interfered with acquiring the second association, whereas the second association had only a minimal impact on the first. That is, odor associations were highly affected by proactive interference, but negligibly influenced by retroactive interference. This was a surprising and important finding because it was well know that in visual, verbal and auditory memory, retroactive interference dominated. This observation also opened the door to further theoretical speculation on the biological significance of olfactory perception – that odors are inherently meaningless but are especially prepared to acquire meaning through experience, and then able to elicit motivational responses accordingly based on that meaning (e.g., approach or avoidance) (see Section 4. The Special Role of Associative Learning, for further discussion).

In comparison to recognition memory, free recall requires one to dredge information from one's memory store independently. Odor recall would therefore be the ability to willfully conjure the perceptual representation of an odor – in other words, odor imagery. Imagery here is defined as a percept-like representation of a stimulus in the absence of sensory receptor stimulation. This definition assumes that perception and imagery share some of the same neural mechanisms (Hebb, 1968; Farah, 1989; Finke 1989). Imagery as measured by overlapping neural activation between perception and imagery is readily shown in vision and audition (see Finke, 1989 for a general discussion), and to a lesser extent in touch (shivers, tingles, phantom limb), and taste (sour salivary reaction). However, in olfaction, similar studies suggest that the degree of overlap between smelling an odor and "imaging" it is much weaker (Djordjevic, Zatorre, Petrides, Boyle, & Jones-Gotman, 2005).

During Engen's career it was very difficult to verifiably or reliably assess the ability to recall odors. Therefore, in spite of various culinary and perfume professionals' claims to the contrary, Engen contended that olfactory imagery was not possible, and that what people were experiencing instead was a conflation of sensations arising from other senses and the cognitive integration of them (Engen, 1982). For example, if one tries to recall the scent of lemon, most people almost instantly experience the conditioned salivary sour response and the visualization of a lemon. Because of the close connection between smell and taste these sensations can be readily confused as feeling like the scent of a lemon, but they are not. Recent advances in methodological technology, however, now suggest that though odor imagery among average people is very rare, expert perfumers may in fact be able to acquire the ability to image odors through training (Plailly, Delon-Martin, & Royet, 2011). Using functional magnetic resonance imagery (fMRI), Plailly and colleagues (2011) found that among professional perfumers, extensive olfactory practice was associated with a functional reorganization of the posterior piriform cortex, the orbitofrontal cortex, and the hippocampus during the creation of mental images of odors, and that activity in these areas was negatively correlated with experience. That is, olfactory areas are engaged when experts attempt to image odors and experience can change the neural organization of olfaction. In addition to showing that the claims of expert smellers may be valid, this finding is a further illustration of how the olfactory system is especially flexible and responsive to learning throughout the lifespan (Poncelet et al., 2010).

Engen's research on odor memory repeatedly pointed to associative learning as the critical mechanism underlying the ability to recognize odors. We recognize an odor because of our past association with it, be it semantic, perceptual or emotional. It was also his observation regarding the importance of associative learning that paved the way for exploring the phenomenon where the association one has to an odor is highly involving and autobiographical.

2. Odor-evoked memory

Recognition memory for odors and odor-evoked memory can overlap. Though Engen never directly studied odor-evoked memory, he began the "Memory" chapter in his seminal book *The Perception of Odors* (1982) with the following anecdote from the Yorkshire veterinarian-writer, James Herriot. In his bestseller, *All Things Wise and Wonderful* (1977), Herriot describes an incident where he used a new bar of "Famous" brand soap after a miserable day in the Royal Air Force (RAF) and how from that night forward he was never able to use the soap again because "…the merest whiff jerks me back to that first night away from my wife, and to the empty feeling I had then. It was a dull, empty ache which never really went away" (p. 2). Herriot's anecdote illustrates how he is able to recognize what a certain scent is (Famous brand soap), and that at the same time this recognition memory unfolds into a personal episodic memory. This is because the identity of the soap and the events of that day are inextricably interwoven. The soap scent *is* the memory and feeling of his first miserable night away from his wife. Engen realized that memory for an odor and the memory evoked by an odor were fundamentally connected.

Most odor-associations are not meaningful and singular but rather inconsequential and redundant. For example, recognizing a certain scent as being "coffee" is unlikely to transport one to a past personal episode because the scent of coffee is so frequently experienced in a variety of mundane situations. Moreover, another important feature of odor-evoked memory is that it usually has little to do with the odor itself. After Herriot's first episode with Famous brand soap, subsequent encounters with the odor did not lead him to contemplate soap or bathing, but rather to the events of a depressing and lonely day in the RAF. Herriot's anecdote further illustrates that odor-evoked memories feel very emotional and evocative. Indeed, odor-evoked memories are claimed to be distinguished from other episodic memory experiences by their: emotionality, evocativeness (being brought back to the original time and place), vividness, rarity (rarely thought of compared to other memories), and age (memories from early life). The following sections examine the experimental evidence for and against these distinguishing features.

2.1 Emotionality and evocativeness

As Herriot's anecdote reveals the emotional sting and evocativeness of odor-evoked memories can be intense; "*a dull, empty ache which never really went away.*" Early experimental studies on autobiographical odor-evoked memories confirmed the claim that odor-evoked memories were markedly emotional (Laird, 1935; Herz & Cupchik, 1992). However, these studies were descriptive and did not compare memories evoked by odors with memories elicited by other stimuli. In order to assess whether odor-evoked memory is verifiably distinct from memories elicited by other stimuli, it is necessary to employ a cross-modal methodological approach where memories elicited by stimuli presented in various sensory modalities are compared. In the first cross-modal odor-evoked memory experiment, Rubin, Groth and Goldsmith (1984) gave participants 15 familiar stimuli (e.g., coffee, band-aids, cigarettes) to assess in either olfactory, verbal or picture form. For each item, the participant described the memory that was evoked, and rated it on several scales: age of memory, vividness, emotionality at time of event, emotionality at time of recall, how many times it had been thought of, and when it was last recalled (prior to the experiment). The results showed a trend for odor-evoked memories to be more emotional, but the effect was statistically marginal. More supportive evidence was reported by Hinton and Henley (1993) who compared participants' free-associations to verbal, olfactory and visual versions of six familiar everyday items (coffee, tobacco, carnations, orange, Ivory soap, pine) and found that those elicited by odors were more emotional, as evaluated by independent judges, than the free associations elicited by words or visual versions of the same items.

To more deeply probe odor-evoked memory, Herz and Cupchik (1995) developed a PA cross-modal paradigm in which a series of familiar source cues were presented to participants in either olfactory, verbal, visual, or tactile form (e.g., smelling the odor of popcorn, hearing the word "popcorn", seeing a picture of popcorn, touching popcorn kernels) while participants' viewed emotionally involving pictures. Two days later these cues were presented again and the participants' memory for the pictures was assessed in several ways pertaining to accuracy-content and emotionality. Across several experiments, we consistently observed that memories recalled to the scent of a particular cue were experienced as significantly more emotional than memories elicited by cues presented in any other sensory modality (Herz, 1998; Herz & Cupchik, 1995). These findings have been replicated by others comparing odors with visual and verbal stimuli (e.g., Bonfigli, Kodilja, & Zanuttini, 2002; Willander & Larsson, 2007).

In order to establish the primacy of odors as emotionally evocative memory cues, my laboratory has also compared odors to music, as music is renowned as an emotionally evocative stimulus and it is often reported that music evokes poignant memories (Royet et al., 2000). In one study, a cross-modal PA paradigm with odors, visual stimuli and music of comparative novelty, pleasantness and complexity as memory cues was implemented and self-report measures of emotion as well as a physiological correlate of emotion – heart-rate – were assessed (Herz, 1998). Results showed that there was a trend for odors to subjectively elicit more emotional memories and this was statistically confirmed by the heart-rate data. Memories elicited by odors were accompanied by greater heart rate increases than those elicited by music (sniffing artifacts were accounted for), thus showing that odors supersede music as emotion evoking memory cues.

A possible explanation for the outstanding emotionality of odor-evoked memories may be found in how they phenomenologically unfold. All descriptive accounts of odor-evoked memory point to the fact that during odor-evoked memory the *feeling* comes first, after which the event to which the emotion is attached is discerned (Herz, 2011). For example, in Proust's madeleine biscuit and linden tea episode, the emotion that the odor elicited came well before awareness of the content to which that emotion was connected. That is, the temporal unfolding of his memory was first emotional and then only after a number of minutes, and pages, did the visual and concrete content to which that memory was attached become apparent (Herz, 2007; Jellinek, 2004). By contrast, in all other memory experiences the content of the event is first recalled, and then it is emotionally appraised. For example, a song evokes the memory of a high-school dance first, and then that event is quickly evaluated as having been embarrassing, joyful, depressing, and so on. In other words, with music and other stimuli the emotion experienced is inherently more analytical and filtered than it is in odor-evoked recollections. The neuroanatomy of olfaction also supports this proposition.

The primary olfactory cortex includes the amygdala, which is where emotional experience in general is processed, and the hippocampus, which is involved in associative learning (Cahill, Babinsky, Markowitsch, & McGaugh, 1995; Turner, Mishkin, & Knapp, 1980; Aggleton & Mishkin, 1986; Eichenbaum, 2001). Classical conditioning of specific cues to emotion is also mediated by the amygdala (LeDoux, 1998), and the olfactory cortex and amygdala have been shown to play a major role in stimulus reinforcement association learning (see Rolls, 1999). Moreover, the orbitofrontal cortex (the secondary olfactory cortex) is also the cortical area responsible for assigning affective value to stimuli and for determining the reinforcement (approach-avoid) value of stimuli in general (Davidson, Putnam, & Larson, 2000; Quirk & Gehlert, 2003). Further, unlike other with other types of sensory processing, olfactory information does not need to be integrated in the thalamus prior to processing in the cortex (Sabri, Radnovich, Li, & Kareken, 2005). Thus, there is a privileged and unique anatomical relationship between the neural substrates of emotion and olfaction, and as such it seems that odors are inherently more emotional and less cognitively analyzed than other stimuli.

My laboratory has also obtained neurobiological evidence for the special affective potency of odor-evoked memory using fMRI. In our study (Herz, Eliassen, Beland, & Souza, 2004) potential female (only) volunteers were interviewed to determine whether a specific perfume could be identified that elicited a positive autobiographical memory. Criteria for participant selection was recalling a positive, personal memory in which both the smell and sight of a perfume bottle figured. That is, the participants needed to have both olfactory and visual recognition of the stimulus. A block design with three blocks of 16 trials was followed where participants were exposed to two experimental stimuli: the smell and sight of the specific perfume they had selected as being emotionally meaningful (Experimental Odor, Experimental Visual); and two control stimuli, the smell and sight of a

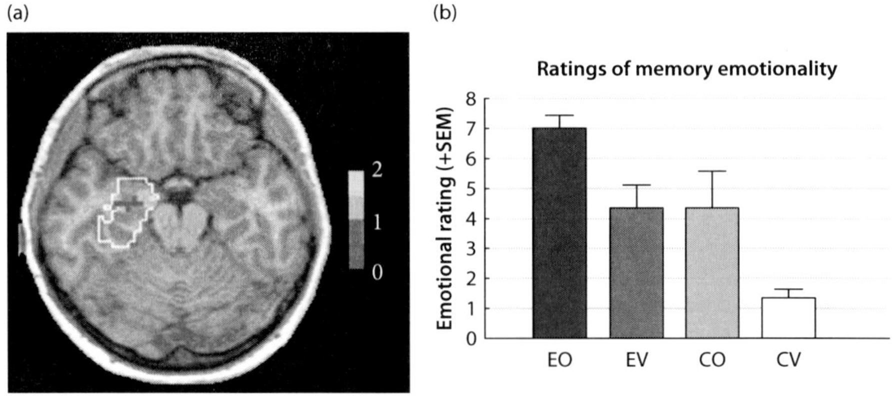

(a) Activation for the Experimental Odor (EO) in the amygdala. The positive activation difference for the comparison EO vs. EV + CO + CV is shown. EO = Experimental Odor; EV = Experimental Visual; CO = Control Odor; CV = Control Visual. The slice shown is at Z = –16 mm inferior to anterior commissure (AC). The maximum intensity difference of 1.65 (MR units) appeared at 14, 8, –16, relative to the AC, corresponding to left hemisphere Brodmann areas 28 and 34. (b) Mean emotion ratings given during memory elicitation to each stimulus.

Figure 1. fMRI analysis of odor versus visually evoked autobiographical memory.

perfume that they were not familiar with (Control Odor, Control Visual), which was an unmarketed perfume that was the same for all participants. The first trial of every block and between each stimulus was an air-only trial.

Analyses of the fMRI data revealed significantly greater activation in the amygdala and hippocampal regions during recall to the odor of participants' personally meaningful perfume than to any other cue. Subsequent behavioral testing also confirmed that participants experienced the strongest feelings of emotion while they were recalling an episode to the scent of their personally selected perfume (see Figure 1). This study therefore confirmed that the subjective emotional potency of odor-evoked memory is specifically correlated with heightened activation in the amygdala-hippocampal region during recall, and demonstrated that odor-evoked memories are neurobiologically distinguished by the activation they produce in the neural substrates of emotion and associative learning. This result is also noteworthy given that odors generally elicit activation in the amygdala-hippocampal complex.

The evocativeness of a memory is defined by the degree to which one has the feeling of "being brought back to the original time and place" of an episode (Herz & Schooler, 2002). Only four studies to date have examined the evocativeness of memories elicited by odors in comparison to memories elicited by other stimuli. Using a basic cross-modal design, in two different experiments Willander and

Larsson (2006, 2007) showed that odor-evoked memories were accompanied by a stronger "feeling of being brought back" than memories elicited by verbal and picture cues of the same items (e.g., tobacco, black-current, soap). Notably, a potential criticism of autobiographical memory research is that memory selection may be confounded with memory recollection. That is, the specific memories one chooses to recall may be inherently tinged with more emotion (for various reasons including demand characteristics) than might be truly representative.

To disambiguate the effects of memory selection with recollection, Herz and Schooler (2002) and Herz (2004) developed a double-cuing method to test autobiographical memory cross-modally. In our experiments, participants were first verbally given an odor name (e.g., "freshly cut grass") and asked to think of a memory from their past that was associated to it, and to rate that memory on a variety of dimensions. They were then given either a visual (a photograph of a freshly mowed lawn), auditory (sound of a lawnmower), or olfactory (the odor of fresh-cut grass) version of the same cue, and asked to recall their memory again. Selecting the memories in response to verbal names prior to the introduction of the sensory cues allowed for matching of the memories recalled in the various cue conditions. Therefore, any differences in the quality of the memories that were subsequently observed could be attributed to the effects of the cues on memory *recollection* rather than memory *selection*. In both studies, we found that autobiographical memories re-cued by olfactory variants led to a significantly greater feeling of being brought back to the original time and place of the event and were felt as more emotional than memories that were re-cued by visual, verbal or auditory versions of the same items.

The feeling of being transported back to the original memory event exemplifies one's emotional connection to the memory experience, and as such evocativeness can be viewed as another dimension of memory emotionality. Together, these findings strongly substantiate the claim that odors elicit uniquely potent emotional memories. Moreover, no experimental studies to date have reported that odor-evoked memories are *not* emotional.

2.2 Vividness

Anecdotal descriptions of odor-evoked memory typically mention their high "vividness". However, in the first cross-modal experiments (Herz & Cupchik, 1995; Herz, 1998; Rubin et al., 1984) no advantage on memory vividness or content detail for odor-evoked memories was found. In fact, none of the memory cues in these experiments differed in the richness of the content details or accuracy that they elicited. By contrast, Aggelton and Waskett (1999) compared relevant odors (e.g., burnt wood, "fish-market") and irrelevant odors (e.g., coconut, peppermint) as cues to autobiographical memory for a Viking museum exhibit. Testing visitors who had been to the Viking exhibit approximately 6.5 years before, they found that the relevant odors (which had actually been present at the exhibit, e.g., burnt wood) elicited better recall for features of the exhibit than irrelevant odors (which had not been present, e.g., coconut). In another study, Chu and Downes (2002) used a double-cuing method where participants first retrieved a memory to a verbal cue (e.g., ginger) and then retrieved it again with either the same verbal cue (label group), or an odor cue that corresponded to the original verbal cue (e.g., ginger odor – congruent group), or did not correspond to the original verbal cue (e.g., cinnamon odor – incongruent group). Participants then evaluated their memories in each condition along various qualitative measures, including vividness. The results from this study revealed that re-recollection with congruent odor cues led to higher ratings of memory vividness than memories re-cued by verbal or incongruent odors did. However, Herz and Schooler (2002) and Herz (2004) also used a recuing method to test autobiographical memories with vividness as a dependent variable and no differences on this measure were obtained.

A problem in empirically determining the strength of memory vividness is that the term "vividness" is colloquially ambiguous and can mean either the clarity and detail of visual imagery in a memory, as well, or in addition to, the emotional vividness (i.e., emotional intensity) of a memory. It appears most likely that the differences in results for memory "vividness" that have been obtained across various studies are due to the meaning of "vividness" that was presumed by the participants. That is, whether the connotation implied emotional vividness or the vividness of visually accessible content. Differences in the methodology of Chu and Downes (2002), in contrast to Rubin et al. (1984), Herz and Schooler (2002) and Herz (2004), may have led participants to presume different meanings in each case; the former invoking more emotional construal and the later experiments suggesting a visual imagery connotation. For example, Chu and Downes (2002) asked participants to rate their memories on a variety of terms: pleasant, painful, anxious, embarrassing, vivid and unique; all of which could be interpreted as relating to emotional quality. By contrast, in the Rubin et al. (1984) and the Herz experiments, participants were asked to rate their memories on a series of dimensions where the emotion-related items and content-related items were more differentiated. Notably, Chu and Downes (2002) reported that odors led to recall of more emotional detail than verbal labels. This further supports the possibility that their vividness measure was interpreted in emotional terms. It should also be noted that there is no neuroanatomical or functional reason why odors should elicit memories of greater visual vividness or content detail than memories evoked by visual or verbal cues, especially given the generally weak to non-existent ability to image odors (Djordjevic et al., 2005).

2.3 Old and rare

Odor-evoked memories are typically characterized as eliciting recollections from early life as well as memories that are rarely thought of in comparison to visual or verbal memories. The clustering of memories from a specific life stage is referred to as a "memory bump" (Rubin & Schulkind, 1997). Traditionally, it has been found that when adults recall episodic memories to verbal and visual cues their recollections center on events which occurred in their second and third decades of life – the period of adolescence and early adulthood (Rubin, Rahhal, & Poon, 1998). The question as to whether odor-evoked memories also cluster in this life phase has only been investigated in a few studies.

Rubin et al. (1984) were unable to find any differences among college students in the age distribution for odor-evoked memories compared to memories elicited by photographs or verbal labels. Likewise, Goddard, Pring, and Felmingham (2005) compared memories elicited by words, pictures and odors in young adults and found no differences in the age distribution of memories. However, in studies where older cohorts have been tested, an earlier bump by approximately ten years has been found for memories elicited by odors compared to memories elicited by words. Chu and Downes (2000) found that among older participants (average age 69 yrs), a bump for odor-evoked memories emerged between the ages of 6–10, compared with the bump for label-cued events which was found between the ages of 11–20. Recently, Willander and Larsson (2006, 2007) have confirmed that the memory bump for odor-evoked memories is earlier than for memories triggered by other cues. Testing adults between the ages of 65–80 years (mean age 74 yrs), they found a significant clustering of memories from the first decade of life elicited by odors compared to verbally evoked memories which peaked in early adulthood.

Why should the temporal distribution of odor-evoked memories be concentrated in the first years of life compared to the memory bump in later development observed for verbal and visual memory? One possibility is that the younger-age shift for odor-evoked recollections is due to the fact that odors evoke the first experiences that have been associated to them, and since most new experiences occur in childhood this could explain why associations to odors predominantly remind one of events from these early years. Delaunay-El Allam and colleagues (Delaunay-El Allam, Soussignan, Marlier, Patris, & Schaal, 2010) recently showed that even infants can establish odor-associations which influence behavior up to at least 21 months later. Newborns who were exposed to the odor of chamomile during breast feeding showed preferences for this odor, as exhibited by several behavioral tasks, when tested at 7 months and 21 months of age compared to infants who did not have this exposure history – an effect which could only have occurred if the exposed infants' had formed an odor-evoked memory to the scent of chamomile.

In addition to the fact that many olfactory associations are acquired during childhood, as previously discussed, a unique feature of odor memory is that proactive interference is especially strong. Consequently, it is difficult to associate an odor to any new experience once an initial association has been made to it (Lawless & Engen, 1977). Therefore, another reason why memories recalled to odors might center on first associations is because proactive interference inhibits later memories from being formed. By contrast, due to the dominance of retroactive interference for verbal and visual cues, newer memories tend to erase first associations and hence recollections are drawn from the more recent past.

Old memories may also be more likely to be rare memories – that is, infrequently relived compared to other episodic events. Indeed, in their pioneering experiment, the only effects that Rubin et al. (1984) found to be statistically reliable were that odor-evoked memories were *talked about* and *thought of* less often than memories evoked by verbal or visual items. Willander and Larsson (2006) have confirmed this effect and shown that odor-evoked memories were thought about less often than memories evoked by pictures, and that fewer memories were elicited over one's lifetime by odors than by verbal cues (Willander & Larsson, 2007). Together these findings support the contention that odors evoke autobiographical memories less frequently than other memory cues, and that there is less rumination on these events than memories triggered by other stimuli. The relatively low frequency of encountering odors that trigger true odor-evoked memories and consequently the reduced rehearsing of these memories may be a further reason for the preservation of first associations and the early memory bump observed. Indeed the low probability of encountering an odor memory cue may shed light on one way in which odors might be considered "the best" cues to memory.

2.4 The best cues to memory

The claim is often made that odors are "better" than other memory cues. However, empirical assessment negates this assertion if better refers to memory accuracy (e.g., Herz, 1998, 2004). It has never been shown that odors bring back more accurate memories than other sensory stimuli. Rather, it seems that the distinctive emotional salience and evocativeness of odor-evoked memories makes them *feel* more real and this then leads to the false impression that these memories are especially accurate (Herz, 1998). The confidence that one's recollections are true, which is so hard to resist when memories are colored by emotional experience, is similar to what often emerges during eyewitness testimony. Eyewitnesses recalling emotionally charged episodes are often extremely confident that their recollections are accurate, but research shows that these memories are often incorrect (Busey, Tunnicliff, Loftus & Loftus, 2000).

There is, however, one way in which odors might be better memory cues. They may be more likely to elicit recall for events which otherwise would never be remembered (Herz, 2007). That is, odors may unlock memories whose only "mental tag" is the odor that was present when the memory was encoded. A specific odor may be explicitly tied to one event in life and no other because that scent was only encountered in that particular situation, whereas visual or verbal versions of the same cue are multiply encountered in various iterations of similar events. For example, in Proust's famous tea and cookie recollection he admits that he has seen numerous madeleine biscuits and taken many teas in the interim, but he had never tasted linden tea and madeleine biscuits together since his childhood in Combray. In other words, due to redundancy and interference with multiple experiences, visual and verbal representations of critical memory cues lose their ability to trigger the memory of a particular episode. By contrast, resistance to being overwritten (proactive interference) and the high distinctiveness of certain scents can combine to make odors faithfully and directly tied to particular events like no other cue. The ability to elicit a memory that might otherwise never be retrieved is possibly the key feature that makes odors "better" than other memory cues (Herz, 2007). However, this claim has not yet been experimentally verified.

2.5 Odor-evoked memory summary

Over the past several decades substantial empirical research has been conducted in the field of odor-evoked memory and major strides have been made in experimentally determining its distinctive characteristics. Table 1 summarizes the findings discussed in this chapter examining the superiority of distinctiveness of odor-evoked memory compared to memories triggered by other cues along these dimensions. Among the distinctive features of odor-evoked memory that have been experimentally verified, the most notable is their uniquely emotional and evocative nature. There is also clear evidence that memories elicited by odors tend to be from an earlier time of life and are more rarely relived than memories elicited by other stimuli. The degree of visual vividness in odor-evoked memory does not appear to be greater than with any other stimuli, although if vividness is

Table 1. Comparison between odor-evoked memory and memories evoked by other stimuli.

	Memory dimension					
Cue type	**Emotionality**	**Evocativeness**	**Vividness**	**Accuracy**	**Old and rare**	**Able to elicit forgotten memories**
Odors			....			
Non-odors			....			

A summary of the dimensions along which odor-evoked memories have been shown to be distinctive/superior (or not) compared to memories evoked by other stimuli. Black = superior; dotted line = ambiguous findings; white = no difference; grey = not yet empirically tested, but anecdotal observation suggests superiority.

interpreted as emotional intensity, then this description may be apt. Odors also do not seem to be the best cues to memory, if "best" refers to accuracy. Here, again, the feeling that odors are superior reminders is based on their emotional potency. However, it may be the case that odors are more likely to trigger memories that otherwise would never be recalled. Whether this is a true defining feature of odor-evoked memory awaits experimental investigation.

3. The uniqueness of odor memory

Trygg Engen had a long-time interest in determining whether odor memory was truly different from other forms of memory. In 1996, he and I examined the proposition of odor memory uniqueness with respect to the extant literature and assessed whether odor-memory fit within a "multiple memory systems" theoretical framework, as compared to a unimodal memory framework (Herz & Engen, 1996). Multiple memory system theories presume that there are specialized types of memory governed by specific and distinct rules and underlying mechanisms. We used the criteria outlined by Roediger, Rajaram, and Srinivas (1990) and Sherry and Schacter (1987) – functional dissociation, stochastic independence, independent neural systems, and functional incompatibility – to assess whether odor memory could be considered a separate memory system. Our analysis led us to conclude that we could not definitively state that odor memory as a whole represented a separate memory system, but that there were sufficient distinctive and special features of olfaction to align it with the principles of a multiple memory systems framework. Notably, in the field of odor-evoked memory there is ample evidence to posit that qualities, such as emotionality, evocativeness and a preponderance of early life events distinguish odor-evoked memory from other memory experiences. Recently, others have also concluded that olfactory cognition and olfactory consciousness are unique from other systems (Zucco, 2003; Wilson & Stevenson, 2006; Stevenson, 2009). Thus, though it is certain that cognitive and perceptual processing systems across the senses share a number of commonalities, the characteristics of olfaction appear to show particular differences.

4. The special role of associative learning

The basis for the distinguishing properties of odor memory can be explained by two central and overlapping features of olfaction. First is the uniquely privileged neuroanatomical relationship between olfactory processing and the structures that govern emotion, memory and associative learning. Second, memory for odors and odor-evoked memory are both predicated on the defining perceptual and cognitive characteristic of olfaction – associative learning.

Engen observed that there are no innate responses to odors. For example, he found that newborns showed the same response to the extract of *asafetida* (described as smelling like rotted garlic or body odor) and anise (smells like licorice) (Engen, 1988). Other research with infants has shown that they often demonstrate opposite responses to those of adults, for example, liking the smell of synthetic sweat and feces (Stein, Ottenberg, & Roulet, 1958). Examining later development, Engen found that even four-year olds still did not show differentiating hedonic reactions to odors that adults designate as very unpleasant (e.g., butyric acid, smells like body odor) or pleasant (e.g., amyl acetate, smells like banana). However, by the age eight, most children's responses to odors had come to mimic adult norms (Engen, 1988). Indeed, there is now a large literature demonstrating that olfactory responses are acquired through associative learning from the earliest moments of life and that they continue to be shaped by culture and experience across the life-span (e.g., Ayabe-Kanamura, Schicker, Laska, Hudson, Distel, Kobayakawa, & Saito, 1998; Beauchamp & Mennella, 2009; Delaunay-El Allam et al., 2010; Haller, Rummel, Henneberg, Pollmer, & Köster, 1999; Mennella & Garcia, 2000; Schaal, Marlier, & Soussignan, 2000).

The reason that the human olfactory system is so sensitive to associative learning is because we, along with rats and cockroaches, are the world's most successful generalists – species that can exploit any habitat. The available resources and potential predators and dangers, however, differ drastically across environments. Therefore, it is not adaptive to have pre-determined olfactory responses to potential prey or predators, but rather to be especially prepared to learn associations on the basis of their significance when encountered. The best natural example of the potency of odor-associative learning is the case of "learned taste aversions". Rats and humans will avoid a flavor that has previously been co-experienced with gastric distress (Garcia & Koelling, 1966; Bernstein, 1978). Bartoshuk (1989) has made it clear that it is the smell, not the taste, of the substance that is responsible for the conditioned aversion response.

The potency of proactive interference and the durability of odor memory further demonstrate the biological significance of odor-associative learning. It confers a significant survival advantage to enduringly remember the first meaning that is learned to an odor, and not to haphazardly re-associate odors with every new experience. One does not want to keep forgetting what the smell of poisonous mushrooms is, or to over-ride a danger association just because a certain scent is later encountered in a benign circumstance. From an evolutionarily perspective, it is adaptive that the olfactory system of generalists not to be predisposed to like or dislike any particular odors, but rather be especially prepared to learn and remember what to approach and what to avoid based on experience (see Herz, 2007 for discussion).

Learned responses to odors not only code an odor as good or bad (approach or avoid), they inform us as to what the odor is. This recognition can range from a vague feeling of familiarity to specific semantic and personal knowledge – *this is the odor of spoiled milk and if I drink it I will get sick*. Moreover, once learned the odor then becomes a proxy for the experiences to which it is associated and itself can elicit both the emotional and motivational states that correspond to the association. For example, both children and adults who have been exposed to a novel odor while performing a frustrating task, later show decreased motivation on a different task when that same odor is present (Epple & Herz, 1999; Herz, Schankler, & Beland, 2004).

5. Conclusion

The study of odor memory began with Trygg Engen's seminal experiments over 40 years ago. From that early work, a field of psychology has evolved that examines the defining characteristics of our ability to remember specific odors, the special qualities of odors to evoke episodic recollections and the mechanism that ties olfactory perception, cognition, memory and emotion together – association learning. Not only have these findings shaped the growing field of odor memory, they have shed important light on the fundamental biological significance of the sense of smell – to elicit emotional meaning and motivate behavior accordingly.

References

Aggleton, J. P., & Mishkin, M. (1986). The amygdala: Sensory gateway to the emotions. In R. Plutchik & H. Kellerman (Eds.), *Emotion: Theory, research and experience. Biological foundations of emotion*, Vol. 3 (pp. 281–299). Orlando: Academic Press.

Aggleton, J. P., & Waskett, L. (1999). The ability of odours to serve as state-dependent cues for real-world memories: Can Viking smells aid the recall of Viking experiences? *British Journal of Psychology, 90*, 1–7.

Ayabe-Kanamura, S., Schicker, I., Laska, M., Hudson, R., Distel, H., Kobayakawa, T., & Saito, S. (1998). Differences in perception of everyday odors: A Japanese-German cross-cultural study. *Chemical Senses, 23*, 31–38.

Bartoshuk, L. M. (1989). The functions of taste and olfaction. *Annals of the New York Academy of Sciences, 575*, 353–362.

Beauchamp, G. K., & Mennella, J. A. (2009). Early flavor learning and its impact on later feeding behavior. *Journal of Pediatric Gastroenterology and Nutrition, 48 Suppl 1*, S25–S30.

Bernstein, I. L. (1978). Learned taste aversions in children receiving chemotherapy. *Science, 200*, 1302–1303.

Bonfigli, L., Kodilja, R., & Zanuttini, L. (2002). Verbal versus olfactory cues: Affect in elicited memories. *Perceptual and Motor Skills, 94*, 9–20.

Busey, T. A., Tunnicliff, J., Loftus, G. R., & Loftus, E. F. (2000). Accounts of the confidence- accuracy relation in recognition memory. *Psychonomic Bulletin and Review, 7*, 26–48.

Cahill, L., Babinsky, R., Markowitsch, H. J., & McGaugh, J. L. (1995). The amygdala and emotional memory. *Nature, 377*, 295–296.

Chu, S., & Downes, J. J. (2000). Long live Proust: The odour-cued autobiographical memory bump. *Cognition, 75*, B41–B50.

Chu, S., & Downes, J. J. (2002). Proust nose best: Odors are better cues of autobiographical memory. *Memory & Cognition, 30*, 511–518.

Davidson, R. J., Putnam, K. M., & Larson, C. L. (2000). Dysfunction in the neural circuitry of emotion regulation – a possible prelude to violence. *Science, 289*, 591–594.

de Araujo, I., Rolls, E. T., Velazco, M.-I., Margot, C., & Cayeux, I. (2005). Cognitive modulation of olfactory processing. *Neuron, 46*, 671–679.

Delaunay-El Allam, M., Soussignan, R., Marlier, L., Patris, B., & Schaal, B. (2010). Long lasting memory for an odor acquired at the mother's breast. *Developmental Science, 13*, 849–863.

Djordjevic, J., Zatorre, R. J., Petrides, M., Boyle, J. A., & Jones-Gotman, M. (2005). Functional neuroimaging of odor imagery. *Neuroimage, 24*, 791–801.

Ebbinghaus, H. (1885/1913). *Memory. A Contribution to Experimental Psychology*. New York: Teachers College, Columbia University.

Eichenbaum, H. (2001). The hippocampus and declarative memory: Cognitive mechanisms and neural codes. *Behavioural Brain Research, 127*, 199–207.

Engen, T. (1982). *The Perception of odors.* Toronto: Academic Press.

Engen, T. (1988). The acquisition of odor hedonics. In S. Van Toller, & G. H. Dodd (Eds.), *Perfumery: The psychology and biology of fragrance*. New York: Chapman & Hall.

Engen, T., & Eaton, J. (1975). *Free recall of odor and color names.* Unpublished manuscript. Brown University.

Engen, T., & Ross, B. M. (1973). Long-term memory odors with and verbal descriptions. *Journal of Experimental Psychology, 100*, 221–227.
Engen, T., Kuisma, J. E., & Eimas, P. D. (1973). Short-term memory of odors. *Journal of Experimental Psychology, 99*, 222–225.
Epple, G., & Herz, R. S. (1999). Ambient odors associated to failure influence cognitive performance in children. *Developmental Psychobiology, 35*, 103–107.
Farah, M. J. (1989). The neuropsychology of mental imagery. In J. W. Brown (Ed.), *Neuropsychology of visual perception* (pp. 183–202). Hillsdale, NJ: Erlbaum.
Finke, R. A. (1989). *Principles of mental imagery*. Cambridge, MA: MIT Press.
Garcia, J., & Koelling, R. A. (1966). Relation of cue to consequence in avoidance learning. *Psychonomic Science, 4*, 123–124.
Goddard, L., Pring, L., & Felmingham, N. (2005). The effects of cue modality on the quality of personal memories retrieved. *Memory, 13*, 79–86.
Haller R., Rummel, C., Henneberg, S., Pollmer, U., & Köster, E. P. (1999). The effect of early experience with vanillin on food preference later in life. *Chemical Senses, 24*, 465–467.
Hebb, D. O. (1968). Concerning imagery. *Psychological Review, 75*, 466–477.
Herriot, J. (1977). *All things wise and wonderful*. New York: St. Martins, Press.
Herz, R. S. (1998). Are odors the best cues to memory? A cross-modal comparison of associative memory stimuli. *Annals of the New York Academy of Sciences, 855*, 670–674.
Herz, R. S. (2004). A comparison of autobiographical memories triggered by olfactory, visual and auditory stimuli. *Chemical Senses, 29*, 217–224.
Herz, R. S. (2005). The unique interaction between language and olfactory perception and cognition. *Trends in Experimental Psychology Research* (pp. 91–109). New York: Nova Science Publishers, Inc.
Herz, R. (2007). *The Scent of Desire: Discovering Our Enigmatic Sense of Smell.* New York: William Morrow/HarperCollins Publishers.
Herz, R. S. (2011, in press). Odor-evoked memory. In J. Decety, & J. Cacioppo (Eds.), *The Oxford Handbook of Social Neuroscience.* New York: Oxford University Press.
Herz, R. S., & Cupchik, G. C. (1992). An experimental characterization of odor-evoked memories in humans. *Chemical Senses, 17*, 519–528.
Herz, R. S., & Cupchik, G. C. (1995). The emotional distinctiveness of odor-evoked memories. *Chemical Senses, 20*, 517–528.
Herz, R. S., & Engen, T. (1996). Odor memory: Review and analysis. *Psychonomic Bulletin and Review, 3*, 300–313.
Herz, R. S., Eliassen, J. C., Beland, S. L., & Souza, T. (2004). Neuroimaging evidence for the emotional potency of odor-evoked memory. *Neuropsychologia, 42*, 371–378.
Herz, R. S., Schankler, C., & Beland, S. (2004). Olfaction, emotion and associative learning: Effects on motivated behavior. *Motivation and Emotion, 28*, 363–383.
Herz, R. S., & Schooler, J. W. (2002). A naturalistic study of autobiographical memories evoked to olfactory versus visual cues. *American Journal of Psychology, 115*, 21–32.
Hinton, P. B., & Henley, T. B. (1993). Cognitive and affective components of stimuli presented in three modes. *Bulletin of the Psychonomic Society, 31*, 595–598.
Jellinek, J. S. (2004). Proust remembered: Has Proust's account of odor-cue autobiographical memory recall really been investigated? *Chemical Senses, 29*, 455–458.
Laird, D. A. (1935). What can you do with your nose? *Scientific Monthly, 41*, 126–130.
Lawless, H., & Engen, T. (1977). Associations to odors: Interference, mnemonics and verbal labeling. *Journal of Experimental Psychology: Human Learning and Memory, 3*, 52–59.

LeDoux, J. (1998). Fear and the brain: Where have we been and where are we going? *Biological Psychiatry, 44,* 1229–1238.

Lorig, T. S. (1999). On the similarity of odor and language perception. *Neuroscience & Biobehavioral Reviews, 23,* 391–398.

Mair, R. G., Capra, C., McEntee, W. J., & Engen, T. (1980). Odor discrimination and memory in Korsakoff's psychosis. *Journal of Experimental Psychology: Human Perception and Performance, 6,* 445–458.

Mennella, J. A., & Garcia, P. L. (2000). Children's hedonic response to the smell of alcohol: Effects of parental drinking habits. *Alcoholism: Clinical and Experimental Research, 24,* 1167–1171.

Plailly, J., Delon-Martin, C., & Royet, J.-P. (2011). Experience induces functional reorganization in brain regions involved in odor imagery in perfumers. *Human Brain Mapping.* doi: 10.1002/hbm.2120.

Poncelet, J., Rinck, F., Bourgeat, F., Schaal, B., Rouby, C., Bensafi, M., & Hummel, T. (2010). The effect of early experience on odor perception in humans: Psychological and physiological correlates. *Behavioural Brain Research, 208,* 458–465.

Proust, M. (1928). *Swann's Way.* New York: Modern Library.

Quirk, G. L., & Gehlert, D. R. (2003). Inhibition in the amygdala: key to pathological states? *Annals of the New York Academy of Sciences, 985,* 263–272.

Roediger, H. L., Rajaram, S., & Srinivas, K. (1990). Specifying criteria for postulating memory systems. *Annals of the New York Academy of Sciences, 608,* 572–595.

Rolls, E. T. (1999). *The Brain and Emotion.* Oxford: Oxford University Press.

Royet, J. P., Zald, D., Versace, R., Costes, N., Lavenne, F., Koenig, O., & Gervais, R. (2000). Emotional responses to pleasant and unpleasant olfactory, visual, and auditory stimuli: a positron emission tomography study. *Journal of Neuroscience, 20,* 7752–7759.

Rubin, D. C., Groth, E., & Goldsmith, D. J. (1984). Olfactory cueing of autobiographical memory. *American Journal of Psychology, 97,* 493–507.

Rubin, D. C., Rahhal, T. A., & Poon, L. W. (1998). Things learned in early adulthood are remembered best. *Memory and Cognition, 26,* 3–19.

Rubin, D. C., & Schulkind, M. D. (1997). Distribution of important and word-cues autobiographical memories in 20–35- and 70-year-old adults. *Psychology and Aging, 12,* 524–535.

Sabri, M., Radnovich, A. J., Li, T. Q., & Kareken, D. A. (2005). Neural correlates of olfactory change detection. *NeuroImage, 25,* 969–974.

Schaal, B., Marlier, L., & Soussignan, R. (2000). Human foetuses learn odours from their pregnant mother's diet. *Chemical Senses, 25,* 729–737.

Sherry, D. F., & Schacter, D. L. (1987). The evolution of multiple memory systems. *Psychological Review, 94,* 439–454.

Stein, M., Ottenberg, M. D., & Roulet, N. (1958). A study of the development of olfactory preferences. *Archives of Neurological Psychiatry,* 80, 264–266.

Stevenson, R. J. (2009). Phenomenal and access consciousness in olfaction. *Consciousness and Cognition, 18,* 1004–1017.

Turner, B. H., Mishkin, M., & Knapp, M. (1980). Organization of the amygdalopetal projections from modality-specific cortical association areas in the monkey. *Journal of Comparative Neurology, 191,* 515–543.

Willander, J., & Larsson, M. (2006). Smell your way back to childhood: Autobiographical odor memory. *Psychonomic Bulletin & Review, 13,* 240–244.

Willander, J., & Larsson, M. (2007). Olfaction and emotion: The case of autobiographical memory. *Memory & Cognition*, *35*, 1659–1663.
Wilson, D. A., & Stevenson, R. J. (2006). *Learning to smell: Olfactory perception from neurobiology to behavior.* Baltimore, MD: The Johns Hopkins University Press.
Zucco, G. M. (2003). Anomalies in cognition: Olfactory memory. *European Psychologist*, *8*, 77–86.

CHAPTER 7

Knowing what we smell

Fredrik U. Jönsson[1] and Mats J. Olsson[2]
[1]Stockholm University / [2]Karolinska Institutet Stockholm, Sweden

In this chapter, we review the nature of, and performance levels for, odor source naming, and the different proposed explanations to the generally low odor naming performance observed in experimental studies. We differentiate between odor naming and odor identification and show that although humans can rarely name more than 50% of common household items, this is not an odor naming problem, but rather reflects the difficulty we have in identifying odors. We investigate two broad accounts of odor identification failures in terms of perceptual and associative processes necessary for correct identification. Additionally, we discuss the *feeling of knowing* and *tip of the nose experience* commonly associated with identification failures. This type of metacognition provides us with odor knowledge in the absence of odor identification. In light of these phenomena, we discuss the importance of odor identification for olfactory functioning.

Introduction

Our ability to name odors is limited and this comes as a surprise to most laypeople when challenged with the task of identifying simple household items by smell alone. If you are visually presented with a lemon you will most certainly be able to name it, as would all healthy observers. Its name is overlearned and is normally triggered quickly and automatically. From the point of view of visual identification the environment is full of items that hardly anyone would fail to identify correctly. However, if you instead smell the odor of lemon without any other sensory input, naming performance drops considerably, although the odor is still considered quite familiar to most people. The ability or perhaps more aptly, the inability, to correctly name odors without the help of visual or other contextual cues has vexed researchers of olfaction for decades. Rachel Herz and Trygg Engen even described it as the "most contentious issue in human olfactory processing" (Herz & Engen, 1996, p. 301). More specifically, how able are we to name odors correctly? It has repeatedly been shown that correct naming performance for a set of common everyday odors rarely exceeds 50% (e.g., Cain, 1979; Cain, de Wijk, Lulejian, Schiet, & See, 1998; Desor & Beauchamp, 1974; de Wijk & Cain, 1994a; de Wijk & Cain, 1994b; de Wijk, Schab, & Cain, 1995; Distel & Hudson, 2001; Lawless & Engen, 1977; Olsson & Fridén, 2001), and this number is considerably lower for more unfamiliar and uncommon odors. In addition, the successful naming rate of *any* single item rarely reaches 100% across a group of participants. It should be noted that these performance levels pertain to our ability to correctly name odors presented in a laboratory environment and in the absence of other contextual information.

Most research about odor identification has consisted of odor naming tasks (e.g., de Wijk et al., 1995). The typical procedure is to let a group of participants sample different odors, one at a time, without visual or other contextual cues and then try to name the object from which the odor emanates, often referred to as "free identification" in the literature. The underlying assumption is that if a person is able to name something, then he/she normally also knows what it is (see e.g., Johnson, Paivio, & Clark, 1996, for similar thoughts in picture naming). However, because an object can be identified even though its name is sometimes not retrieved or even learned, it is important to distinguish between *odor naming* and *odor identification*. In fact, picture naming (McCauley, Parmalee, Sperber, & Carr, 1980; Johnson et al., 1996), as well as person naming (Burton & Bruce, 1992) are thought to occur in three broad stages, namely: (1) object identification, (2) name activation, and (3) response generation.

First an object has to be identified as a member of a particular class of objects. Following identification, one or more candidate names are retrieved from memory. Finally, in the third stage the best candidate name is articulated in the form of an overt response (e.g., saying it aloud or writing it down). According to Johnson and colleagues (1996) these stages are usually assumed to occur more or less sequentially. Hence, an object can be identified without its name being activated, and a name can be activated without it being overtly expressed. An olfactory example of identification without naming would be if you smelled juniper and could imagine the tree on which it grows, how the juniper berries look, that it is an ingredient in gin, and so forth, but you could not retrieve its name. Obviously, if you can name something, the object will normally also be identified, but the opposite may not be true.

Holley (2002) argued that odors are nothing but "attributes of objects and substances whose natural function is to reveal the presence of those objects and substances in the environment. As a consequence, odor naming turns out to be odor-source naming" (p. 19). In line with this, in this chapter we utilize the term *odor naming* as the ability to correctly name an odor source. *Odor identification* on the other hand, is used to denote identification ability, independent of naming ability. As will be shown below, inability to name odors rarely depends on the observer's failure to come up with a proper name for an already identified odor (as would be the case if you can picture the juniper berry on the tree but not name it). The inability to name an odor is more likely a consequence of not knowing exactly what it is that smells this way, that is, on a failure to identify the odor source (Jönsson, Tchekhova, Lönner, & Olsson, 2005). Because we will likely be able to name the odor source when we have identified it, in particular if it is very familiar, odor naming performance turns out to be a reasonable approximation of odor identification performance.

1. Accounts of odor identification failure

Visual agnosia is the case when a neurologically compromised person cannot identify a common object by sight alone, whereas he or she would be able to understand which object is referred to when hearing a description of it, or after haptic investigation. Visual agnosia can serve as a visual parallel to our (in)ability to identify odors, because in comparison with visual identification, olfactory identification performance suggests "olfactory agnosia" to be the rule. Of course, the aim here is not to pathologize the olfactory cognition of healthy observers but it is interesting to note that for visual agnosia there is an apperceptive and an associative type. The former diagnosis is applicable when a patient cannot veridically represent the object perceptually. For example, it could reveal itself as an inability to draw the object of interest. The latter is applicable when failure to identify the object is present without a perceptual distortion. It could be instructive to use these processes as models for understanding how olfactory identification works. In other words, which processes fall short when we fail to identify an odor source? With this in mind we categorize below the several competing, complementary and overlapping views on the inability to identify odors into either *associative* or *perceptual* accounts.

1.1 Associative accounts

1.1.1 *Odor names lack ecological importance*

Köster (2002) stressed that, in real life, rather than being able to identify odors and judge their intensity, it is much more important to be able to detect their presence, discriminate between them, and recognize them as either familiar or unfamiliar. Hence, our partial inability to name odors may just be a consequence of it not being so important for us to do so. However, one might argue that there are occasions when accurate identification of odors is advantageous, if not essential. For example, accurate identification of a kitchen gas leak, spoiled food, or fire – all which could have detrimental consequences if not understood (White & Kurtz, 2003). It should also be noted that positive identification of odors seems to enhance processing at the perceptual level. When odors are identified, we are better at discriminating between them than when they are not (Jönsson, Møller, & Olsson, 2011).

1.1.2 *Odor names are non-consensual*

De Wijk et al. (1995) argued that in everyday life, odors are typically experienced in different contexts and that what we smell is to a large extent interpreted based on the contextual information available. Sulmont-Rossé, Issanchou, and Köster (2005) argued that some of the poor performance found in odor naming tasks may be due to the lack of social consensus about odor names. That is, people encounter certain odors in different contexts or in different products, which means that which odor is associated with which objects differs between persons and cultures. For example, in their experiment several participants consistently labeled artificial flower odors as "cleaning supply" or "bathroom freshener". This was not the veridical name expected by the experimenters, but they may still be considered as correct descriptions of the odor sources, because these products are often fragranced with such odors. However, this account cannot explain why it is so difficult to name even common household objects by their smell alone since these have names that are generally agreed upon within a culture.

1.1.3 *The poor-link view*

Another explanation for why it is so difficult to name odors is that there is a "poor link" between the odor and its name. This is consistent with the previous argument that odor identification is not as important as other olfactory abilities (i.e., odor names are poorly learned, because learning the name is not important for us). But more specifically the poor-link hypothesis suggests that the link is *inherently* weak in that the verbal areas of the brain are poorly associated with the olfactory processing areas and more so than for other modalities (Engen, 1991; Engen, 1987; Herz & Engen, 1996). In a similar vein, Lorig (1999) argued that odor information processing and language processing share some of the same cortical resources. Due to this, simultaneous processing of olfactory and language information leads to interference, which in turn explains why it is so difficult to name odors. De Wijk et al. (1995) on the other hand proposed that our inability to name odors is due to a combination of a poor odor-name association (either inherent or due to less learning) and poor odor discrimination ability. The implications of odor quality discrimination failures on identification performance will be further developed below.

To conclude, there is some confusion in the literature about associative accounts of odor identification and naming with respect to why there should be a poor link, as well as the exact nature of this link. The latter confusion has to do with the fact that there is one association from the perceptual representation to the identity and another from the identity to the proper name, which only a few researchers have taken into account (Cleary, Konkel, Nomi, & McCabe, 2010; Jönsson & Olsson, 2003; Jönsson et al., 2005).

1.2 Perceptual accounts of odor identification failure

The perceptual accounts of odor identification failures more or less all concern *odor quality discrimination* (OQD) as an underlying ability-limiting function. We agree with Wise, Olsson and Cain (2000) who pointed out that almost all tests of olfactory performance are limited to some degree by the property of discrimination (see also Cain, 1979; Schab & Cain, 1991). Similarly, de Wijk et al. (1995) stated that "all subsequent higher order processing, including recognition memory and identification, can only be as accurate as the resolution of the sensory system" (p. 24). Eskenazi, Cain, and Friend (1986) were of the same opinion, namely that "the ability to discriminate must underlie the ability to identify" (p. 204) an odor.

1.2.1 *Odor quality discrimination (OQD)*

What then characterizes OQD, per se? OQD is often regarded as very good in humans. The arguments behind this view are based on the fact that we have a large number of olfactory receptors that could be combined to code for an almost infinite number of odorants. Another argument is the finding that very "small" changes of a molecular structure can result in clear differences in odor quality (as found by Laska, Ayabe-Kanamura, Hubener, & Saito, 2000; Laska & Freyer, 1997). Moreover, odor discrimination can also improve with repeated exposure (Jehl, Royet, & Holley, 1995; Rabin, 1988; but see Olsson, Faxbrink, & Jönsson, 2000), or through conditioning (Li, Moallem, Paller, & Gottfried, 2007). In a recent paper, Yeshurun and Sobel (2010) argued that OQD is "outstanding", and perhaps even more so for ecologically valid odors as judged by kin recognition studies. However, a closer look at the kin recognition literature reveals moderate performance levels. For instance, in one experiment 16 mothers out of 20 could, by smell alone, identify their own offspring in a binary choice (i.e., chance equals 10 correct identifications). In a second experiment 13 out of 17 mothers (chance being 8.5) made a correct choice (Porter, Cernoch, & McLaughlin, 1983). Another study showed that participants could identify their own T-shirt among nine other participants' shirts (all worn for 24 hours) in only 75% of the cases (Lord & Kasprzak, 1989). In the case of non-body odors, Olsson and Cain (2000) attempted to define a method to quantify an OQD threshold. They assessed the degree to which an odor (A) needed to be substituted (in liquid phase) with another fairly dissimilar odorant (B) in order to reach a just noticeable difference in perceived odor quality. The substitution needed averaged as high as 30% (see also Wise et al., 2000). In other words, within this experimental protocol, an odor A could barely be discriminated from a mixture of 70% of odorant A and 30% of odorant B. In our view, OQD has a better reputation than behavioral studies can support. OQD may therefore play a major role in identification failures.

1.2.2 *Does odor quality discrimination underlie odor naming?*

A number of studies have indicated that discrimination is indeed a problem in odor identification and naming. In the first experiment by Cain and colleagues (1998) participants tried to name the same set of odors in four sessions (each separated by about two days). The overall naming accuracy increased over time, but the naming performance varied across the sessions and an odor that was correctly named on one day was sometimes incorrectly named on another. The hit rate varied between 42% and 47%, across the sessions. However, only about 33% of the odors were correctly named in all sessions. The authors suggested that these results were indicative of errors of discrimination. Further support for the idea that odor identification and naming ability are dependent on OQD comes from several correlational studies. For example, in another study by Cain and colleagues (1998), it was found that participants' discrimination ability for one odor set was highly correlated with their naming ability for another set (see also Eskenazi, Cain, & Novelly, 1983; Rabin, 1988, for similar findings). De Wijk and Cain (1994a) further showed that odor naming performance follows an inverted U-shaped function with age, reflecting best performance in young adults and worse performance in children and the elderly. Discrimination ability showed the same pattern and was correlated with naming ability. In another study, de Wijk and Cain (1994b) found significant correlations between discrimination and naming performance for both free and cued naming (see also Eskenazi et al., 1986) as well as with consistency in applying the same label to a given odor with different intensity. In these studies the correlations varied from moderate (.48) to high (.80), clearly tying OQD with odor identification ability.

The odor naming errors people make often reveal errors of discrimination (e.g., Engen, 1987), such as when the name lemon or lime is used to describe the smell of orange. Researchers of olfaction have long been aware of this and often find it meaningful to separate two measures of naming performance, a strict measure of accuracy (the odor label has to be fully correct) and a measure that includes so called "near misses". An example of the latter would be to name orange as either lemon or lime or the more generic label "citrus fruit". Such responses may, at least in some cases, reveal some level of knowledge about the odor. To conclude, our ability to discriminate between odors seems closely connected to our ability to identify and consequently name odors.

1.2.3 *How does odor quality discrimination affect odor naming?*

Odor presentation will often fail to activate the odor name, but if the name is provided, odor perception instantly becomes unambiguous (see de Wijk et al., 1995, for similar reasoning). The clarity with which an identified odor is perceived can be contrasted with the more fuzzy (unclear) perception of the odor before correct (or incorrect) identification has occurred. We call this the *fuzzy perception account* of odor identification. To clarify, the basic idea of this perceptual account is that as long as an odor is not identified, the perception of it is *fuzzy*, like a blurred image that is not clearly perceived until you get some contextual clues or are told what it represents. The consequence of this lack of clarity is that it becomes harder to discriminate between unidentified odors, than between identified odors (e.g., de Wijk & Cain, 1994b; Jönsson et al., 2011), as seen in the correlation between OQD and odor naming ability. However, when additional information is available, the correct association between the perceptual representation and the identity can be made (Jönsson et al., 2005).

Support for a perceptual account of odor identification failures can be inferred from Wilson and Stevenson (2006). They argued that odor quality has little to do with a stimulus-based model of how chemical components of an odorant act upon olfactory receptors. Rather odor quality is the result of olfactory learning. For instance, if an odorant has been perceived in conjunction with a sweet tastant, the odor will acquire some of the tastant's sweetness when perceived alone. That is, the sweetness is synthesized with the original odor quality. This odor-taste learning is also paralleled by odor-odor learning. In our view, the lack of a stable one-to-one correspondence between odor quality and a specific stimulus may contribute to non-consensual labeling of odors.

Yet another perceptual explanation of why we often fail to identify even familiar odors was recently introduced by Yeshurun and Sobel (2010). They suggested that odor quality is limited to a single dimension, pleasantness. That is, olfactory object perception is really a matter of precisely representing the pleasantness of the odor that is then linked to the odor object considered to be the best match for that specific level of pleasantness. Since it is possible that several odor objects can be comparable in odor pleasantness, exact identification for a given odor is difficult. From this it follows that if a person were to rate the pleasantness of coffee and lemon as exactly the same, he/she would not be able to tell them apart.

2. Cued identification

Whereas free identification (i.e., by smell alone) of odors is difficult for most people, cued identification is considerably easier. Notably most clinical tests of olfactory ability are given in the format of cued identification tests, such as the Scandinavian Odor Identification Test (Nordin, Brämerson, Lidén, & Bende, 1998; Nordin, Nyroos, Maunuksela, Niskanen, & Tuorila, 2002), the University of Pennsylvania Smell Identification Test (Doty, Shaman, Kimmelman, & Dann, 1984), the Connecticut Chemosensory Clinical Research Center test (Cain, Goodspeed, Gent, & Leonard, 1988), and the Sniffin' Sticks test (Kobal et al., 1996). If an observer smells a target odor and is presented with four names, one of which is correct, their performance level is remarkably higher than in a free identification task (i.e., where naming is used as a measure of identification ability). For instance, de Wijk and Cain showed that identification of odors changed from around 40% correct when not cued (i.e., free) to above 80% correct when cued in a young population. For a group of middle aged participants, performance improved from approximately 30% to above 80% correct responses. Moreover, as the chance level for a correct response was 17% for the cued identification test, this increase is statistically substantial.

2.1 Cued identification versus discrimination

As shown above, cued identification performance is considerably better than free identification. With this background, de Wijk and Cain (1994b) compared odor quality discrimination to cued identification. They showed that when performance measures were corrected for chance, cued identification (chance level = .17) fell just 10% below discrimination (chance level = .50) among young adults and only 1% below discrimination in a middle-aged group. In sum, cued identification seems to be about as good as our discrimination ability.

2.2 The reciprocity of the odor-name association

The marked difficulty of odor naming has led to the idea that there is some asymmetry in the odor-name association (de Wijk et al., 1995). To test this hypothesis, Olsson and Jönsson (2008) asked participants to either smell an odor and then choose the proper name among three alternatives (of which one was correct) or to read an odor name and then choose the matching odorant among three alternatives. Although instances of an asymmetry were suggested, in both directions, depending on which odors or odor names were involved, the overall conclusion was that no general asymmetry is evident. Overall, it was as difficult to match a name to an odor as it was to match an odor to a name. Moreover, this was true for odorants of both high and low familiarity and identifiability.

3. Metacognitive aspects on odor naming

Although we are sometimes unable to name an odor, we can still have knowledge about it. That there is knowledge in the absence of proper odor identification is seen in metacognitive experiments. In these studies, participants can feel quite strongly that they know the odor and have some semantic information about it – e.g., a "fruit", despite the fact that they cannot come up with a name or other description momentarily. These feelings are somewhat predictive of actual knowledge about the odor, or more precisely the ability to correctly pick the correct name of an odor amongst lures. This aspect of odor knowledge will be discussed below.

3.1 The feeling of knowing the odor

It is intriguing that we can, with some accuracy, monitor the availability of a memory, even though we might fail to retrieve it, (Hart, 1965; Koriat, 1993, 1994, 1995; Koriat & Levy-Sadot, 2001; Metcalfe, 2000; Nelson, 1996; Nelson & Narens, 1990). This subjective state – that a currently unretrieved memory is stored in memory is referred to as a feeling of knowing (FOK; Hart, 1965). Indeed, we are able to tell whether an answer to a general information question is in memory (or not) faster than we can articulate the answer (Nhouyvanisvong & Reder, 1998; Reder, 1987). Reder and Ritter (1992) argued that FOK is a general process that operates whenever memory is queried and it can vary from a strong feeling of *not* knowing, to a strong feeling of knowing. A special case of FOK is the strong and imminent feeling of retrieval people can have when a word they cannot remember is on the tip of their tongue (see next section).

The most common way of investigating FOK judgments is to present participants with general information questions (e.g., what is the capital of Australia?). For all questions where the answer is not remembered the participants judge how strong their FOK is on a scale and are then given a multiple alternative forced choice recognition test (e.g., four alternatives: Sidney, Melbourne, Canberra, Perth[1]). FOK ratings are then correlated with actual recognition performance. An extensive body of FOK data has shown that people are, in general, moderately accurate in monitoring their knowledge (see Metcalfe, 2000, for a review). However, few studies have examined FOK in olfaction (Cain et al., 1998; Jönsson & Olsson, 2003; Jönsson et al., 2005). In fact, a FOK state can inform us about the degree of knowledge people have about odors, even when they fail to identify them properly. Cain and colleagues compared the predictive validity of FOK for verbal and olfactory material. They replicated the common finding that FOKs for answers to semantic general knowledge questions were predictive of later multiple choice recognition performance, but found that this was not the case for FOK judgments about odor names. In a follow-up study by Jönsson and Olsson (2003; Experiment 1), participants were asked to name a set of common odors, and if they thought that they could name an odor correctly they attempted to do so. If they could not name the odor, they instead rated how sure they were that they would recognize the correct odor name if it were shown to them. When presented with a name, participants accepted a correct name on 72% of trials and rejected an incorrect name on 86% of trials. The participants FOK judgments were moderately, and significantly, correlated with their recognition of the correct odor name ($G = .34$), as measured by the Goodman-Kruskal gamma correlation (Nelson, 1984). For other modalities, the gamma correlation typically ranges between .45 and .55 (Metcalfe, 2000). In a later study, Jönsson et al. (2005) directly compared the accuracy (i.e., predictive validity) of FOK for the labels of common odors with FOK for names of famous persons. The data showed that although both were reliably related to actual ability to retrieve an accurate label for the items, FOK responses for person names were superior. Hence, even when we fail to immediately name an odor in a free-recall identification task, we still have enough knowledge of the odor to somewhat accurately say whether we can later recognize or retrieve it from memory, but to a lesser extent than we are capable of in other modalities. Two other metacognitive experiences that relate to feeling of knowing are described below.

3.2 The tip of the tongue and the tip of the nose experience

Brown and McNeill (1966; see Brown, 1991, 2011, for reviews) were the first to perform a systematic investigation of the tip of the tongue (TOT) experience, and described it as follows: "If you are unable to think of the word, but feel sure

1. The correct answer is Canberra.

that you know it and that it is on the verge of coming back to you, then you are in a TOT state" (p. 327). Most subsequent studies have relied on this description or some version of it. Brown (1991) pointed out that a definition of TOT almost always includes: (i) that a missing word is known but at the moment unavailable and (ii) that recall of the word feels imminent. Typical ways of eliciting TOT states in the laboratory are to present participants with word definitions, or to show pictures of famous persons. The TOT state seems to occur universally across different languages (Schwartz, 1999; see Schwartz, 2002, for a review). The TOT state is also closely related to the FOK experience and most probably, any FOK study includes some TOT states which are defined as strong FOK experiences. However, all strong FOK states are not TOT experiences. For example, Yaniv and Meyer (1987) found instances of high FOK ratings for items not reported to be on the tip of the tongue (see Brown, 2011, for a review). In addition, the research methodology typically differs between FOK and TOT experiments. Whereas a person in a FOK experiment is merely asked about the likelihood of being able to recognize (or retrieve) the missing answer at a later time, independent of the strength of the FOK, TOT experiments only target the stronger feelings of imminent retrieval people sometimes have following retrieval failures (Nhouyvanisvong & Reder, 1998).

When we fail to retrieve the name of an object (e.g., a person or an odor), we often have partial access to both relevant and irrelevant information that appears in our consciousness as a byproduct of our memory search. For example, a distinctive feature of TOT experiences is the frequent access to structural-phonological information of relevance to the unretrieved target word (Brown, 1991, 2011). This may be the first letter of the sought-for word (e.g., the missing word begins with an A, as in "apple"; note that actually it is the phonemes not the letters that are retrieved), other letters, similar sounding words, or other information disclosing partial access to the missing word. According to Koriat, Levy-Sadot, Edry, and Marcas (2003) the partial clues we have when failing to retrieve a sought-for word can be divided into two categories; First, the just mentioned structural-phonological clues and secondly, other semantic clues. Examples of the latter would be "it is yellow" and "it is a fruit" (for lemon). Although the availability of structural-phonological clues has attracted the most attention when investigating TOT experiences, they also trigger a multitude of other semantic clues (Brown, 1991; Brown, 2011; Koriat et al., 2003; Lovelace, 1987; Yarmey, 1973). Access to structural-phonological clues clearly indicates failures to fully activate the sought-for word, that is, name-activation failures. This can be viewed as a (possibly temporary) "poor link" between the stimulus that cues the memory and the memory for the name of the object. Hence, if odor naming failures would be the symptom of a poor link between the odor and its name, TOT experiences should be frequent, and those experiences should be frequently associated with structural-phonological information. However, this is not the case.

In one of Trygg Engen's most cited papers (Lawless & Engen, 1977; Experiment 3), the authors triggered TOT experiences for the veridical names of odor sources. In this study, participants made familiarity ratings and attempted to name a set of 48 odors. If the participants were unable to name an odor, but gave a high familiarity rating and felt that they had the odor name on the tip of the tongue, they filled out a questionnaire asking for partial information about the sought-for odor name (e.g., first letter, number of syllables) and other associated information (e.g., if they could form a visual image of the odorous object or give a plausible odor category, for example, "spice" for black pepper). In contrast to verbal TOTs, where partial access to such information is frequent (Brown, 1991, 2011), Lawless and Engen (1977) found that the participants had virtually no partial information about the odor name. Therefore, they coined the separate term "tip of the nose" (TON) experience as an olfactory parallel to the TOT experience. The key difference between the otherwise highly related TOT and TON experiences, the almost complete lack of structural-phonological information when having a very strong feeling of knowing for an odor that you cannot name, was also later replicated by us (Jönsson & Olsson, 2003). Notably, odors do trigger some semantic information even when odor name retrieval fails. Participants can sometimes categorize the odor correctly (e.g. spice for black pepper), or have access to other semantic information (Lawless & Engen, 1977). For example, we found that when reporting a TON experience, 20% of the participants could name a place from which the odor might have come, 13% could name an object from which it might have come, 17% could name a general category for the odor, and 8% could name a similar odor (Jönsson & Olsson, 2003). Note that the latter responses were not scored for accuracy, rather they indicate what kind of semantic information the participants have about the odor.

Lawless and Engen's (1977) findings concerning TON experiences have recieved very frequent citation in the olfactory literature, but very little empirical attention. Up to 2005 the scientific literature still consisted of the single article by Lawless and Engen (1977; but see also Engen, 1987), which is surprising, given the attention it has received. The absence of partial information about the missing word when having a TON experience, though typical of the TOT experience, has been taken as evidence for a poor link between the odor and its name (Engen, 1991; Engen, 1987; Herz & Engen, 1996). However, later studies have contradicted this assertion (Jönsson & Olsson, 2003; Jönsson et al., 2005). If odor naming failures would be mere name activation failures we would expect odors to trigger as much structural-phonological information as other modalities do (Brown, 1991, 2011), which they do not. In addition, as TON experiences are typically triggered by very familiar, well-encoded, everyday items (e.g., juniper, banana, or coffee), name activation failures seem implausible.

However, the lack of structural-phonological information (i.e., partial access to the sought for name) when having a TON experience also explains why the resolution rate is generally lower than that for TOT experiences for other types of semantic stimuli (e.g., person names; Jönsson et al., 2005), as well as why we observe a somewhat lower predictive validity of the more general FOK judgments (Cain et al., 1998; Jönsson et al., 2005). The lower resolution rate is simply symptomatic of the fact that the TON state (and FOK for odors) is not about the name of the odor; rather participants in the TON state still search for the exact identity of the odor (i.e., knowledge of what it is). As the naming fails earlier in the naming process, in the identification stage, the resolution rate (measured as correct naming) is expected to be lower because successful resolution calls for two successful activations, first of the identity and then the name. Jönsson et al. (2005; Experiment 1) further investigated why odor naming sometimes fails or, more precisely, where in the naming process it fails. In their first experiment they let participants try to name a set of common odors and a set of famous people, of comparable difficulty levels. If the participants could immediately name the item they were told to do so, but if the name was not immediately retrieved, the participants were asked to indicate which of two states they were in: (i) they lacked the name and the identity of the odor or person, or (ii) they lacked the name but had identified the stimulus (i.e., they knew who or what it was). The latter is referred to as "subjective identification". If the participants thought they knew the identity of the odorous object, they were asked whether they were having a TOT experience or not, that is, whether resolution of the sought-for name felt imminent. If they did not know the identity, they instead made a judgment about how certain they were that they would be able to retrieve the exact name or properly describe the item (i.e., a FOK judgment), if given more time to search their memory.

If odor naming failures, in particular those associated with very strong feelings of imminent retrieval, were due to name-activation failures we would expect (i) the rate of subjective identifications to be high and (ii) the odor-elicited TOT rate to be high or at the very least comparable to the TOT rate for the names of famous people. In addition, (iii) if the subjective identification rate is high, the odors should also activate a high degree of semantic information other than the name during the memory search (i.e., the amount of semantic information triggered by an odor should follow the subjective identification rate). However, these predictions failed on all accounts. Only 8% of the naming failures for odors were associated with subjective identifications; significantly lower than for the person names (43%). Further, whereas 24% of the person naming failures led to a TOT state, this was the case for only 7% of the odor naming failures. Instead, odor naming failures often led to strong or very strong FOK judgments (i.e., the participants had not identified the odors but felt that that they were very familiar). Finally, the participants' general knowledge of the unsuccessfully named odors and pictures showed the same pattern as the subjective identifications, namely, they had reliably less semantic information about the odors than the people.

This line of research shows that people typically do not have an odor name on the tip of their tongue (as also indicated by the lack of structural-phonological information); instead, they generally fail to name odors due to failures to identify them (i.e., to know what the odor is). Nevertheless, the subjective feeling of knowing an odor does to some extent reflect knowledge about the odor in that it predicts future identification performance (Jönsson & Olsson, 2003; but see Cain et al., 1998), but less so than it does for other modalities (Jönsson et al., 2005, Experiment 2). For instance, the stronger the feeling of knowing, the more probable is it that a correct odor name will be recognized as correct and an incorrect name will be rejected.

4. Discussion

Starting from the observation that humans in a laboratory setting, without the aid of other sensory or contextual cues, fail to name most common odors, we have reviewed the different attempts to understand why odor naming is so difficult. We show that failures to name an odor very rarely depend on failures to retrieve the particular name, but rather are due to the fact that most of the time we do not know what we are smelling. This lack of odor source knowledge is an overlooked fact in the literature on odor naming.

In this chapter, we have also discussed perceptual and associative accounts of odor identification failures, referring to the associative and perceptual processes that are necessary for successful identification. The perceptual processes lead to the odor being reliably perceived in a manner such that it can be discriminated from other similar odors. The associative processes link the percept of the odor to the identity and then, if this is successful, most likely to a proper name. Association failures have often been suggested as an explanation for identification failures (not based on evidence but rather on observation); possibly because participants often have a strong feeling that they know the odor well, but that its identity will just not manifest to them at that moment.

The discrepancy between cued and free identification can be taken as support for the idea that associative rather than perceptual problems underlie odor identifications failures. The knowledge about the odor source is retained but needs to be cued in order to be retrieved. However, the fact that it becomes easier to map the odor percept to the proper name when cued is also consistent with the fuzzy perception account. If perception is fuzzy, mapping the percept to the proper semantic information (e.g., the odor name), should be more difficult, than when semantic information is given, which helps in the identification process. Hence, fuzzy perception (which is associated with poorer discrimination performance for unidentified odors) might in itself hinder the activation of the proper association between an odor and its identity. Several studies have suggested that discrimination failures are behind identification failures. However, several authors argue that odor quality discrimination is excellent. Others, including us, are less impressed with human odor discrimination abilities. In the normal case, when perception of the environment is multisensory and full of other cues, the perception of the odor may become more complete, there by leading to an association with the identity of an odor. According to Gottfried (2009), new findings from both rodents and humans suggest that pattern-based odor representations in the piriform cortex (a site for odor quality representations) are involved first in binding odorant parts into perceptual wholes, second, in overcoming fragmentation of stimuli, and third, in defining perceptual boundaries of odor qualities and categories. There is, therefore, reasonable evidence to suggest that not only will cues strengthen the association between a percept and its identity but that the cues available will also shape the quality representation of the odor. The latter is in line with the idea that odor names sharpen odor quality discrimination, as described above. To conclude, it is not easy to discern the exact reason for why odor identification and naming fail, as the associative and perceptual processes are likely interdependent. The answer, therefore, is likely that identification failures are due to a combination of perceptual and associative processes.

In light of these issues, it is interesting to note that in the absence of immediate identification, we do possess some semantic knowledge about an odor. This knowledge is reflected by metacognitive states such as feeling of knowing and tip of the nose experiences, and by associated semantic information that predict future identification performance. For instance, the stronger the feeling of knowing, the more probable is it that a given correct odor name can be recognized as correct and that an incorrect one can be rejected (Jönsson & Olsson, 2003; Jönsson et al., 2005; but see Cain et al., 1998).

From another angle, if we consider olfactory functioning, where does odor identification make a difference? As suggested by Stevenson (2010) the most significant categories of human olfactory function concern ingestion, avoiding environmental hazards, and social communication. How crucial is successful identification in serving these functions? Since an unidentified odor evokes feelings of, for example, pleasantness, disgust, familiarity and so forth, most possible functions within these categories seem not to depend heavily on proper identification of the source. But, if the information is ambiguous, identification could resolve it. For example, if the smell of a food violates your expectancy of a peculiar odor note, identifying it as parmesan cheese and not vomit (for example) would be advantageous. Similarly, the ability to reject a negative alternative (vomit) in the absence of a proper identification can also be informative. Hence, the immediate feelings of familiarity and pleasantness (or disgust) we have when smelling an odor can be viewed as very fast and generally functional heuristic rules on which we can base our immediate reaction to an odor – whether we should approach the odorous object or avoid it. However, as with all heuristic rules, they may not be right all of the time, which is the price paid for the speed gained in using them. Hence, the outcome of a slower identification process allows us to reevaluate the stimulus properly and, as noted, it might very well change the perception of the odor.

Although there are very little data elucidating the role of odor identification in basic human functioning, a few studies can be mentioned that reflect the importance of identifying a food source. De Araujo, Rolls, Velazco, Margot, and Cayeux (2005; see also Herz & von Clef, 2001) presented participants with a mixture of isovaleric acid and cheddar cheese flavor and labeled it as either "body odor" or "cheddar cheese". They found that both ratings of pleasantness and activations of its neural substrates varied as a function of the given label. This suggests that our ability to identify odors correctly could support successful foraging. On the other hand, actual data have indicated the dispensability of odor identification in a task where it could be expected to matter: edibility judgments. Data from Olsson and Fridén (2001) and Olsson and Cain (2003) suggest that subvocal edibility judgments are considerably faster than subvocal identification, indicating that edibility judgments do not rely on a previous identification. Moreover, results suggested that correct identification of odors only explains < 1% of the variance across odors in how much edibility judgments could be sped up as a function of a previous judgment (so-called edibility priming; Koenig, Bourron, & Royet, 2000). In other words, the correct identification of odors seems to be of only marginal importance for knowing what to eat and not to eat.

In contrast to the literature which touts the importance of odor identification, some authors have argued that awareness of odor identity actually distorts rather than enhances olfactory functioning. Köster and colleagues (Degel & Köster, 1999; Degel, Piper, & Köster, 2001) found that that people tend to implicitly link a nonconscious odor (target) with the place where that odor was present: Participants often matched the target odor (presented in suprathreshold concentration) to a picture of the room in which the odor had been previously presented, but only if the target odor could not be identified. The authors, therefore, argued that awareness of odor identity disrupted the link between odor and place. In a similar vein, Li and colleagues (2007) found that participants rated the likeability of neutral faces differently depending on whether they simultaneously smelled pleasant, neutral, or unpleasant odors delivered below detection thresholds (for example, pleasant odors leading to higher likability). Importantly, however, odor hedonics significantly shifted likeability ratings only for those participants who were not consciously aware of the presence of the odors. These results indicate that social preferences are subject to influences from odors that escape awareness, whereas the availability of conscious odor information (which is a prerequisite for identification) may disrupt such effects. These observations further put the necessity of proper odor identification in question.

Knowledge about the odor source, both in terms of explicit knowledge about the odor source and in terms of metacognitive experience reflecting the knowledge we think we possess but cannot momentarily retrieve, is substantial but also clearly fallible. In parallel, we can more readily report the perception of pleasantness, edibility, emotionality, and familiarity of odors. These features may be the primary information that leads us to knowing what to eat, who or what to approach, or avoid.

Acknowledgements

We would like to thank Rachel Herz, Amy Gordon, and Stephanie Juran for their valuable feedback during the writing of this chapter. Last but not least we would like to express our sincere gratitude to the late Trygg Engen whose scientific work has always been a source of inspiration for us.

References

Brown, R., & McNeill, D. (1966). The "tip of the tongue" phenomenon. *Journal of Verbal Learning and Verbal Behavior, 5*, 325–337.

Brown, A. S. (1991). A review of the tip-of-the-tongue experience. *Psychological Bulletin, 109*, 204–223.

Brown, A. S. (2011). *Tip of the tongue state*. Hove, UK: Psychology Press.

Burton, A. M., & Bruce V. (1992). I recognize your face but I can't remember your name: a simple explanation? *British Journal of Psychology, 83*, 45–60.

Cain, W. S. (1979). To know with the nose: Keys to odor identification. *Science, 203*, 467–470.

Cain, W. S., Goodspeed, R. B., Gent, J. F., & Leonard, G. (1988). Evaluation of olfactory dysfunction in the Connecticut Chemosensory Clinical Research Center. *Laryngoscope, 98*, 83–88.

Cain, W. S., de Wijk, R., Lulejian, C., Schiet, F., & See, L. C. (1998). Odor identification: Perceptual and semantic dimensions. *Chemical Senses, 23*, 309–326.

Cleary, A. M., Konkel, K. E., Nomi, J. S., & McCabe, D. P. (2010). Odor recognition without identification. *Memory & Cognition, 38*, 452–460.
De Araujo, I. E., Rolls, E. T., Velazco, M. I., Margot, C., & Cayeux, I. (2006). *Neuron, 46*, 671–679.
De Wijk, R. A., & Cain, W. S. (1994a). Odor identification by name and edibility: Life-span development and safety. *Human factors, 36*, 182–187.
De Wijk, R. A., & Cain, W. S. (1994b). Odor quality: Discrimination versus free and cued identification. *Perception and Psychophysics, 56*, 12–18.
De Wijk, R. A., Schab, F. R., & Cain, W. S. (1995). Odor identification. In F. R. Schab, & R. G. Crowder (Eds.), *Memory for odors* (pp. 21–37). Mahwah, NJ: Lawrence Erlbaum associates.
Degel, J., & Köster, E. P. (1999). Odors: Implicit memory and performance effects. *Chemical Senses, 24*, 317–325.
Degel, J., Piper, D., & Köster, E. P. (2001). Implicit learning and implicit memory for odors: The Influence of odor identification and retention time. *Chemical Senses, 26*, 267–280.
Desor, J. A., & Beauchamp, G. K. (1974). The human capacity to transmit olfactory information. *Perception & Psychophysics, 16*, 551–556.
Distel, H., & Hudson, R. (2001). Judgment of odor intensity is influenced by subjects' knowledge of the odor source. *Chemical Senses, 26*, 247–251.
Doty, R. L., Shaman, P., Kimmelman, C. P., Dann, M. S. (1984). University of Pennsylvania Smell Identification Test: a rapid quantitative olfactory function test for the clinic. *Laryngoscope, 94*, 176–178.
Engen, T. (1987). Remembering odors and their names. *American Scientist, 75*, 497–503.
Engen, T. (1991). *Odor sensation and memory*. New York: Praeger Publishers.
Eskenazi, B., Cain, W. S., & Friend, K. B. (1986). Exploration of olfactory aptitude. *Bulletin of the Psychonomic Society, 24*, 203–206.
Eskenazi, B., Cain, W. S., Novelly, R. A., & Friend, K. B. (1983). Olfactory functioning in temporal lobectomy patients. *Neuropsychologia, 21*, 365–374.
Gottfried, J. A. (2009). Function follows form: Ecological constrains on odor codes and olfactory percepts. *Current opinion in neurobiology, 19*, 422–429.
Hart, J. T. (1965). Memory and the feeling-of-knowing experience. *Journal of Educational Psychology, 56*, 208–216.
Herz, R. S., & Engen, T. (1996). Odor memory: Review and analysis. *Psychonomic Bulletin and Review, 3*, 300–313.
Herz R. S., & von Clef, J. (2001). The influence of verbal labeling on the perception of odors: Evidence for olfactory illusions? *Perception, 30*, 381–391.
Holley, A. (2002). Cognitive aspects of olfaction in perfumer practice. In C. Rouby, B. Schaal, D. Dubois, R. Gervais, & A. Holley (Eds.), *Olfaction, taste and cognition* (pp. 16–26). Cambridge University Press.
Jehl, C., Royet, J. P., & Holley, A. (1995). Odor discrimination and recognition memory as a function of familiarization. *Perception and Psychophysics, 57*, 1002–1011.
Johnson, C. J., Paivio, A., & Clark, J. M. (1996). Cognitive components of picture naming. *Psychological Bulletin, 120*, 113–139.
Jönsson, F. U., Möller, P., & Olsson, M. J. (2011). Olfactory working memory: Effects of verbalization on the 2-back task. *Memory and Cognition, 39*, 1023–1032.
Jönsson, F. U., & Olsson, M. J. (2003). Olfactory metacognition. *Chemical Senses, 28*, 651–658.

Jönsson, F. U., Tcheckova, A., Lönner, P., & Olsson, M. J. (2005). A metamemory perspective on odor naming and identification. *Chemical Senses, 30*, 353–365.

Kobal, G., Hummel, T., Sekinger, B., Barz, S., Roscher, S., & Wolf, S. (1996). "Sniffin' Sticks": screening of olfactory performance. *Rhinology, 34*, 222–226.

Koenig, O., Bourron, G., & Royet, J.-P. (2000). Evidence for separate perceptive and semantic memories for odours: a priming experiment. *Chemical Senses, 25*, 703–708.

Koriat, A. (1993). How do we know that we know? The accessibility model of the feeling of knowing. *Psychological Review, 100*, 609–639.

Koriat, A. (1994). Memory's knowledge about its own knowledge: The accessibility account of the feeling-of-knowing. In J. Metcalfe, & A. P. Shimamura (Eds.), *Metacognition: Knowing about knowing* (pp. 115–135). Cambridge, MA: Bradford.

Koriat, A. (1995). Dissociating knowing and the feeling of knowing: Further evidence for the accessibility model. *Journal of Experimental Psychology: General, 124*, 311–333.

Koriat, A., & Levy-Sadot, R. (2001). The combined contributions of the cue familiarity and accessibility heuristics to feelings of knowing. *Journal of Experimental Psychology: Learning, Memory, and Cognition, 27*, 34–53.

Koriat, A., Levy-Sadot, R., Edry, E., & de Marcas, S. (2003). What do we know about what we cannot remember? Accessing the semantic attributes of words that cannot be recalled. *Journal of Experimental Psychology: Learning, Memory, and Cognition, 29*, 1095–1105.

Köster, E. P. (2002). The specific characteristics of the sense of smell. In C. Rouby, B. Schaal, D. Dubois, R. Gervais, & A. Holley (Eds.), *Olfaction, Taste, and Cognition* (pp. 27–43). New York: Cambridge University Press.

Laska, M., Ayabe-Kanamura S., Hubener, F., & Saito, S. (2000). Olfactory discrimination ability for aliphatic odorants as a function of oxygen moiety. *Chemical Senses, 25*, 189–197.

Laska, M., & Freyer, D. (1997). Olfactory discrimination ability for aliphatic esters in squirrel monkeys and humans. *Chemical Senses*, 22, 457–465.

Lawless, H., & Engen, T. (1977). Associations to odors: Interference, mnemonics, and verbal labeling. *Journal of Experimental Psychology: Human Learning and Memory, 3*, 52–59.

Li, W., Moallem, I., Paller K. A., & Gottfried, J. A. (2007). Subliminal Smells can Guide Social Preferences. *Psychological Science, 18*, 1044–1049.

Lord, T., & Kasprzak, M. (1989). Identification of self through olfaction. *Perceptual and Motor Skills, 69*, 219–224.

Lorig, T. (1999). On the similiarity of odor and language perception. *Neuroscience and Biobehavioral Reviews, 23*, 391–389.

Lovelace, E. (1987). Attributes that come to mind in the TOT state. *Bulletin of the Psychonomic Society, 25*, 370–372.

Metcalfe, J. (2000). Metamemory: Theory and data. In E. Tulving, & F. I. M. Craik (Eds.), *The Oxford handbook of memory* (pp. 197–211). New York: Oxford University Press.

McCauley, C., Parmelee, C. M., Sperber, R. D., & Carr, T. H. (1980). Early extraction of meaning from pictures and its relation to conscious identification. Journal of Experimental psychology. *Human Perception and Performance, 6*, 265–276.

Nelson, T. O., & Narens, L. (1990). Metamemory: A theoretical framework and some new findings. In G. H. Bower (Ed.), *The psychology of learning and motivation* (pp. 1–45). New York: Academic Press.

Nelson, T. O. (1996). Consciousness and metacognition. *American Psychologist, 51*, 102–116.

Nelson, T. O. (1984). A comparison of current measures of the accuracy of feeling-of-knowing predictions. *Psychological Bulletin, 95*, 109–133.

Nordin, S., Brämerson, A., Lidén, E., & Bende, M. (1998) .The Scandinavian odor-identification test: Development, reliability, validity and normative data. *Acta Otolaryngology, 118,* 226–234.

Nordin, S., Nyroos, M., Maunuksela, T., Niskanen, T., & Tuorila, H. (2002). Applicability of the Scandinavian odor identification test: A Finnish–Swedish comparison. *Acta Otolaryngology, 122,* 294–297.

Nhouyvanisvong, A., & Reder, L. M. (1998). Rapid feeling-of-knowing: A strategy selection mechanism. In V. Y. Yzerbyt, G. Lories, & B. Dardenne (Eds.), *Metacognition: Cognitive and social dimensions* (pp. 35–52). Thousand Oaks: Sage Publications.

Olsson, M. J., & Cain, W. S. (2000). Psychometrics of odor quality discrimination: Method for threshold determination. *Chemical Senses, 25,* 493–499.

Olsson, M. J., & Cain, W. S. (2003). Implicit and explicit memory for odors: Hemispheric differences. *Memory & Cognition, 31,* 44–50.

Olsson, M. J., Faxbrink, M., & Jönsson, F. U. (2002). *Repetition priming in odor memory.* Rouby, Catherine; Schaal, Benoist; Dubois, Danièle; Gervais, Rémi; Holley, André (2002). Olfaction, taste, and cognition (pp. 246–260). New York, NY, US: Cambridge University Press.

Olsson M. J., & Fridén M. (2001). Evidence of odor priming: edibility judgments are primed differently betwen the hemispheres. *Chemical Senses, 26, 117–123.*

Olsson, M. J., & Jonsson, F. U. (2008). Is it easier to match a name to an odor than vice versa? *Chemosensory perception, 1,* 184–189.

Porter, R. H., Cernoch, J. M., & McLaughlin, F. J. (1983). Maternal recognition of neonates through olfactory cues. *Physiology & Behavior, 30,* 151–154.

Rabin, M. D. (1988). Experience facilitates olfactory quality discrimination. *Perceptionand Psychophysics, 44,* 532–540.

Reder, L. M. (1987). Strategy selection in question answering. *Cognitive Psychology,* 19, 90–138.

Reder, L. M., & Ritter, F. E. (1992). What determines initial feeling of knowing? Familiarity with question terms, not with the answer. *Journal of Experimental Psychology: Learning, Memory and Cognition, 18,* 435–451.

Schwartz, B. L. (1999). Sparkling at the end of the tongue: The etiology of tip-ofthe-tongue states: A diary study. *Psychonomic Bulletin & Review, 6,* 379–393.

Schwartz, B. L. (2002). *Tip-of-the-tongue states: Phenomenology, mechanism, and lexical retrieval.* Mahwah, NJ, US: Lawrence Erlbaum Associates Publishers.

Schab, F. R., & Cain, W. S. (1991). Memory for odors. In D. G. Laing, R. L. Dory, & W. Breipohl (Eds.), *The human sense of smell* (pp. 217–240). New York: Springer.

Stevenson, R. J. (2010). An Initial Evaluation of the Functions of Human Olfaction. *Chemical Senses, 35,* 3–20.

Sulmont-Rossé, C., Issanchou, S., & Köster, E. P. (2005). Odor naming methodology: Correct identification with multiple-choice versus repeatable identification in a free task. *Chemical Senses, 30,* 23–27.

White, T. L., & Kurtz, D. B. (2003). The relationship between metacognitive awareness of olfactory ability and age in people reporting chemosensory disturbances. *American journal of psychology, 116,* 99–110.

Wilson, D. A., & Stevenson, R. J. (2006). *Learning to smell: Olfactory perception from neurobiology to behavior.* Baltimore, MD, USA: John Hopkins University Press.

Wise, P. M., Olsson, M. J., & Cain, W. S. (2000). Quantification of odor quality. *Chemical Senses, 25,* 429–443.

Yaniv, I., & Meyer, D. E. (1987). Activation and metacognition of inaccessible stored information: Potential bases for incubation effects in problem solving. *Journal of Experimental Psychology: Learning, Memory, and Cognition, 13,* 187–205.
Yarmey, A. D. (1973). I recognize your face but I can't remember your name: Further evidence on the tip-of-the-tongue phenomenon. *Memory & Cognition, 1,* 287–290.
Yeshurun, Y., & Sobel, N. (2010). An odor is not worth a thousand words: From multidimensional odor to unidimensional objects. *Annual Review of Psychology, 61,* 219–241.

CHAPTER 8

Attending to olfactory short-term memory

Theresa L. White
Le Moyne College and SUNY Upstate Medical University, Syracuse, USA

Short-term memory generally refers to the temporary internal representation of a limited amount of information. Trygg Engen's early results on olfactory short-term memory were so dissimilar to those found in other sensory systems that they led him to question the very existence of this type of memory (Engen, 1987). His later work, however, suggests that he came to support the concept of olfactory short-term memory (Engen, Gilmore, & Mair, 1991), though he felt that it functioned quite differently from other memory systems (Herz & Engen, 1996). Although a variety of general models regarding the nature of short-term memory exist (e.g., Baddeley & Hitch, 1974; Jonides et al., 2008), most recognize the role of attention as an integral part of the memory process. Likewise, general models of attention include short-term memory as a fundamental component (Knudsen, 2007). So, in other sensory systems, attention has been shown to both determine the contents of memory (Broadbent, 1958) and be guided by short-term memory in perceptual tasks (Downing, 2000). The question remains, however, as to whether olfactory short-term memory enjoys the same two-way relationship with attention. The most recent model of olfactory memory (Wilson & Stevens, 2006) does not include attention as an aspect of the memory process. An examination of the literature gives evidence that although olfactory short-term memory has some unique characteristics when compared to memory for other types of stimuli, attention is still a highly relevant component.

Introduction

Although earlier scientists dabbled in the area (e.g., Heywood & Vortriede, 1905; Laird, 1935), Trygg Engen was the first to systematically study olfactory memory and to grapple with the idea of its structure (e.g., Engen, Kuisma, & Eimas, 1973; Engen & Ross, 1973). One of his central questions was whether memory for olfactory stimuli was stored and processed in the same way as information from other sensory systems. Engen's experimentation led him to answer that question as "no", and he viewed olfactory cognition as unique, functioning quite differently from that of other sensory systems, though he recognized that similarities also existed (Herz & Engen, 1996).

Everyday life experience bears out Engen's scientific observations as to the uniqueness of olfactory cognition. Consider the following fairly normal exchange between two friends:

> One: "Do you smell that?"
> Two: "Smell what?"
> One: "Fresh bread, baking."
> Two: "Oh, I didn't notice until you mentioned it, but now I smell it. The bread smells great!"

This sort of discussion is common regarding olfactory stimuli, but it is difficult to imagine it occurring with supra-threshold levels of visual or auditory stimuli. Yet, this ordinary conversation illustrates both the difficulty in the detection of olfactory stimuli (e.g., Cain, 1977) and the difficulty in identification of an odorant (e.g., Cain, 1979), as well as the influence of existing knowledge on odor processing (e.g., Rabin & Cain, 1984; Herz & von Clef, 2001). This exchange also shows that even the simple act of detecting and identifying an odorant based on a stored memory requires a certain amount of attention.

1. Attention

Although a variety of general models regarding the nature of memory exist (e.g., Baddeley & Hitch, 1974; Nee & Jonides, 2011), most recognize attention as an integral part of the process. Attention is awareness, or "taking possession by the mind" (James, 1890); it is the name given to cognitive phenomena that can broadly be divided into two categories, exogenous and endogenous. In *exogenous attention*, the attentional focus is reflexively drawn toward objects in the environment, such as those that are novel (Shiffrin & Schneider, 1977) or salient (Park, Shin, & Lee, 2002). The object requiring attention typically elicits an automatic orienting response, coupled with a redirection of sensory receptors. An example would be a door that is slammed loudly in an otherwise quiet hallway; the sound can capture the attentional focus and result in head turning toward the door. It is possible, however, to use *endogenous attention* in such a situation and thus, to control the focus of attention intentionally. One might choose, for example, not to turn to look toward the sound of the slamming door, but to focus attention on the computer instead. Endogenous attention is an effortful process related to self-control

(Schmeichel & Baumeister, 2010), in which one portion of the environment garners attention while the other aspects are ignored.

Both exogenous and endogenous categories of attention seem to share the same limited cognitive resource that influences how incoming information is processed once it reaches the receptors. Neurally, attention is represented as enhanced firing (Brefczynski & DeYoe, 1999) in sensory cortices (Coull, Frith, Büchell, & Nobre, 2000), fronto-parietal cortical networks (Buschman & Miller, 2007), and in thalamus (Rees, 2009). Endogenous attention (as an aspect of self-control) has been shown to be related to blood glucose level (Gailliot et al., 2007).

2. General attentional model

Attention in other sensory systems has been conceptualized as object-based (e.g., Shinn-Cunningham, 2008), essentially a process that when focused increases neural processing and enhances recognition. A recent model conceptualizing attention (Knudsen, 2007) identifies the involvement of four fundamental processes: Working memory, competitive selection, top-down control, and bottom-up saliency filters (similarly to Figure 1). *Working memory* (Baddeley & Hitch, 1974) or short-term memory, a temporary store that must be refreshed regularly in order to maintain information, is at the center of the model. The higher cognitive processes that regulate the signal intensity of information in the competitive selection process are called *top-down sensitivity control*. The sensitivity regulation achieved through top-down control processes influences goal direction and the selection of new information for working memory; this regulation also allows for the voluntary allocation of endogenous attention. Particularly salient stimuli, such as those that occur rarely or have biological relevance, capture access to working memory through automatic enhancement from *bottom-up saliency filters* that command exogenous attention. The process of *competitive selection* determines which pieces of information enter the system and gain access to working memory. In this model, working memory and attention are inexorably intertwined: Objects that garner attention are represented in working memory, working memory representations require attention in order to be maintained, and attention in perceptual tasks is guided by memory.

This model has been fairly robust, applying not only to visual but also to auditory perception (Shinn-Cunningham, 2008). Indeed, some have suggested that attentional resources are independent of specific stimulus modality, and constitute a single shared resource based on frontal lobe activity (Dade, Zatorre, Evans, & Jones-Gotman, 2001; Spence, Kettenman, Kobal, & McGone, 2001).

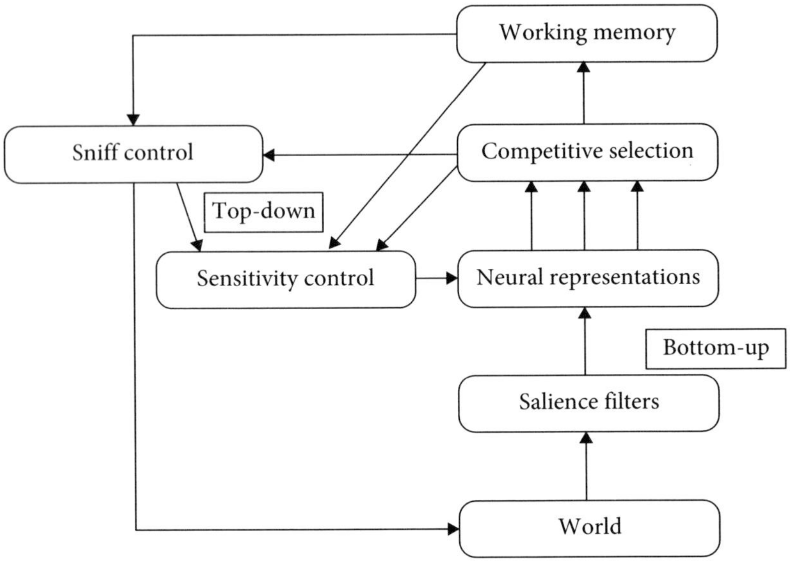

Figure 1. Modified version of the model of functional attentional components described by Knudsen (2007, p. 59). Information about the world is transmitted by the receptors and nervous system to salience filters that respond differentially to unusual or salient stimuli. Neural representations code the information about the world, stored memories, internal state, etc. Representations with the highest signal strength compete for entry to working memory. Working memory can guide top-down signal bias that alters the sensitivity of the representations that it is processing and can direct sniff control and other orienting behaviors. Voluntary attention involves working memory, sensitivity control, and competitive selection working in a recurrent loop.

3. Olfactory attention may be unique

While there has been considerable interest in the way that central factors such as attention affect the processing of visual and auditory information (e.g., Schmeichel & Baumeister, 2010), only a modest amount of research has focused on these aspects of olfactory cognition. Yet, the neural architecture of the olfactory system differs considerably from that of other sensory systems (Tham, Stevenson, & Miller, 2009), and olfactory cognition has been described as unique (Zucco, 2003; Wilson & Stevenson, 2006), as has olfactory consciousness (Stevenson, 2009). Thus, olfactory attention may be somewhat dissimilar to that of other sensory systems.

The sense of smell is unique in its neuroanatomy (Tham et al., 2009) and its stimulus acquisition process. Incoming information is received either orthonasally through sniffing (Mainland & Sobel, 2006; Sela & Sobel, 2010) or retronasally through the oral cavity (Pierce & Halpern, 1996). The sensory receptors are neurons located at the top of the nasal cavity in the nasal epithelium, and projections from the receptor neurons proceed mostly ipsilaterally through the cribiform plate to synapse in the olfactory bulb (Gottfried, 2010). Primary olfactory pathways do not include a thalamic relay per se, as do the primary pathways of other sensory systems, such as vision, but instead project directly to piriform cortex (Shepherd, 2005). The pathway can then either proceed directly to orbitofrontal cortex or project to mediodorsal thalamus before continuing to orbitofrontal cortex, though the neural fibers in this secondary thalamic pathway are sparse (Price, 1985). The thalamus has been shown to be important for mediating arousal and conscious attention (John, 2001; Guillery & Sherman, 2002), as well as for gating information to higher cortical regions from sensory systems other than olfaction (McCormick & Bal, 1994). Although some fMRI evidence seems to suggest thalamic activity in response to an olfactory selective attention task (Plailly, Howard, Gitelman, & Gottfried, 2008), other studies have failed to find an association (Sabri, Radnovich, Li, & Kareken, 2005; Zelano et al., 2005), suggesting that olfactory attentional processing may be different from other sensory systems.

Olfactory cognition also differs from other sensory systems, particularly in terms of memory performance (Zucco, 2003). The rate of acquisition of olfactory information is low compared to other sensory systems (Lawless, 1978), possibly because odors are encoded with relatively few features (Engen, Gilmore, & Mair, 1991). In contrast to memory for other types of stimuli, odors show little decrease in retention over periods from 30 sec (Engen, Kuisma, & Eimas, 1973) to 3 months (Engen & Ross, 1973). Also in contrast to visual or verbal memory, serial position effects in olfactory memory are not always observed (Lawless & Cain, 1975; Gabassi & Zanuttini, 1983), and when they are, the shape of the curve varies (e.g., White & Treisman, 1997; Reed, 2000; Miles & Hodder, 2007). Lastly, a model of olfactory memory has been proposed that does not include a functional dissociation between short- and long-term memory stores (Wilson & Stevenson, 2006; Engen, 1982). While the concept of a unitary memory is consistent with some general models of memory, these models typically posit variations in states of memory accompanied by a focus of attention (e.g., Nee & Jonides, 2011), which is missing from the olfactory memory model (White, 2009).

The consciousness associated with olfactory stimuli has been proposed to differ from that of other sensory systems (Zucco, 2003; Stevenson, 2009). At least one paper has suggested that odors lack conscious representation in memory, and are instead represented implicitly or automatically (Zucco, 2003). Another review of the literature found no evidence of a distinction between phenomenonal consciousness (experience of sensation) and access consciousness (experience available for other cognitive processes) as has been reported in other sensory systems (Stevenson, 2009), though it is worth mentioning that the distinction between these two types of consciousness has been generally disputed (Kouider, de Gardelle, Sackur, & Dupoux, 2010).

As discussed above, the olfactory system has a unique neural architecture that results in cognitive and memory processes that differ from that of other sensory systems. Although similarities exist, such as the ability to remember information over relatively brief intervals, the differences are remarkable. Further, the conscious processing of olfactory stimuli has been questioned via a hypothesis that proposed that odors do not give rise to a conscious representation as visual and verbal stimuli do (Zucco, 2003), which suggests that the awareness and attentional focus for odors may be distinctive. With this in mind, how might a general model of attention (Knudsen, 2007) account for olfactory stimuli?

4. Fitting olfactory information into an attentional model

4.1 Working memory

Although Engen first questioned the existence of short-term olfactory memory (Lawless & Engen, 1977; Engen, 1987), his later work suggests that he came to support the concept (Engen et al., 1991). Essentially, Engen recognized that olfaction is naturally a serial sense; all information must be processed sequentially. Therefore, to do any sort of comparison at all (e.g. "this one is stronger"), one must temporarily maintain representation in some type of short-term store or working memory. In the model of attention under evaluation (Knudsen, 2007), a limited amount of information in working memory is maintained flexibly for brief periods of time while it is evaluated and manipulated.

Like visual or verbal stimuli, odors can be maintained in a temporary memory state (White, 1998; Andarade & Donaldson, 2007). The information contained in olfactory memory seems to be primarily perceptually coded (White, Hornung, Kurtz, Treisman, & Sheehe, 1998) and subject to interference from other odors (Walk & Johns, 1984; Andarade & Donaldson, 2007), though verbal re-coding undoubtedly occurs (Wilson & Stevenson, 2006). Neuroimaging data (fMRI) further supports the concept of a temporary memory, in that effects that continue after the termination of the presentation of an odorant may be observed (Rolls, Grabenhorst, Margot, da Silva, & Velazco, 2008). Sustained activity in primary olfactory (piriform) cortex may reflect a specialized temporary storage for cognitive representations of odors that are not easily named, while activity in areas associated with verbal information is associated with namable odors (Zelano, Montag, Khan, & Sobel, 2009).

Evidence as to whether the temporary representation of olfactory information may be manipulated, a key requirement for the concept of working memory, is somewhat equivocal. For example, the primacy portion of the serial position curve is often taken to indicate evidence of rehearsal of the early portion of a list. However, primacy is not reliably observed with olfactory stimuli, though it can be found under circumstances in which verbal elaboration occurs (Annett & Lorimer, 1995). The more common olfactory serial position finding is that of a curve characterized by recency only (Miles & Hodder, 2005). Support for an olfactory working memory store that is able to manipulate information comes from studies involving a two-back task, in which participants have to determine whether a probe item is the same as a comparison item that was presented two trials before. Since the comparison item is constantly changing, the information in working memory must be continually manipulated. Memory performance on the two-back task with familiar odors was similar to performance with visual stimuli, suggesting similar types of working memory systems for the two types of stimuli (Dade et al., 2001). Familiar, easily named odors were also remembered better than with those that were difficult to name in the two back task, though both types of odors were remembered well above chance (Jönsson, Møller, & Olsson, 2011). This finding indicates that verbal information can be co-opted in processing olfactory stimuli, but also supports the presence of an olfactory working memory system that is able to manipulate information.

Another way that olfactory information can be manipulated is in terms of parsing out particular odors in odor-taste mixtures. Participants who could memorize large numbers of odors were better able to detect a target odor of benzaldehyde in a flavored soft drink than people with poorer olfactory memory spans, suggesting that larger olfactory memory capacity leaves more resources available for discrimination (Dacremont & Valentin, 2004). It is worth mentioning that people are worse at identifying the olfactory components of an odor-taste mixture than the taste components (Laing, Link, Jinks, & Hutchinson, 2001), which suggests that highly limited working memory resources may be shared among some sensory systems, such as taste. However, it is also possible that this finding reflects the limited number of identification possibilities available for basic tastes in comparison to the many possible odorants that could be identified in a chemical mixture. Collectively, these studies suggest that working memory capacity is associated with the ability to focus attention, and support the presence of an olfactory working memory that meets the requirement of the general model of attention (Knudsen, 2007).

4.2 Top-down sensitivity control

In the model under consideration, working memory not only stores and manipulates information, but also sharpens and enhances the quality of that information via several different mechanisms (Knudsen, 2007). One mechanism is to instigate an orienting response toward an object, such as sniffing. Though people continually breathe through their noses, sniffing represents an active attempt to sample and process olfactory stimuli. Sniffing is necessary for the perception of an odorant and affects both the intensity and the identification of an incoming olfactory object (Mainland & Sobel, 2006). Functional MRI evidence shows that airflow from a sniff is represented differently in the brain when it is attended to, activating frontal piriform cortex and the olfactory tubercle as opposed to only temporal piriform cortex when attention is not directed towards an odor that is present (Zelano et al., 2005).

Another mechanism for improving the quality of information in working memory is to modulate the responses of neural circuits. In the case of olfaction, olfactory cortex sends considerable feedback to the olfactory bulb, modulating inputs (Wilson, Kadohisa, & Fletcher, 2006). The anterior portion of primary olfactory cortex habituates rapidly to specific odors; this process relegates the habituated odors to the background, enhancing the discrimination of target odors (Linster, Henry, Kadohisha, & Wilson, 2007). Conversely, however, when people are told that an odor may be harmful, habituation can also be resisted in order to maintain focus on a potential danger (Dalton, 1996; Kobayashi et al., 2008).

People can easily shift endogenous attention between two different odors, ignoring one and attending to the other (Krauel, Pause, Sojka, Schott, & Ferstl, 1998). Electroencephalogram (EEG) studies suggest that attentional selection is performed early in the process for olfaction on the basis of physical characteristics (Krauel et al., 1998). Mononaral and binaral rivalry is also possible, with olfactory percepts shifting relative to attention when different odorants are presented to each nostril (Zhou & Chen, 2009).

Voluntary attentional shifts between olfaction and other sensory systems result in different levels of brain activation (Spence, Kettenmann, Kobal, & McGlone, 2000, 2001). In a functional magnetic resonance imaging (fMRI) study selective attention toward an odor and away from a tone led to an increase in response in the mediodorsal thalamic to orbitofrontal cortex pathway, while response to this area was weakened when the tone was selectively attended and the olfactory stimuli ignored (Plailly et al., 2008). In other words, subjects were able to direct their attention to either the odor or the tone during the same experimental paradigm based on instructions, and selective attention to the odor produced specific brain activity (Plailly et al., 2008) in an area typically associated with conscious attention in other sensory systems (e.g. John, 2001; Guillery & Sherman, 2002).

The ability to attend selectively to olfaction or another sensory system may be more difficult when the competing sense is taste, particularly when the odor is detected both ortho-nasally and retro-nasally, as occurs in normal food consumption. Selective attention to the odor of vanillin did not enhance its detectability when it was compared to water, sucrose, or citric acid when the odor/taste combination was sipped by subjects (Ashkenazi & Marks, 2004). Although attention to taste stimuli results in fMRI activity in primary taste cortex and attention to orthonasally presented odorants actives primary olfactory cortex, an area that is active during the presentation of both types of stimuli has been observed in a region of rostral anterior insula (Veldhuizen & Small, 2011), underscoring a high level of integration between these two sensory systems.

Top-down alternations of neural responses are specific to the judgments that will be made about an olfactory stimulus, in that the pattern of brain activity induced in one task may not be present during other tasks. When people were asked to perform a task that involved pleasantness, fMRI revealed that the medial orbito-frontal and pregenual cingulate cortex were active, while an intensity task resulted in inferior frontal gyrus activity (Rolls et al., 2008). These differences in brain activity began before odor delivery, in an anticipatory fashion following the instructions, suggesting the attentional biasing of incoming information in a way that matched the instructions (Rolls et al., 2008). These findings support previous work indicating that subjective affective judgments activate secondary olfactory (orbitofronal) cortex, while sensory information is processed in primary olfactory (piriform) cortex (Rolls, Kringelbach, & de Araujo, 2003).

4.3 Bottom-up saliency filtering

Trygg Engen once noted the "attention-getting" aspect of odors (Engen et al., 1991), by observing that a salient odor can intrude upon awareness. People continuously monitor the environment for odors, even when attention is engaged in other tasks, which means that people can have their attention captured by a salient odor (Stevenson, Mahmut, & Oaten, 2011). This supports the idea that strong neural responses can gain access to working memory (Knudsen, 2007), including neural responses that are generated by olfactory stimuli. However, the olfactory stimuli necessary to elicit an exogenous attentional response must be relatively strong in order to draw attentional focus. The absence of a spatial dimension for odors, coupled with the necessity of discrete sampling of the environment by sniffing, has led to the argument that people essentially live in a state of constant olfactory change blindness that minimizes attentional capture (Sela & Sobel, 2010). In other words, because of the amount of time that lapses between sniffs, people tend to "fill in" their representation of the olfactory environment cognitively, and are generally unaware of changes unless the newly introduced stimuli are highly salient.

Salience filters for stimuli may be created from a variety of neural mechanisms, including adaptation. Adaptation is defined here as a peripheral phenomenon, that is distinct from cortical habituation and reflects a tendency for receptors to decrease their firing in response to prolonged stimulation (Ekman, Berglund, Berglund, & Lindvall, 1967). Practically speaking, both adaptation and habituation may contribute to both daily odor processing and phenomena such as binaral rivalry (Zhou & Chen, 2009). Nevertheless, peripheral adaptation serves as one way to alter the incoming information seeking access to working memory.

The sense of smell also has a salience filter that is inherently biased toward false alarms – signaling the presence of an odor when none is actually present – possibly because one of the major functions of the sense of smell is to alert the organism to potential hazards (Engen et al., 1991). In an early demonstration of this bias in humans, students in a course were told that a flask containing distilled water contained a strong odor, and were asked to raise their hands when they smelled it (Slosson, 1899). According to Slosson's report, within 15 seconds most of the first row had their hands up and before one minute was over, three quarters of the class had their hands raised. In a modern version of the same manipulation, listeners to a radio program were told that odors were being transmitted via radio waves, and were asked to call the station when they smelled the odor (O'Mahoney, 1978). Again many people reported that they smelled an odor when none were present – some listeners even complained of having an allergic reaction to the *scent*. This strong response bias likely reflects the high variability associated with olfactory stimuli (Cain, 1977). A myriad of environmental factors, such as temperature or humidity, can interfere with the perception of the quality or strength of an odor, and a bias toward false alarm responses allows the system to be maximally effective in odor detection. So, a person trying to detect amyl acetate, an odorant that smells somewhat like nail polish remover at a high concentration and like banana at a lower concentration (De Rovia, 1999), is best served by a system that has a tendency to say "yes, something's there" at any hint of a scent, so as not to miss any potentially meaningful signal.

In vision, salient stimuli are given an advantage in the competition for access to working memory through bottom-up filters, and have a tendency to "pop out" from the background (Egeth & Yantis, 1997). The same has been shown to be true of olfactory information in a study that utilized an "oddball" paradigm, where the task was either to attend to an infrequently occurring odor or a sound (Sabri et al., 2005). Significant fMRI activity in the subgenual cingulate and in central posterior orbitofrontal cortex was observed during the shift from frequent to infrequent odors, but only in the condition where the participant was supposed to be ignoring the odors. These findings suggest that attention can be automatically drawn to olfactory novelty, even when much attention is directed elsewhere.

4.4 Competitive selection

In the attentional model's competitive selection component, the signal strengths of information arising from top-down and bottom-up biases are compared for access to working memory (Knudsen, 2007). The representations of external stimuli compete, both with each other and with representations arising from internal sources. The competition helps to eliminate distractions and to select relevant stimuli.

Competition of strength between basic stimulus parameters occurs with olfactory stimuli. Exposure to one olfactory stimulus may impair detection of another olfactory stimulus, though under some circumstances one stimulus may instead enhance a person's sensitivity to another odor (Engen & Bosack, 1969). For example, the number of responses to ethanol is greater when it closely follows octanol in time, possibly due to differences in the solubility of the odorants that may sharpen the contrast between the two. Cross-adaptation, in which adaptation to one odorant decreases the perception of another odorant, also occurs (Cain, 1970), particularly between odorants that are perceptually similar to each other (Kurtz, Lawless, & Acree, 2010). It is worth noting that these cross-adaptive relationships are not necessarily reciprocal. An odorant, such as pentanol, that decreases the perception of another, such as propanol, may not be affected itself; so, propanol induces only limited cross-adaptation in pentanol (Engen, 1963). These findings support the idea of competition at early levels in the olfactory system, either in the periphery (Kurtz et al., 2010) or at the level of anterior piriform cortex (Gottfried, Winston, & Dolan, 2006).

Higher levels of processing, in brain areas where features associated with stimuli are processed, are additional sites of competition. In contrast to visual stimuli, olfactory stimuli have been described as "unitary" percepts, with few features (Engen & Ross, 1973), other than hedonic tone (Yeshurun & Sobel, 2010) or intensity (Rolls et al., 2008). Yet, most smells are actually comprised of a myriad of odorants that combine to form a perceptual whole. The combination of these odorants may be synergistic (response is more than the sum of responses to the parts), or suppressive (response is less than responses to the parts), and likely takes place at the level of the posterior piriform cortex (Gottfried et al., 2006).

In addition to the competition found at higher and lower stimulus processing levels, competition can also take place at the interface with working memory, where different types of information, such as verbal, visual, taste, or olfactory, compete for entry to the limited capacity working memory system. When incoming information from other sensory systems conflicts with olfactory stimuli, memory is impaired. Inappropriate labels for olfactory stimuli impair olfactory memory, such as "root beer" for licorice (Engen & Ross, 1973), as do pictures of inappropriate source locations, such as seeing a picture of a grill when attempting to remember the smell of soap (Streeter & White, 2011). These findings suggest that competition for access to working memory resources occurs for olfactory stimuli.

5. Conclusion

In olfactory perception, as in perception of all types of stimuli, the initial challenge is to dissociate the relevant stimuli from the barrage of incoming information from the environment and existing representations. This remarkable feat is achieved through attentional focus. Although the bulk of the research on attention has focused on visual (Schmeichel & Baumeister, 2010) and auditory (Shinn-Cunningham, 2008) stimulation, similar principles seem to govern olfactory stimuli. Evidence suggests that olfactory cognition fits reasonably well into a general attentional framework that is consistent across the senses (Knudsen, 2007). The four processes fundamental to attention, (1) working memory, (2) top-down sensitivity control, (3) bottom-up salience filtering, and (4) competitive selection, are all shown with studies of olfactory stimuli. This is not to suggest that many aspects of olfactory attention are not unique, but instead points out that despite these differences, attention, as a general property of cognition, enhances the neural processing of olfactory stimuli and allows the focus of cognitive resources on olfactory information.

References

Andrade, J., & Donaldson, L. (2007). Evidence for an olfactory store in working memory? *Psychologia, 50*, 76–89.

Annett, J. A., & Lorimer, A. W. (1995). Primacy and recency in recognition of odors and recall of odor names. *Perceptual & Motor Skills, 81*, 787–794.

Ashkenazi, A., & Marks, L. E. (2004). Effect of endogenous attention on detection of weak gustatory and olfactory flavors. *Perception & Psychophysics, 66*, 596–608.

Baddeley, A. D., & Hitch, G. J. (1974). Working memory. In G. Bower (Ed.), *Recent advances in learning and motivation,* Vol. 8 (pp. 47–90). New York, NY: Academic Press.
Brefczynski, J. A., & DeYoe, E. A. (1999). A physiological correlate of the 'spotlight' of visual attention. *Nature Neuroscience, 2*, 370–374.
Broadbent, D. E. (1958). *Perception and Communication*. London: Pergamon.
Buschman, T. J., & Miller, E. K. (2007). Top-down versus bottom-up control of attention in the prefrontal and posterior parietal cortices. *Science, 315*, 1860–1862.
Cain, W. S. (1970). Odor intensity after self-adaptation and cross-adaptation. *Perception & Psychophysics, 7*, 271–275.
Cain, W. S. (1977). Differential sensitivity for smell: 'Noise' at the nose. *Science, 195*, 796–798.
Cain, W. S. (1979). To know with the nose: Keys to odor identification. *Science, 203*, 467–469.
Coull, J. T., Frith, C. D., Büchel C., & Nobre, A. C. (2000). Orienting attention in time: Behavioural and neuroanatomical distinction between exogenous and endogenous shifts. *Neuropsychologia, 38*, 808–819.
Dacremont, C., & Valentin, D. (2004). Mobilization of short-term memory capacity for odors in discriminative tests: Implication for assessors' selection. *Food Quality and Preference, 15*, 735–742.
Dade, L. A., Zatorre R. J., Evans, A. C., & Jones-Gotman, M. (2001). Working memory in another dimension: Functional imaging of human olfactory working memory. *NeuroImage, 14*, 650–660.
Dalton, P. (1996). Odor perception and beliefs about risks. *Chemical Senses, 21*, 447–458.
De Rovira, D. A. (1999). *The dictionary of flavors and general guide for those training in the art and science of flavor chemistry.* Trumbull, CT: Food & Nutrition Press, Inc.
Downing, P. E. (2000). Interactions between visual working memory and selective attention. *Psychological Sciences, 11*(6), 467–473.
Egeth, H. E., & Yantis, S. (1997). Visual attention: Control, representation, and time course. *Annual Review of Psychology, 48*, 269–297.
Ekman, G., Berglund, B., Berglund, V., & Lindval, T. (1967). Perceived intensity of odor as a function of time of adaptation. *Scandinavian Journal of Psychology, 8*, 177–186.
Engen, T. (1963). Cross-adaptation to the aliphatic alcohols. *American Journal of Psychology, 76*, 96–102.
Engen, T. (1982). *The perception of odors*. New York: Academic Press.
Engen, T. (1987). Remembering odors and their names. *American Scientist, 75*, 497–503.
Engen, T., & Bosack, T. N. (1969). Facilitation in olfactory detection. *Journal of Comparative and Physiological Psychology, 68*, 320–326.
Engen, T., Gilmore, M., & Mair, R. G. (1991). Odor memory. In T. V. Getchell, R. L. Doty, L. M. Bartoshuk, & J. B. Snow (Eds.), *Smell and taste in health and disease* (pp. 315–328). New York: Raven.
Engen, T., Kuisma, J., & Eimas, P. (1973). Short-term memory of odors. *Journal of Experimental Psychology, 99*, 222–225.
Engen, T., & Ross, B. (1973). Long-term memory of odors with and without verbal descriptions. *Journal of Experimental Psychology, 100*, 221–227.
Gabassi, P. G., & Zanuttini, L. (1983). Ricinosimento di stimol: olfattivi nella memoria a brevtermine. *Giornale Italiano di Psicologia, 10*, 51–60.

Gailliot, M. T., Baumeister, R. F., DeWall, C. N., Maner, J. K., Plant, E. A., Tice, D. M., Brewer, L. E., & Schmeichel, B. J. (2007). Self-control relies on glucose as a limited energy source: Willpower is more than a metaphor. *Journal of Personality and Social Psychology, 92*, 325–336.

Gottfried, J. A. (2010). Central mechanisms of odour object perception. *Nature Reviews Neuroscience, 11*, 628–641.

Gottfried, J. A., Winston, J. S., & Dolan, R. J. (2006). Dissociable codes of odor quality and odorant structure in human piriform cortex. *Neuron, 49*, 467–479.

Guillery, R. W., & Sherman, S. M. (2002). Thalamic relay functions and their role in corticocortical communication: Generalizations from the visual system. *Neuron, 33*, 163–175.

Herz, R., & Engen, T. (1996). Odor memory: Review and analysis. *Psychonomic Bulletin & Review, 3*, 300–313.

Herz, R. S., & von Clef, J. (2001). The influence of verbal labeling on the perception of odors: Evidence for olfactory illusions? *Perception, 30*, 381–391.

Heywood, A., & Vortriede, H. A. (1905). Some experiments on the associative power of smells. *American Journal of Psychology, 16*, 537–541.

James, W. (1890). *Principles of psychology*. New York: Holt.

John, E. R. (2001). A field theory of consciousness. *Conscious Cognition, 10*, 184–213.

Jonides, J., Lewis, R. L., Nee, D. E., Lustig, C. A., Berman, M. G., & Moore, K. S. (2008). The mind and brain of short-term memory. *Annual Review of Psychology, 59*, 193–224.

Jönsson, F. U., Møller, P., & Olsson, M. J. (2011). Olfactory working memory: Effects of verbalization on the 2-back task. *Memory and Cognition*. doi: 10.3758/s13421-011-0080-5.

Knudsen, E. I. (2007). Fundamental components of attention. *Annual Review of Neuroscience, 30*, 57–78.

Kobayashi, T., Sakai, N., Kobayakawa, T., Akiyama, S., Toda, H., & Saito, S. (2008). Effects of cognitive factors on perceived odor intensity in adaptation/habituation processes: From 2 different odor presentation methods. *Chemical Senses, 33*, 163–171.

Kouider, S., de Gardelle, V., Sackur, J., & Dupoux, E. (2010). How rich is consciousness? The partial awareness hypothesis. *Trends in Cognitive Sciences, 14*, 301–307. doi: 10.1016/j.tics.2010.04.006.

Krauel, K., Pause, B. M., Sojka, B., Schott, P., & Ferstl, R. (1998). Attentional modulation of central odor processing. *Chemical Senses, 23*, 423–432.

Kurtz, A. J., Lawless, H. T., & Acree, T. E. (2010). The cross-adaptation of green and citrus odorants. *Chemosensory Perception, 3*, 149–155.

Laing, D. G., Link, C., Jinks, A. L., & Hutchinson, I. (2002). The limited capacity of humans to identify the components of taste mixtures and taste-odour mixtures. *Perception, 31*, 617–635.

Laird, D. A. (1935). What can you do with your nose? *The Scientific Monthly, 41*, 126–130.

Lawless, H. T. (1978). Recognition of common odors, pictures, and simple shapes. *Perception & Psychophysics, 24*, 493–495.

Lawless, H. T., & Cain, W. S. (1975). Recognition memory for odors. *Chemical Senses & Flavor, 1*, 331–337.

Lawless, H. T., & Engen, T. (1977). Associations to odors: Interference, mnemonics, and verbal labeling. *Journal of Experimental Psychology, 3*, 52–59.

Linster, C., Henry, L., Kadohisha, M., & Wilson, D. A. (2007). Synaptic adaptation and odor-background segmentation. *Neurobiology of Learning and Memory, 87*, 352–360.

Mainland, J., & Sobel, N. (2006). The sniff is part of the olfactory percept. *Chemical Senses, 31*, 181–196.

McCormick, D. A., & Bal, T. (1994). Sensory gating mechanisms of the thalamus. *Current Opinions in Neurobiology, 4*, 550–556.

Miles, C., & Hodder, K. (2005). Serial position effects in recognition memory for odors: A reexamination. *Memory and Cognition, 33*, 1303–1314.

Nee, D. E., & Jonides, J. (2011). Dissociable contributions of prefrontal cortex and the hippocampus to short-term memory: Evidence for a 3-state model of memory. *NeuroImage, 54*, 1540–1548.

O'Mahony, M. (1978). Smell illusions and suggestions: Reports of smells contingent on tones played on television and radio. *Chemical Senses and Flavor, 3*, 183–189.

Park, S., Shin, J., & Lee, M. (2002). Biologically inspired saliency map model for bottom-up visual attention. *Lecture Notes in Computer Science, 2525*, 113–145. doi: 10.1007/3-540-36181-2_42.

Pierce, J., & Halpern, B. P. (1996). Orthonasal and retronasal odorant identification based upon vapor phase input from common substances. *Chemical Senses, 21*, 529–543.

Plailly, J., Howard, J. D., Gitelman, D. R., & Gottfried, J. A. (2008). Attention to odor modulates thalamocortical connectivity in the human brain. *Journal of Neuroscience, 28*, 5257–5267.

Price, J. L. (1985). Beyond the primary olfactory cortex: Olfactory-related areas in the neocortex, thalamus, and hypothalamus. *Chemical Senses, 10*, 239–258.

Rabin, M. D., & Cain, W. S. (1984). Odor recognition: Familiarity, identifiability, and encoding consistency. *Journal of Experimental Psychology: Learning, Memory, and Cognition, 10*, 316–325.

Reed, P. (2000). Serial position effects in recognition memory for odors. *Journal of Experimental Psychology: Learning, Memory, and Cognition, 26*, 411–422.

Rees, G. (2009). Visual attention: The thalamus at the centre? *Current Biology, 19*, R213–R214. doi: 10.1016/j.cub.2009.01.011.

Rolls, E. T., Grabenhorst, F., Margot, C., da Silva, M. A. A. P., & Velazco, M. I. (2008). Selective attention to affective value alters how the brain processes olfactory stimuli. *Journal of Cognitive Neuroscience, 20*, 1–12.

Rolls, E. T., Kringelbach, M. L., & de Araujo, I. E. T. (2003). Different representations of pleasant and unpleasant odors in the human brain. *European Journal of Neuroscience, 18*, 695–703.

Sabri, M., Radnovich, A. J., Li, T. Q., & Kareken, D. A. (2005). Neural correlates of olfactory change detection. *NeuroImage, 25*, 969–974.

Schmeichel, B. J., & Baumeister, R. F. (2010). Effortful attention control. In B. Bruya (Ed.), *Effortless attention: A new perspective in the cognitive science of attention and action* (pp. 29–49). Cambridge, MA: MIT Press.

Sela, L., & Sobel, N. (2010). Human olfaction: A constant state of change-blindness. *Experimental Brain Research, 205*, 13–29.

Shepherd, G. M. (2005). Perception without a thalamus how does olfaction do it? *Neuron, 46*, 166–168.

Shiffrin, W., & Schneider, R. M. (1977). Controlled and automatic human information processing: I. Detection, search, and attention. *Psychological Review, 84*, 1–66.

Shinn-Cunningham, B. G. (2008). Object-based auditory and visual attention. *Trends in Cognitive Science, 12*, 182–186.

Slosson, E. E. (1899). A lecture experiment in hallucinations. *Psychological Review, 6*, 407–408.
Spence, C., Kettenmann, B., Kobal, G., & McGlone, F. (2000). Selective attention to the chemosensory modality. *Perception & Psychophysics, 62*, 1265–1271.
Spence, C., Kettenmann, B., Kobal, G., & McGlone, F. (2001). Shared attentional resources for processing visual and chemosensory information. *Quarterly Journal of Experimental Psychology, 54A*, 775–783.
Streeter, N. L., & White, T. L. (2011). Incongruent contextual information intrudes on short-term olfactory memory. *Chemosensory Perception, 4*, 1–8. doi: 10.1007/s12078-010-9082-0.
Stevenson, R. J. (2009). Phenomenal and access consciousness in olfaction. *Consciousness and Cognition, 18*, 1004–1017.
Stevenson, R. J., Mahmut, M., & Oaten, M. J. (2011). The role of attention in the localization of odors to the mouth. *Attention, Perception, & Psychophysics, 73*, 247–258.
Tham, W. W. P., Stevenson, R. J., & Miller, L. A. (2009). The functional role of the medio dorsal thalamic nucleus in olfaction. *Brain Research Reviews, 62*, 109–126.
Veldhuizen, M. G., & Small, D. M. (2011). Modality-specific neural effects of selective attention to taste and odor. *Chemical Senses*. doi: 10.1093/chemse/bjr043.
Walk, H. A., & Johns, E. E. (1984). Interference and facilitation in short term memory for odors. *Perception & Psychophysics, 36*, 508–514.
White, T. L. (1998). Olfactory memory: The long and short of it. *Chemical Senses, 23*, 433–441.
White, T. L. (2009). A second look at the structure of human olfactory memory. *Annals of the New York Academy of Science, 1170*, 338–342.
White, T. L., Hornung, D. E., Kurtz, D. B., Treisman, M., & Sheehe, P. (1998). Phonological and perceptual components of short-term memory for odors. *American Journal of Psychology, 111*, 411–441.
White, T. L., & Treisman, M. (1997). A comparison of the encoding of content and order in olfactory memory and in memory for visually presented verbal materials. *British Journal of Psychology, 88*, 459–472.
Wilson, D. A., Kadohisa, M., & Fletcher, M. L. (2006). Cortical contributions to olfaction: Plasticity and perception. *Seminars in Cell & Developmental Biology, 17*, 462–470.
Wilson, D. A., & Stevenson, R. J. (2006). *Learning to smell: Olfactory perception from neurobiology to behavior.* Baltimore: The Johns Hopkins University Press.
Yeshurun, Y., & Sobel, N. (2010). An odor is not worth a thousand words: From multidimensional odors to unidimensional odor objects. *Annual Review of Psychology, 61*, 219–41.
Zelano, C., Bensafi, M., Porter, J., Mainland, J., Johnson, B., Bremner, E., Telles, C., Khan, R., & Sobel, N. (2005). Attentional modulation in human primary olfactory cortex. *Nature Neuroscience, 8*, 114–120.
Zelano, C., Montag, J., Khan, R., & Sobel, N. (2009). A specialized odor memory buffer in primary olfactory cortex. *Public Library of Science (PLoS) ONE, 4*(3), 1–11.
Zhou, W., & Chen, D. (2009). Binaral rivalry between the nostrils and in the cortex. *Current Biology, 19*, 1561–1565.
Zucco, G. M. (2003). Anomalies in cognition: Olfactory memory. *European Psychologist, 8*(2), 77–86.

PART III

Neuropsychology and olfactory dysfunctions

CHAPTER 9

Olfactory function in Parkinson's disease

Richard L. Doty and Hakan Tekeli
University of Pennsylvania School of Medicine, Philadelphia, USA / Kasimpasa Military Hospital, Istanbul, Turkey

Olfactory dysfunction is an early 'pre-clinical' sign of Parkinson's disease (PD). In this chapter we briefly review what is known about such dysfunction in PD and related diseases. In general, the dysfunction is robust and bilateral, anosmia is not the norm, women exhibit less loss than men, and some monogenetic forms of the disease exhibit the dysfunction. The loss is unrelated to disease stage or magnitude of the motor symptoms, save perhaps in the earliest stages of the disorder. Lewy body pathology, a hallmark of PD, appears to be correlated with the olfactory dysfunction, although it is not known whether this reflects a causal relationship. Dopamine is over-expressed in the olfactory bulbs of PD patients. Interestingly, deficiencies in other neurotransmitters, most notably acetylcholine, are correlated with the olfactory loss. A number of diseases often misdiagnosed as PD have little or no smell loss, including essential tremor, vascular parkinsonism, and progressive supranuclear palsy, making olfactory testing of value in differential diagnosis. The limited data suggest that the physiological basis of PD-related olfactory dysfunction is likely multifactorial, representing a combination of environmentally and genetically determined pathologies that involve multiple transmitter systems.

Introduction

Parkinson's disease (PD) is an insidious progressive movement disorder that afflicts 6.3 million people worldwide. The median age at disease onset is 60 years, although 10% of cases occur before the age of 45 years (Lees, Hardy, & Revesz, 2009). Prevalence markedly increases with age. In one Dutch study, for example, such prevalence was 0.6% in those between 65 and 69 years of age and 2.6% in those between 85 and 90 years of age (de Rijk et al., 2000). The average life expectancy from the time of diagnosis is 15 years, with the most common cause of death being pneumonia (Katzenschlager et al., 2008).

Although traditionally considered solely a motor system disease with the cardinal features of bradykinesia, resting tremor, rigidity, and postural instability, it is now widely recognized that PD is accompanied by a range of non-motor symptoms. Among such symptoms are aberrations in olfaction, taste, vision, cardiovascular function, sleep, balance, gastric and bowel function, mood, and cognitive function (Halliday, Barker, & Rowe, 2011). A number of these symptoms precede the onset of the classic motor phenotype by years, being early components of the so-called 'pre-clinical', 'pre-motor', or 'pre-symptomatic' stage of the disease (Gaig & Tolosa, 2009).

1. The olfactory phenotype

Among the most salient of the non-motor elements of PD is olfactory dysfunction. Since first described in 1975, more than 80 published studies have appeared in the peer-reviewed literature demonstrating an altered ability to smell in patients with PD. These studies have shown that the PD-related smell loss is marked and can be detected by numerous olfactory tests, including those of odor detection, identification, and memory. Most studies have employed the University of Pennsylvania Smell Identification Test (UPSIT), a 40-item microencapsulated ("scratch & sniff") test developed at our center that correlates highly with detection threshold tests (Doty, Shaman, & Dann, 1984). Anosmia is not the norm and the olfactory loss is somewhat greater in men than in women (Doty, Deems, & Stellar, 1988a). The dysfunction is typically bilateral (Doty, Stern, Pfeiffer, Gollomp, & Hurtig, 1992), robust (Mesholam, Moberg, Mahr, & Doty, 1998), and only related to disease development or motor dysfunction at the earliest phases of the disorder (Doty, Deems, & Stellar, 1988b; Doty, Riklan, Deems, Reynolds, & Stellar, 1989; Barz et al., 1997; Siderowf et al., 2005; Berendse, Roos, Raijmakers, & Doty, 2011). Correlations exist between olfactory test scores and measures of cardiac sympathetic function, implying associations with some other 'pre-clinical' PD-related symptoms (Lee, Yeo, Kim, & Youm, 2006). Interestingly, the smell loss of early

Table 1. Scores on the University of Pennsylvania Smell Identification Test (UPSIT) and Picture Identification Test (PIT) in early stage patients with three major neurodegenerative diseases (n = 24 per disease).*

	Parkinson's disease	Alzheimer's disease	AD-PD Guam complex
UPSIT Score (SD)	20.2 (7.1)	18.4 (7.1)	20.5 (7.3)
PIT Score (SD)	39.5 (0.9)	38.3 (1.8)	36.2 (3.4)

* Data from (Doty et al., 1991b). The PIT is identical to the UPSIT except that pictures, not odorants, serve as stimuli.

PD is indistinguishable from that observed in early Alzheimer's disease and the Parkinson-Dementia Complex of Guam (Table 1).

2. Physiological underpinnings

Despite being well characterized phenotypically, the physiological basis of the olfactory dysfunction of PD is poorly understood. Repletion of dopamine clearly has no effect on the PD-related dysfunction, implying its independence from synaptic levels of this neurotransmitter (Doty et al., 1992; Quinn, Rossor, & Marsden, 1987; Roth, Radil, Ruzicka, Jech, & Tichy, 1998). Moreover, dopamine is overexpressed in the olfactory bulbs of PD patients. Thus, the expression of tyrosine hydroxylase, the rate-limiting enzyme involved in the synthesis of dopamine, is markedly *increased* in the olfactory bulbs of most patients with PD (Huisman, Uylings, & Hoogland, 2004). Similar increases in tyrosine hydroxylase expression have been found in the olfactory bulbs of mice and Macaca monkeys injected with the proneurotoxin 1-methyl-4-phenyl-1,2,3,6-tetrahydropyride (MPTP) (Belzunegui, Sebastian, & Garrido-Gil, 2007; Yamada, Onodera, Mizuno, & Mochizuki, 2004), a synthetic heroine-like agent that, when injected intravenously, induces a syndrome very similar to that of PD. However, Lewy bodies, a hallmark of PD-related pathology comprised of sphere-like abnormal intraneuronal aggregations of the protein α-synuclein, are not present in MPTP-related parkinsonism, nor is olfactory dysfunction (Doty, Singh, Tetrud, & Langston, 1992). The increased dopamine may well reflect increased migration of dopamine-secreting cells from the subventricular zone along the rostral migratory stream into the olfactory bulb, compensating for the loss of a dopamine-responsive substrate (Bedard & Parent, 2004).

The olfactory receptor neurons of PD patients exhibit a number of pathological elements generally associated with the disease [e.g., dystrophic neurites without Lewy bodies, amyloid precursor protein fragments, and all varieties of synuclein (α, β, γ)] (Crino et al., 1995). However, such aberrations are not specific to PD, being also found within the olfactory receptor neurons of elderly controls and patients with Alzheimer's disease, Lewy body disease, and multiple system atrophy (Duda, Shah, Arnold, Lee, & Trojanowski, 1999). More specific disease-related alterations are present, however, in the olfactory bulbs and tracts of patients with PD. For example, Lewy bodies have been found in the mitral cells (the major projection neurons of the bulb), as well as within the anterior olfactory nucleus (AON) – a structure that receives projections from mitral cells (Daniel & Hawkes, 1992). The number of Lewy bodies within the AON correlates with the loss of neurons, as well as with disease duration (Pearce, Hawkes, & Daniel, 1995).

Alterations in numerous brain regions beyond the olfactory receptor cells and olfactory bulb likely contribute to the olfactory alterations seen in PD. A recent diffusion tensor imaging study of olfaction-related brain structures found an association between the degree of microstructural reduction in white matter and the degree of smell dysfunction (Ibarretxe-Bilbao et al., 2010). Post-mortem studies have identified Lewy bodies and neurofibrillary tangles in a range of central brain structures associated with olfaction, including the amygdala, hippocampus, and entorhinal cortex (Mattila, Roytta, Torikka, Dickson, & Rinne, 1998; Braak & Braak, 1990). Based upon α-synuclein immunostaining of brain tissue, Braak et al. have proposed that the pathology of PD advances in a predictable sequence (Figure 1), with the olfactory bulbs, the associated AON, and the dorsal motor nuclear complex of the glossopharyngeal and vagus nerves (DMC) exhibiting the earlier changes (Braak, Rub, Gai, & Del, 2003). When Lewy bodies were identified in the substantia nigra, they were also always present in the olfactory bulb, the anterior olfactory nucleus, and the dorsal medulla, implying earlier involvement of these structures and providing one possible explanation of why many patients with PD experience smell loss years before the onset of their motor symptoms. However, this important theory is not without its critics (Attems & Jellinger, 2008; Kalaitzakis, Graeber, Gentleman, & Pearce, 2008; Kingsbury et al., 2010) and individual variation likely occurs in the temporal susceptibility of certain pathways to the disease process. One potential explanation of the pattern of pathology described by Braak et al., albeit based largely on circumstantial evidence, is that yet-to-be identified pathogens enter the brain via the olfactory fila and the enteric

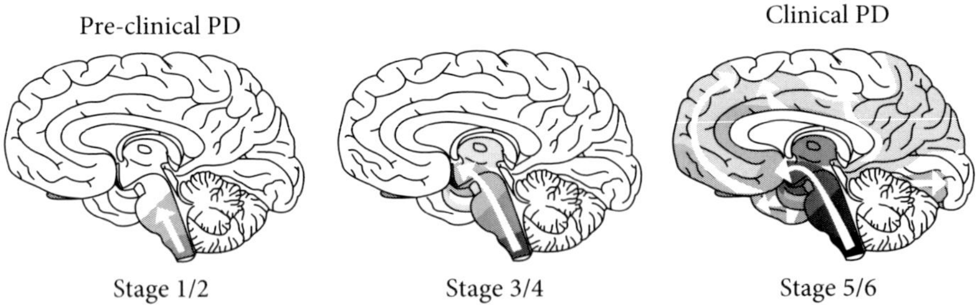

Figure 1. PD-related Lewy body pathology evolves in predictable stages. According to the staging system of Braak, Lewy bodies (LB) first form within in the olfactory bulb and dorsal motor nucleus of the vagal nerve (Stage 1). In Stages 2 and 3, LB pathology expands from these induction sites into additional brain stem nuclei (e.g., locus coeruleus and substantia nigra) and then into the amygdala. In Stages 5 to 6, the pathology extends into the cerebral cortex. Clinical symptoms arise during Stages 4 to 6 when the pathology involves significant regions of the substantia nigra and related brain areas. From Thal, Del, and Braak (2004).

nervous system of the stomach, inducing neural damage in the bulb and dorsal medulla and somehow catalyzing the disease process (Hawkes, Del Tredici, & Braak, 2007; Doty, 2008).

While dopamine is generally unrelated to the olfactory dysfunction of PD, another major neurotransmitter, acetylcholine, is intimately associated with such dysfunction. Cholinergic neurons make up major projections to a number of olfaction-related brain regions. Thus, there is a strong cholinergic projection from the horizontal limb of the diagonal band of Broca to the anterior olfactory nucleus, the olfactory tubercle, and the olfactory bulb's glomerular layer. Recent studies suggest that mouse strains that differ in the expression of α7-nicotinic acetylcholine receptors correspondingly differ in their ability to perform on an odor detection/discrimination task (Hellier et al., 2010). Injections of physostigmine, an acetycholinesterase (AChE) inhibitor, increases the ability of rats to detect low concentrations of butanol within a background of pentyl acetate (Doty, Bagla, & Kim, 1999) and tightens odor generalization gradients in mice (Mandairon et al., 2011).

Perhaps the most compelling evidence for an association between acetylcholine and olfactory function in PD comes from a study of 58 patients employing [^{11}C] methyl-4-piperidinyl propionate acetylcholinesterase brain positron emission tomography (PET) (Bohnen et al., 2010). This radioactively labeled ligand provides a measure of the integrity of the forebrain cholinergic pathway. Bohnen et al. (2010) found relatively strong correlations between scores on the University of Pennsylvania Smell Identification (UPSIT) and this PET imaging measure for the hippocampus, amygdala, and neocortex (respective *r*s = 0.63, 0.55, and 0.57; all ps < 0.001).[1] Similar correlations with dopaminergic activity in the striatum, as measured by (+)-[^{11}C]dihydrotetrabenazine vasicular monoamine type 2 (VMAT2) binding, were non-significant. The relationship between the hippocampal acetylcholinesterase PET activity and the original UPSIT scores is shown in Figure 2. Weak associations were noted between UPSIT scores and measures of episodic verbal learning ($r = 0.30$, $p = 0.02$) and Mini-Mental State Examination scores ($r = 0.25$, $p = 0.055$), but not with composite z-scores for measures of visual non-verbal memory ($r = 0.18$, $p = 0.17$), visuospatial function ($r = 0.001$, $p = 0.99$), attention ($r = -0.05$, $p = 0.80$), or executive function ($r = -0.10$, $p = 0.46$). In other studies, lower cognitive function of PD patients has been associated with

1. These correlations were computed after UPSIT scores with zeros were omitted from the analysis. The author inadvertently included zero scores for persons who were clearly anosmic on other grounds, but because the UPSIT is a 4-alternative 40-odorant forced-choice test, chance performance would be, on average, 10, not 0. The original correlations with the UPSIT scores were 0.56 for the hippocampus, 0.50 for the amygdala, and 0.46 for the neocortex (all ps < 0.001).

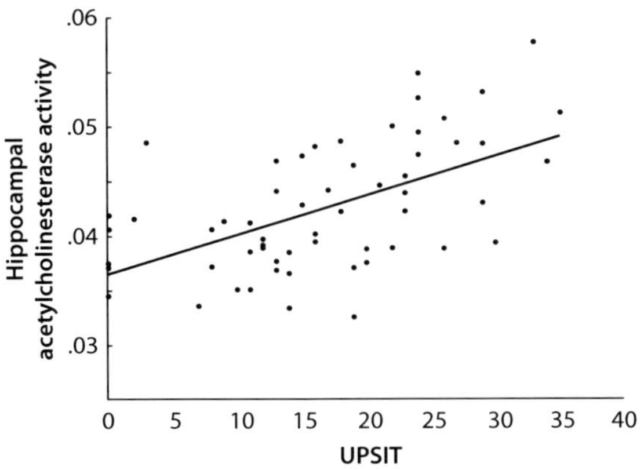

Figure 2. Scatter plot of the relationship between hippocampal AChE k3 hydrolysis (min^{-1}) rates and scores on the University of Pennsylvania Smell Identification Test (UPSIT) obtained from the PD subjects. Modified from Bohnen et al. (2010).

lower choline acetyltransferase (ChAT) levels, an enzyme involved in the production of acetylcholine (Mattila et al., 2001). Interesting, this relationship is independent of the presence of neuritic plaques or neurofibrillary tangles – structures classically associated with cognitive dysfunction.

Despite the fact that AChE activity is markedly reduced in the cerebral cortex of even *de novo* PD patients with no dementia, no further reduction appears to occur in non-demented patients with advanced PD (Shimada et al., 2009). If true, this implies that the associated cholinergic deficits, like the olfactory deficits, may not progress during the course of the disease, even at advanced stages. In a phenomenon seemingly analogous to the increased levels of tyrosine hydroxylase reported in the bulbs of PD patients (Huisman et al., 2004), the olfactory bulbs of aging mice contain increased levels of acetylcholine (Mandairon et al., 2010), likely reflecting compensatory mechanisms for the decline in the density of muscarinic receptors.

3. Environmental and genetic determinants

It is well established from epidemiologic studies that a host of environmental factors likely play a role in the development of sporadic PD. Among such factors are agricultural chemicals, dairy products, heavy metals, and viruses (Doty, 2012). A case in point is a recent study in which lifelong occupational and job task histories were obtained, including exposures to specific pesticides, for 519 PD cases and 511 controls (Tanner et al., 2009). Pesticide use of any sort was associated with the risk of having PD (Odds Ratio, 1.90; 95% Confidence Interval, 1.12–3.21). Such risk was greater for any one of eight pesticides known to induce PD-like syndromes in experimental animals (OR 2.20, CI 1.02–4.75). Three agents, the insecticide Permethrin, the herbicide Paraquat, and the organochlorine 2,4-dichlorophenyoxy-acetic acid were associated with an approximately 3-fold increased risk of PD [respective OR's & CI's = 3.21 (0.65–15.80), 2.80 (0.81–9.72), & 2.59 (1.03–6.48)]. Interestingly, animal studies support the concept that toxic exposures early in life may predispose the organism for later-life development of PD induced by stress or additional toxic exposures (Thiruchelvam, Richfield, Goodman, Baggs, & Cory-Slechta, 2002; Barlow, Richfield, Cory-Slechta, & Thiruchelvam, 2004). How such exposures interact with genetic factors is largely unknown and unexplored.

Although twin studies have failed to find evidence of strong heritability for PD (Tanner et al., 1999), as is also the case for olfactory function in older non-PD cohorts (Doty, Petersen, Mensah, & Christensen, 2011), mutations have been identified that account for some forms of PD (Table 2). Many of these forms differ from classical PD in subtle ways. For example, a number exhibit an earlier onset and more rapid disease progression associated with dementia (Tanner, 2008).

Table 2. Genes that have been identified that are associated with familial forms of Parkinson's Disease (PD).

Locus (map position)	Protein	Inheritance pattern	Mutations	Clinical features
PARK1/ PARK4 (4q21–q23)	α-synuclein	Autosomal Dominant	A30P, E46K, A53T duplications and triplications	Early onset PD, Dementia, autonomic dysfunction. Rare and not yet observed in sporadic PD cases
PARK2 (6q25.2–q27)	Parkin	Autosomal Recessive	A wide variety of mutations, exonic deletions, duplications and triplications	Juvenile and early onset, slow progression, good response to L-dopa
PARK3 (2p13)	Unknown	Autosomal Dominant	–	Late Onset PD, Typical PD symptoms
PARK5 (4p14)	UCHL1	Autosomal Dominant	I93M, S18Y	LOPD, Typical PD symptoms
PARK6 (1p35–p36)	Pink-1	Autosomal Recessive	G309D, exonic deletions	EOPD, slow progression
PARK7 (1p36)	DJ-1	Autosomal Recessive	Homozigot exonic deletion, L166P	EOPD, slow progression

Table 2. (*continued*)

Locus (map position)	Protein	Inheritance pattern	Mutations	Clinical features
PARK8 (12q12)	LRRK2	Autosomal Dominant (incomplete penetrance)	G2019S (most common), R1441C/G/H, Y1699C, I2020T, G2385R, others	LOPD, PD symptoms as the dominant
PARK9 (1p36)	ATP13A2	Autosomal Recessive	Loss-of-function mutations	Kufor-Rakeb syndrome; Juvenile and early onset, dementia, pyramidal degeneration, spasticity, paralysis of gaze
PARK10 (1p32)	Unknown	Not clear	–	LOPD, Typical PD symptoms
PARK11 (2q36–q37)	GIGYF2	Autosomal Dominant (incomplete penetrance)	–	LOPD, Typical PD symptoms
PARK12 (Xq21–25)	Unknown	Not clear	–	Not clear
PARK13 (2p12)	OMI/ HTAR2	Not clear	A141S, G399S	Not clear
PARK14 (22q13.1)	PLA2G6	Autosomal Recessive	–	Adult onset dystonia-parkinsonism
PARK 15 (22q12–q13)	FBXO7	Autosomal Recessive	–	Early onset parkinsonian-pyramidal syndrome
PARK 16 (1q32)	Unknown	Autosomal Domiant?	–	–
PARK 17 (4P16)	GAK	–	–	–
PARK 18 (6P21.3)	HLA-DRA	–	–	–

In a pioneering study, Markopoulou et al. (1997) tested olfaction in six members of three kindreds who exhibited the classic PD phenotype (i.e., tremor, rigidity, bradykinesia, postural instability, and positive response to levodopa). Of these six individuals, four were anosmic and two were microsmic on the UPSIT. Of four asymptomatic relatives, three were microsmic and one was normosmic.

One individual, who showed only a single cardinal sign of PD, also proved to be hyposmic. Additionally, six individuals from three families with a Parkinson-plus syndrome (PPS) phenotype were tested. PPS was defined as having at least three cardinal parkinsonian signs and additional features such as dystonia, pyramidal signs, dementia, amyotrophy, respiratory and autonomic abnormalities. Of the six PPS patients, five were anosmic and one was normosmic. Of the eight at-risk asymptomatic persons from these families that were tested, two were microsmic and six were normosmic. Of two monosymptomatic persons who were tested, one was microsmic and the other normosmic. The mean UPSIT scores of the affected PPS individuals did not differ from those of the PD group. The authors suggested that the olfactory loss most like reflects a heritable defect and that olfactory testing unlikely differentiates between sporadic and inherited forms of PD.

More recently, a number of so-called monogenetic forms of PD have been identified, although a thorough understanding of their olfactory function is yet to be achieved. This reflects several factors, including the relatively small numbers of subjects available for testing, failure in some instances to assess age- and sex-matched contemporaneous controls, and technical difficulties, including statistical issues, in assessing multiple gene interactions. Described below are studies that have shed at least some light on this complex issue, as organized by the type of PARK mutation.

3.1 The PARK1/PARK4 (α-synuclein) mutation

Most patients with PARK1/PARK4 – an autosomal-dominant form of PD – usually exhibit the motor symptoms before the age of 40 years, although later onset may occur in some with the A30P mutation (Kruger et al., 1998). The encoded protein, α-synuclein, is the major fibrillary component of Lewy bodies and neurites, the pathological hallmarks of sporadic and familial PD. Bostantjopoulou et al (2001) evaluated the olfactory function of 7 of 8 individuals with PD who carried the G209A PARK1/PARK4 mutation. Two were anosmic, with UPSIT scores of 12 and 13. The other five subjects were reportedly normal, although no test scores were provided. More recently (Tijero et al., 2010) administered the Brief Smell Identification Test (B-SIT), a 12-item version of the UPSIT, to an asymptomatic carrier of the E46K substitution in the α-synuclein gene and found no evidence of smell loss, with the test score being 10. Severe sympathetic myocardial denervation was present, however, as documented by cardiac scintigraphy.

3.2 The PARK2 (Parkin) mutation

Mutations in the Parkin gene (PARK2) result in the most common form of *early-onset* parkinsonism. Over 100 such mutations have been reported in the literature. The inheritance of this form of parkinsonism is autosomal recessive, unlike the dominant form of PARK1/PARK4. The Parkin protein is largely expressed in cytoplasm and neuronal processes and co-localizes with actin filaments. Cell death commonly occurs via autophagocytosis of mitrochondria (Narendra, Tanaka, Suen, & Youle, 2009).

Khan et al. (2004) administered the UPSIT to 27 patients with PARK2 and found the scores to be essentially equivalent to those of normal controls. A deficit in UPSIT scores was noted in the PARK2-negative group and in PD patients without the PARK2 mutation. Alcalay et al. (2011), in an extensive study assessing the relationship between the Parkin genotype and olfaction in PD probands and their unaffected relatives, found that among the PD probands, compound heterozygotes – i.e., individuals who had PARK2 mutations on both the paternal and material alleles – had UPSIT scores similar to those expected from controls (mean = 31.9). Non-PARK2 mutation carriers and heterozygotes exhibited average UPSIT scores analogous to those observed previously in cases of sporadic PD (respective mean UPSIT scores = 20.1 & 19.9). Both of these UPSIT values were lower than those of the compound heterozygotes ($ps < 0.001$). These differences persisted after adjustment for age, gender, disease duration, and smoking. Among relatives without PD, UPSIT performance was similar in heterozygotes (32.5) and noncarriers (32.4), and was better than that of heterozygotes with PD (20.1) ($p = 0.001$). The latter observations suggest that the olfactory impairment was primarily associated with PD, rather than the PARK2 genotype.

3.3 The PARK6 (Pink-1) mutation

PARK6 PD patients exhibit clinical features similar to those of PARK2. This autosomal-recessive disorder is caused by mutations of PTEN-induced putative kinase 1 (PINK1). Eggers et al., (2010) investigated mild motor signs and nonmotor signs such as altered olfaction and color discrimination as early indicators of a dopaminergic deficit among four homozygous PARK6 patients with PD and ten heterozygous PARK6 mutation carriers. None of the four homozygous PARK6 PD patients exhibited UPSIT scores outside of the 10th percentile of age- and sex-matched normative controls. Four of the ten heterozygotes did so, but only one of the patients was at or below the 5th percentile, suggesting only modest decrements in smell function. Ferraris et al. (2009) used the 'Sniffin' Sticks' test to assess olfactory function in five groups of subjects: sporadic PD (n = 19), PARK6 homozygous PD (n = 7), heterozygous PARK6 PD (n = 6), asymptomatic PARK6 heterozygous carriers (n = 12), and healthy subjects (n = 67). Odor identification was impaired in all PD and PINK1 cases and preserved in healthy heterozygotes. In contrast, detection thresholds were said to be more preserved and discrimination more impaired in the PINK1 patients than in PD cases.

3.4 The PARK 8 (LRRK2) mutation

The PARK 8 mutation (LRRK2/Dardarin) is the most prevalent cause of dominantly inherited PD, accounting for 2–5% of all PD in Europe with a much higher level in Spain, Portugal and North Africa (Ferreira et al., 2007). It is also the form of familial parkinsonism that has received the most attention in terms of olfactory testing. A summary of the general findings of olfactory studies with PARK8 mutations is presented in Table 3. A key aspect of this table is the observation that a number of the reports of abnormality actually are not abnormal if one takes into account the age of the patient. It is also apparent that, with rare exception, too few subjects have been examined to allow for conclusive findings.

Several studies have shown associations between UPSIT scores and cardiovascular function in patients with PARK8 mutations. Valldeoriola et al. (2011) tested ^{123}I-metaiodobenzylguanidine (MIBG) cardiac uptake and smell identification in 14 parkinsonian patients with PARK8 mutations. A positive correlation was found between the MIBG uptake and UPSIT scores ($r = 0.801$, $p = 0.001$). Interestingly, MIBG cardiac uptake was less impaired in the PARK8 patients than in patients with idiopathic PD. Ross et al. (2009) evaluated olfaction using the Brief Smell Identification, a 12-item version of the UPSIT, in 44 PD patients carrying the R1441G and G2019S mutations of PARK8, as well as in 146 PD patients with no known PD-related mutations. The patients were classified as PARK8 mutation carriers or noncarriers. MIBG scintigraphy was performed on 27 of the PARK8 mutation carriers and on 63 of the non-mutation carriers. Thirty-six percent of the PARK8 patients exhibited hyposmia, as compared to 75% of the noncarrier PD patients ($p < .001$). Sixty-six percent of the PARK8 mutation carriers had low early MIBG uptake, compared to 86% of the noncarriers ($p = .048$). These data indicate that olfactory and cardiac impairment is less prevalent in PARK8 PD than in non-PARK8 PD, although the underlying mechanisms are unclear.

Table 3. Olfactory test results of PD patients with various PARK8 (LRRK2) mutation.

Mutation	N	Age	UPSIT score	Relative to age	Author
Y1699C	4	67, 68, 67, 49	24, 25, 34, 36	1/2 Normal	Khan et al. (2005)
I2020T	1	60	35*	Normal	Berg et al. (2005)
R793M	1	67	10*	Abnormal	Berg et al. (2005)
R1441C	3	87, 83, 65	25, 28, 34	2/3 Abnormal	Markopoulou et al. (1997)
R1441H	2	56, 81	10, 24	Abnormal	Ferreira et al. (2007)
R1441G	39	70 (9.6)	NA	1/3 Abnormal**	Ruiz-Martinez et al. (2011)
S1096C	1	78	35	Normal	Berg et al. (2005)
S1228T	2	57, 61	25*, 35*	1/2 Abnormal	Berg et al. (2005)
3342A>G	1	81	35*	Normal	Berg et al. (2005)
G2385R	2	50, 73	15, 15	Abnormal	Lin et al. (2008)
G2019S	19	66.09 (10.08)	16.83 (SD: 7.12)	Abnormal	Ferreira et al. (2007)
G2019S	19	NA	"51% w/abnormal scores"	1/2 Abnormal	Healy et al. (2008)
G2019S	3	80.6 (9.9)	2 anosmic, 1 microsmic	Abnormal	Lohmann et al. (2008)
G2019S	14	62.3 (21.3)	17.5 (SD: 6.8)	Abnormal	Silveira-Moriyama et al. (2008)
G2019S	14	66 (14.6)	23.5*	Abnormal	Silveira-Moriyama et al. (2010)
G2019S	14	NA	27.7 (SD: 7.2)	Abnormal	Kertelge et al. (2010)
G2019S	25	72 (11)	33rd percentile	Abnormal	Marras et al. (2011)
G2019S	31	64.7 (9.8)	24.8 (SD: 7.08)	Abnormal	Saunders-Pullman (2011)
G2019S	14	61.9 (12.6)	21.5 (SD: 7.3)	Abnormal	Valldeoriola et al. (2011)

Abbreviations: CI = 95% confidence interval; NA = Not available; SD = Standard Deviation; UPSIT = University of Pennsylvania Smell Identification Test.
* UPSIT score estimated from ID score on Sniffin' Sticks by multiplying percent correct times 40.
** Based on B-SIT and a combination of 39 R1441G and 5 G2019S mutation carriers.

3.5 Color vision and olfactory function in genetic forms of PD

A number of genetic forms of PD have been associated with deficits in both olfaction and color vision. Kertelge et al. (2010) administered the UPSIT and the Farnsworth-Munsell (FM) color discrimination test to 100 patients with idiopathic PD, 27 mutation carriers (MC), 20 non-mutation carriers (NMC), and 110 controls. The monogenic group consisted of three PARK1/PARK4, fifteen PARK2, seventeen PARK6, eight PARK8, and four PARK9 carriers. Olfaction was most impaired in IPD [mean (SD) UPSIT scores = 10.1 ± 13.5] compared with all other groups (MC 13.8 ± 11.9, NMC 19.6 ± 13.0, Controls 33.8 ± 22.4). Within the mutation carriers those with two mutations in PARK2 and PARK6 showed higher UPSIT scores than PARK1/PARK4 and PARK8 carriers. Olfactory and color discrimination impairment did not evolve in parallel.

4. Other forms of parkinsonism and related disorders

Classic PD is but one sector of a continuum of disorders associated with similar, but distinct, motor dysfunction and neuropathology. Among diseases classified as "parkinsonism" or "parkinson plus syndromes" are Lewy body disease, multiple system atrophy (MSA), progressive supranuclear palsy (PSP), corticobasal degeneration (CBD), drug induced PD (DPD), the PD-dementia complex of Guam (PDC), X-linked dystonia-parkinsonism ('Lubag'), and vascular parkinsonism (VP).

4.1 Lewy body disease

Lewy body disease is a general term applied to diseases in which Lewy bodies are widely deposited in the brain. Classical PD is technically a form of Lewy body disease, given that wide-spread Lewy body deposition largely defines its pathology. Some forms of Lewy body disease are associated with dementia. Dementia with Lewy bodies is used to describe the syndrome when the dementia occurs before, during, or within a year of the onset of the motor symptoms (McKeith et al., 2005). If the dementia occurs more than a year after the onset of PD, the disorder is termed Parkinson disease dementia. It is difficult to differentiate between the latter two diseases on either clinical or pathological grounds (Ballard et al., 2006; Galvin, Pollack, & Morris, 2006). The only published histological study of olfactory bulbs from patients with dementia with Lewy bodies found 9 of 10 cases had tau pathology, Lewy bodies, and α-synuclein deposits (Tsuboi, Wszolek, Graff-Radford, Cookson, & Dickson, 2003).

Several studies have noted severe olfactory impairment in patients with Lewy body disease, as distinguished from PD (McShane et al., 2001; Liberini, Parola, Spano, & Antonini, 2000; Olichney et al., 2005; Wilson et al., 2011). In the most extensive of these studies, autopsies were performed on 201 persons who had completed the B-SIT, on average, 3.5 years before death (Wilson et al., 2011). The distribution of Lewy bodies was examined and segregated into the mutually exclusive categories of nigral (substantial nigra), limbic (entorhinal cortex and/or anterior cingulate cortex, but not neocortex), and neocortical (inferior partietal, midfrontal, superior, or temporal cortices). While no olfactory dysfunction was noted in the nigral group, considerable dysfunction was present in the limbic and neocortical groups even when the influences of age, sex, education, time from olfactory testing to death, and dementia were taken into account. The greatest dysfunction was observed in the neocortical group.

4.2 Vascular parkinsonism

Vascular parkinsonism is a syndrome that mimics PD. This disorder, which has a variable response to L-DOPA and no resting tremor, occurs in patients with extensive cerebrovascular disease involving the basal ganglia, particularly the putamen and striatum. Usually striatal dopamine transporter imaging is normal in this disorder. Cases with acute onset tend to have lesions in the subcortical grey nuclei (striatum, globus pallidus and thalamus), whereas those with insidious onset have more diffusely distributed lesions (Zijlmans et al., 1995). One study of 14 patients fulfilling strictly defined clinical criteria for vascular parkinsonism exhibited UPSIT scores significantly higher than those of PD patients and essentially equivalent to those of age-matched controls (respective means = 17.1, 25.5 & 27.6) (Katzenschlager, Zijlmans, Evans, Watt, & Lees, 2004). This finding suggested to the authors that olfactory testing may be of value in differentiating VP from idiopathic PD, which exhibit, on average, much lower UPSIT scores.

4.3 Drug-induced Parkinson's disease

Drug-induced Parkinson's disease (DPD) can be clinically indistinguishable from classic PD and can arise from a number of sources. DPD was most prevalent when broad-spectrum dopamine antagonists were used to treat psychotic disorders before selective D2 dopamine receptor blockers became available.

Lee, Yeo, Yong, and Kim (2007) compared B-SIT scores of 15 patients with DPD to those of 15 controls and 24 patients with sporadic PD. The B-SIT scores of the DPD patients did not differ from those of controls and were significantly higher than the scores of the PD patients (6.9 vs. 4.4; $p < 0.001$). The DPD had been induced by use of Levosulpiride, Haloperidone, Flunarizine, Perphenazine, Metoclopramide, or Risperidone. Hensiek, Bhatia, and Hawkes (2000) administered the UPSIT to 10 patients with DPD induced by a range of phenothiazine preparations that had been administered for at least 2 weeks. Five had abnormal UPSIT scores for their age and none made a complete recovery from DPD even when the offending medication was changed or stopped. Of the remaining five who did regain normal motor function after treatment adjustment, all but one had normal smell function. Unfortunately, a number of these patients had a psychotic disorder which may have contributed or even caused their smell problem. Bovi et al. (2010) recently assessed olfactory function in sixteen DPA patients (7 Haloperidol, 5 Amisulpride, 2 Perphenazine, 1 Fluphenazine, 1 Clomipramine), 13 PD patients, and 19 age- and sex-matched normal controls. Olfactory function was determined using the "Sniffin' Sticks" test. DPD patients were divided based on normal (n = 9) and abnormal (n = 7) putamen dopamine transporter binding as determined from [(123)I] FP-CIT SPECT imaging. Patients with DPA and a pathological putamen uptake had abnormal olfactory test scores which correlated significantly with putamen uptake values, as observed in PD patients. In contrast, DPA patients with normal putamen uptake exhibited generally normal olfactory function that did not correlate with the putamen uptake values. The authors suggested that the presence of smell deficits in DPA patients may be more closely associated with dopaminergic loss than with drug-mediated dopamine receptor blockade.

The most famous drug known to induce parkinsonism is 1-methyl-4-phenyl-1, 2, 3, 6-tetrahydropyridine (MPTP). MPTP is metabolically converted to MPP+, a neurotoxin that selectively damages dopamine neurons within the striatum. In the only study on this topic, six of the original young MPTP-induced parkinsonism patients were administered the UPSIT and a phenyl ethyl alcohol detection threshold test (Doty et al., 1992). Comparison groups consisted of 13 young PD patients and ten normal subjects. The UPSIT scores of the MPTP-P patients did not differ significantly from normal controls. The young PD patients, on the other hand, had olfactory dysfunction, differing significantly from controls.

4.4 X-linked Recessive Dystonia-parkinsonism

X-linked Recessive Dystonia-parkinsonism, also termed 'Lubag', is found among adult male Filipinos with maternal roots from the Philippine Island of Panay. Evidente et al. (2004) administered a culturally modified 25-item UPSIT to 20 affected males and 20 controls and found moderate impairment even early in the disorder (respective mean scores = 18/25 & 20.5/25, p = 0.003). The smell loss was unrelated to disease duration, severity, and the degree of dystonia.

4.5 Parkinsonism Dementia Complex of Guam

The Parkinsonism Dementia Complex (PDC) of Guam – the largest island of Micronesia – is a disorder in which Alzheimer's disease (AD), parkinsonism, and motor neuron disease (amyotrophic lateral sclerosis; ALS) co-exist, either individually or in combination. The smell loss of patients with this disorder does not differ significantly from that of early AD and PD.

In the first of two studies of PDC, 24 patients were administered the UPSIT (Doty et al., 1991a). Half were from Umatac and Merizo, two southern villages associated with a high prevalence of PDC. The others were from Guamanian villages with lower PDC prevalence rates. For comparison, UPSIT data from 24 AD and 24 PD North American patients matched to the PDC data on the basis of smoking behavior, gender and age were used. The UPSIT scores of the three groups, which were lower than those of normal controls, did not differ significantly from one another (see Table 1). Prior to testing, the patients were generally unaware of their deficit, with 87% of the PDC reporting that they had no smell problems. Corresponding values for the AD and PD patients were 92% and 87%, respectively. In the second study, Ahlskog et al. (1998) administered an abbreviated version of the UPSIT to 9 Guamanians with symptoms of ALS, 9 with symptoms of pure parkinsonism, 11 patients with pure dementia, and 31 patients with PDC, as well as to neurologically normal Guamanians and 25 North American controls. The UPSIT scores were markedly depressed in the four disease groups relative to the controls, and did not differentiate among the four phenotypes.

4.6 Essential tremor

Although not formally a form of PD, essential tremor (ET) is sometimes mistaken for PD. However, it clearly differs from PD in a number of ways, most notably that the tremor occurs when the hands are being moved or used, rather than in the resting state as occurs in PD. Unlike PD, ET is not associated with stooped posture, slow movement, or a shuffling gait. However, the tremor of ET can involve legs, hands, voice and the head, unlike PD tremors that typically affect mainly the hands.

In contrast to PD, patients with ET have no meaningful olfactory function. As early as 1992, Busenbark, Huber, Greer, Pahwa, and Koller (1992) found 15 ET patients to score within the normal range on the UPSIT. Although a later study reported that a significant proportion of ET patients had mild impairment on the UPSIT (Louis, Bromley, Jurewicz, & Watner, 2002), all subsequent studies have found normal UPSIT scores in ET patients (Shah, Findley, Muhammed, & Hawkes, 2005; Shah, Muhammed, Findley, & Hawkes, 2008). For example, Shah, Muhammed, Findley, and Hawkes (2008) compared UPSIT scores of 59 healthy persons with ET to those from 64 persons with tremor-dominant PD. Nearly complete separation of the two groups was made on the basis of UPSIT scores and to lesser degree on measures from odor event-related potentials (OERPs). Surprisingly, when ET subjects were separated by family history of tremor in a first degree relative (FET), this group scored significantly *better* than age- and gender-matched controls, a finding that, if replicated, is very important.

4.7 Progressive supranuclear palsy

Progressive supranuclear palsy (PSP), which is often mistaken for PD, is characterized by an inability to voluntarily look up or down, as well as a rapid progression of motor dysfunction, imbalance, and cognitive decline. Although, like PD, PSP is associated with widespread accumulation of tau protein in degenerating nerves, the olfactory bulbs appear to be spared (Tsuboi et al., 2003). In accord with such sparing, Doty et al. (1993) found equivalent UPSIT scores in 21 PSP patients and 21 matched normal controls, although there was a trend towards higher threshold values in the PSP group ($p = 0.085$). Similar UPSIT findings were found in a subsequent study of 15 cases of PSP by Wenning et al. (1995).

5. Summary and conclusion

The studies reviewed in this chapter clearly demonstrate that smell dysfunction is a key element of PD. The basis of the disorder is not entirely clear, although tau pathology within olfactory eloquent structures may be involved, at least in some cases. Additionally, alterations in certain neurotransmitter systems, most notably those related to acetylcholine, appear to be at the forefront of the causes of the olfactory disorder. Recent genetic studies suggest that some genes associated with familial forms of PD may influence olfactory function. Interestingly, some genetic influences may be of a protective, rather than a denigrative, nature. To what degree environmental agents, some of which are known to enter the brain via the olfactory pathways, may initiate or catalyze the olfactory disfunction or the PD process per se is unknown. Most likely numerous factors are responsible for the various non-motor components of PD, including the involvement of the olfactory system. Only the future will tell whether classical PD, as diagnosed clinically, actually represents a number of distinct disease entities.

References

Ahlskog, J. E., Waring, S. C., Petersen, R. C., Esteban-Santillan, C., Craig, U. K., O'Brien, P. C., et al. (1998). Olfactory dysfunction in Guamanian ALS, parkinsonism, and dementia. *Neurology, 51,* 1672–1677.

Alcalay, R. N., Siderowf, A., Ottman, R., Caccappolo, E., Mejia-Santana, H., Tang, M. X., et al. (2011). Olfaction in Parkin heterozygotes and compound heterozygotes: The CORE-PD study. *Neurology, 76,* 319–326.

Attems, J., & Jellinger, K. A. (2008). The dorsal motor nucleus of the vagus is not an obligatory trigger site of Parkinson's disease. *Neuropathology & Applied Neurobiology, 34,* 466–467.

Ballard, C., Ziabreva, I., Perry, R., Larsen, J. P., O'Brien, J., McKeith, I., et al. (2006). Differences in neuropathologic characteristics across the Lewy body dementia spectrum. *Neurology, 67,* 1931–1934.

Barlow, B. K., Richfield, E. K., Cory-Slechta, D. A., & Thiruchelvam, M. (2004). A fetal risk factor for Parkinson's disease. *Developmental Neuroscience, 26,* 11–23.

Barz, S., Hummel, T., Pauli, E., Majer, M., Lang, C. J., & Kobal, G. (1997). Chemosensory event-related potentials in response to trigeminal and olfactory stimulation in idiopathic Parkinson's disease. *Neurology, 49,* 1424–1431.

Bedard, A., & Parent, A. (2004). Evidence of newly generated neurons in the human olfactory bulb. *Developmental Brain Research, 151,* 159–168.

Belzunegui, S., Sebastian, W. S., & Garrido-Gil, P. (2007). The number of dopaminergic cells is increased in the olfactory bulb of monkeys chronically exposed to MPTP. *Synapse, 61,* 1006–1012.

Berendse, H. W., Roos, D. S., Raijmakers, P., & Doty, R. L. (2011). Motor and non-motor correlates of olfactory dysfunction in Parkinson's disease. *Journal of Neurological Sciences, 31,* 596–605.

Berg, D., Schweitzer, K., Leitner, P., Zimprich, A., Lichtner, P., Belcredi, P., et al. (2005). Type and frequency of mutations in the LRRK2 gene in familial and sporadic Parkinson's disease. *Brain, 128,* 3000–3011.

Bohnen, N. I., Muller, M. L., Kotagal, V., Koeppe, R. A., Kilbourn, M. A., Albin, R. L., et al. (2010). Olfactory dysfunction, central cholinergic integrity and cognitive impairment in Parkinson's disease. *Brain, 133,* 1747–1754.

Boudadi, K., Boxhorn, C. E., Narla, V. A., Suffis, S. D., et al. (2010). Compensatory responses to age-related decline in odor quality acuity: Cholinergic neuromodulation and olfactory enrichment. *Neurobiology of Aging,* in press.

Bovi, T., Antonini, A., Ottaviani, S., Antonioli, A., Cecchini, M. P., Di, F. V., et al. (2010). The status of olfactory function and the striatal dopaminergic system in drug-induced parkinsonism. *Journal of Neurology, 257,* 1882–1889.

Braak, H., & Braak, E. (1990). Cognitive impairment in Parkinson's disease: Amyloid plaques, neurofibrillary tangles, and neuropil threads in the cerebral cortex. *Journal of Neural Transmission – Parkinson's Disease & Dementia Section, 2,* 45–57.

Braak, H., Rub, U., Gai, W. P., & Del, T. K. (2003). Idiopathic Parkinson's disease: Possible routes by which vulnerable neuronal types may be subject to neuroinvasion by an unknown pathogen. *Journal of Neural Transmission, 110,* 517–536.

Busenbark, K. L., Huber, S. I., Greer, G., Pahwa, R., & Koller, W. C. (1992). Olfactory function in essential tremor. *Neurology, 42,* 1631–1632.

Crino, P. B., Martin, J. A., Hill, W. D., Greenberg, B., Lee, V. M., & Trojanowski, J. Q. (1995). Beta-Amyloid peptide and amyloid precursor proteins in olfactory mucosa of patients with Alzheimer's disease, Parkinson's disease, and Down syndrome. *Annals of Otology, Rhinology & Laryngology, 104,* 655–661.

Daniel, S. E., & Hawkes, C. H. (1992). Preliminary diagnosis of Parkinson's disease by olfactory bulb pathology [letter]. *Lancet, 340,* 186.

de Rijk, M. C., Launer, L. J., Berger, K., Breteler, M. M., Dartigues, J. F., Baldereschi, M., et al. (2000). Prevalence of Parkinson's disease in Europe: A collaborative study of population-based cohorts. Neurologic Diseases in the Elderly Research Group. *Neurology, 54,* S21–S23.

Doty, R. L. (2008). The olfactory vector hypothesis of neurodegenerative disease: Is it viable? *Annals of Neurology, 63,* 7–15.

Doty, R. L. (2012). Environmental Influences on the Development of Parkinson's Disease. In R. D. Prediger, & R. Raisman-Vozari (Eds.), *Frontiers in Parkinson's Disease Research.* New York: Nova Science Publishers, in press.

Doty, R. L., Bagla, R., & Kim, N. (1999). Physostigmine enhances performance on an odor mixture discrimination test. *Physiology & Behavior, 65,* 801–804.

Doty, R. L., Deems, D. A., & Stellar, S. (1988a). Olfactory dysfunction in parkinsonism: A general deficit unrelated to neurologic signs, disease stage, or disease duration. *Neurology, 38,* 1237–1244.

Doty, R. L., Deems, D. A., & Stellar, S. (1988b). Olfactory dysfunction in parkinsonism: A general deficit unrelated to neurologic signs, disease stage, or disease duration. *Neurology, 38,* 1237–1244.

Doty, R. L., Golbe, L. I., McKeown, D. A., Stern, M. B., Lehrach, C. M., & Crawford, D. (1993). Olfactory testing differentiates between progressive supranuclear palsy and idiopathic Parkinson's disease. *Neurology, 43,* 962–965.

Doty, R. L., Perl, D. P., Steele, J. C., Chen, K. M., Pierce, J. D., Jr., Reyes, P., et al. (1991a). Odor identification deficit of the parkinsonism-dementia complex of Guam: Equivalence to that of Alzheimer's and idiopathic Parkinson's disease. *Neurology, 41,* 77–80.

Doty, R. L., Perl, D. P., Steele, J. C., Chen, K. M., Pierce, J. D., Jr., Reyes, P., et al. (1991b). Olfactory dysfunction in three neurodegenerative diseases. *Geriatrics, 46* (Suppl. 1), 47–51.

Doty, R. L., Petersen, I., Mensah, N., & Christensen, K. (2011). Genetic and environmental influences on odor identification ability in the very old. *Psychology and Aging, 26,* 864–871.

Doty, R. L., Riklan, M., Deems, D. A., Reynolds, C., & Stellar, S. (1989). The olfactory and cognitive deficits of Parkinson's disease: evidence for independence. *Annals of Neurology, 25,* 166–171.

Doty, R. L., Shaman, P., & Dann, M. (1984). Development of the University of Pennsylvania Smell Identification Test: a standardized microencapsulated test of olfactory function. *Physiology & Behavior, 32,* 489–502.

Doty, R. L., Singh, A., Tetrud, J., & Langston, J. W. (1992). Lack of Major Olfactory Dysfunction in MPTP-Induced Parkinsonism. *Annals of Neurology, 32,* 97–100.

Doty, R. L., Stern, M. B., Pfeiffer, C., Gollomp, S. M., & Hurtig, H. I. (1992). Bilateral olfactory dysfunction in early stage treated and untreated idiopathic Parkinson's disease. *Journal of Neurology, Neurosurgery & Psychiatry, 55,* 138–142.

Duda, J. E., Shah, U., Arnold, S. E., Lee, V. M., & Trojanowski, J. Q. (1999). The expression of alpha-, beta-, and gamma-synucleins in olfactory mucosa from patients with and without neurodegenerative diseases. *Experimental Neurology, 160,* 515–522.

Eggers, C., Schmidt, A., Hagenah, J., Bruggemann, N., Klein, J. C., Tadic, V., et al. (2010). Progression of subtle motor signs in PINK1 mutation carriers with mild dopaminergic deficit. *Neurology, 74,* 1798–1805.

Evidente, V. G., Esteban, R. P., Hernandez, J. L., Natividad, F. F., Advincula, J., Gwinn-Hardy, K., et al. (2004). Smell testing is abnormal in 'lubag' or X-linked dystonia-parkinsonism: A pilot study. *Parkinsonism & Related Disorders, 10,* 407–410.

Ferraris, A., Ialongo, T., Passali, G. C., Pellecchia, M. T., Brusa, L., Laruffa, M., et al. (2009). Olfactory dysfunction in Parkinsonism caused by PINK1 mutations. *Movement Disorders, 24,* 2350–2357.

Ferreira, J. J., Guedes, L. C., Rosa, M. M., Coelho, M., van Doeselaar, M., Schweiger, D., et al. (2007). High prevalence of LRRK2 mutations in familial and sporadic Parkinson's disease in Portugal. *Movement Disorders, 22,* 1194–1201.

Gaig, C., & Tolosa, E. (2009). When does Parkinson's disease begin? *Movement Disorders, 24 Suppl 2,* S656–S664.

Galvin, J. E., Pollack, J., & Morris, J. C. (2006). Clinical phenotype of Parkinson disease dementia. *Neurology, 67,* 1605–1611.

Halliday, G. M., Barker, R. A., & Rowe, D. B. (2011). *Non-dopamine lesions in Parkinson's disease.* Oxford: Oxford University Press.

Hawkes, C. H., Del Tredici, K., & Braak, H. (2007). Parkinson's disease: A dual-hit hypothesis. *Neuropathology and Applied Neurobiology, 33,* 599–614.

Healy, D. G., Falchi, M., O'Sullivan, S. S., Bonifati, V., Durr, A., Bressman, S., et al. (2008). Phenotype, genotype, and worldwide genetic penetrance of LRRK2-associated Parkinson's disease: A case-control study. *Lancet Neurology, 7,* 583–590.

Hellier, J. L., Arevalo, N. L., Blatner, M. J., Dang, A. K., Clevenger, A. C., Adams, C. E., et al. (2010). Olfactory discrimination varies in mice with different levels of alpha7-nicotinic acetylcholine receptor expression. *Brain Research, 1358,* 140–150.

Hensiek, A. E., Bhatia, K., & Hawkes, C. H. (2000). Olfactory function in drug induced parkinsonism. *Journal of Neurology, 247* (Suppl. 3), 82.

Huisman, E., Uylings, H. B. M., & Hoogland, P. V. (2004). A 100% increase of 687 dopaminergic cells in the olfactory bulb may explain hyposmia in Parkinson's disease. *Movement Disorders, 19,* 687–692.

Ibarretxe-Bilbao, N., Junque, C., Marti, M. J., Valldeoriola, F., Vendrell, P., Bargallo, N., et al. (2010). Olfactory impairment in Parkinson's disease and white matter abnormalities in central olfactory areas: A voxel-based diffusion tensor imaging study. *Movement Disorders, 25,* 1888–1894.

Kalaitzakis, M. E., Graeber, M. B., Gentleman, S. M., & Pearce, R. K. B. (2008). The dorsal motor nucleus of the vagus is not an obligatory trigger site of Parkinson's disease: A critical analysis of alpha-synuclein staging. *Neuropathology and Applied Neurobiology, 34,* 284–295.

Katzenschlager, R., Head, J., Schrag, A., Ben-Shlomo, Y., Evans, A., & Lees, A. J. (2008). Fourteen-year final report of the randomized PDRG-UK trial comparing three initial treatments in PD. *Neurology, 71,* 474–480.

Katzenschlager, R., Zijlmans, J., Evans, A., Watt, H., & Lees, A. J. (2004). Olfactory function distinguishes vascular parkinsonism from Parkinson's disease. *Journal of Neurology Neurosurgery & Psychiatry, 75,* 1749–1752.

Kertelge, L., Bruggemann, N., Schmidt, A., Tadic, V., Wisse, C., Dankert, S., et al. (2010). Impaired sense of smell and color discrimination in monogenic and idiopathic Parkinson's disease. *Movement Disorders, 25,* 2665–2669.

Khan, N. L., Jain, S., Lynch, J. M., Pavese, N., bou-Sleiman, P., Holton, J. L., et al. (2005). Mutations in the gene LRRK2 encoding dardarin (PARK8) cause familial Parkinson's disease: Clinical, pathological, olfactory and functional imaging and genetic data. *Brain, 128,* 2786–2796.

Khan, N. L., Katzenschlager, R., Watt, H., Bhatia, K. P., Wood, N. W., Quinn, N., et al. (2004). Olfaction differentiates parkin disease from early-onset parkinsonism and Parkinson disease. *Neurology, 62,* 1224–1226.

Kingsbury, A. E., Bandopadhyay, R., Silveira-Moriyama, L., Ayling, H., Kallis, C., Sterlacci, W., et al. (2010). Brain stem pathology in Parkinson's disease: An evaluation of the Braak staging model. *Movement Disorders, 25,* 2508–2515.

Kruger, R., Kuhn, W., Muller, T., Woitalla, D., Graeber, M., Kosel, S., et al. (1998). Ala30Pro mutation in the gene encoding alpha-synuclein in Parkinson's disease. *Nature Genetics, 18,* 106–108.

Lee, P. H., Yeo, S. H., Kim, H. J., & Youm, H. Y. (2006). Correlation between cardiac 123I-MIBG and odor identification in patients with Parkinson's disease and multiple system atrophy. *Movement Disorders, 21,* 1975–1977.

Lee, P. H., Yeo, S. H., Yong, S. W., & Kim, Y. J. (2007). Odour identification test and its relation to cardiac 123I-metaiodobenzylguanidine in patients with drug induced parkinsonism. *Journal of Neurology, Neurosurgery and Psychiatry, 78,* 1250–1252.

Lees, A. J., Hardy, J., & Revesz, T. (2009). Parkinson's disease. *Lancet, 373,* 2055–2066.

Liberini, P., Parola, S., Spano, P. F., & Antonini, L. (2000). Olfaction in Parkinson's disease: methods of assessment and clinical relevance. *Journal of Neurology, 247,* 88–96.

Lin, C. H., Tzen, K. Y., Yu, C. Y., Tai, C. H., Farrer, M. J., & Wu, R. M. (2008). LRRK2 mutation in familial Parkinson's disease in a Taiwanese population: Clinical, PET, and functional studies. *Journal of Biomedial Sciences, 15,* 661–667.

Lohmann, E., Leclere, L., de Anna, F., Lesage, S., Dubois, B., Agid, Y., Dürr, A., & Brice, A. (2008). A clinical, neuropsychological and olfactory evalution of a large family with LRRK2 mutations. *Parkinsonism and Related Disorders, 15,* 273–276.

Louis, E. D., Bromley, S. M., Jurewicz, E. C., & Watner, D. (2002). Olfactory dysfunction in essential tremor: A deficit unrelated to disease duration or severity. *Neurology, 59,* 1631–1633.

Mandairon, N., Peace, S. T., Boudadi, K., Boxhorn, C. E., Narla, V. A., Suffis, S. D., & Cleland, T. A. (2011). Compensatory responses to age-related decline in odor quality activity: Cholinergic neuromodulation and olfactory enrichment. *Neurobiology of Aging, 32,* 2254–2265.

Markopoulou, K., Larsen, K. W., Wszolek, E. K., Denson, M. A., Lang, A. E., Pfeiffer, R. F., et al. (1997). Olfactory dysfunction in familial parkinsonism. *Neurology, 49,* 1262–1267.

Marras, C., Schuele, B., Munhoz, R. P., Rogaeva, E., Langston, J. W., Kasten, M., Meaney, C., Klein, C., Wadia, P. M., Lim, S.-Y., Chuang, R. S.-I., Zadikof, C., Steeves, T., Prakash, K. M., de Bie, R. M. A., Adeli, G., Thomsen, T., Johansen, K. K., Teive, H. A., Asante, A., Reginold, W., & Lang, A. E. (2011). Phenotype in parkinsonian and nonparkinsonia *LRRK2* G2019S mutation carriers. *Neurology, 77,* 325–333.

Mattila, P. M., Roytta, M., Lonnberg, P., Marjamaki, P., Helenius, H., & Rinne, J. O. (2001). Choline acetytransferase activity and striatal dopamine receptors in Parkinson's disease in relation to cognitive impairment. *Acta Neuropatholgica, 102,* 160–166.

Mattila, P. M., Roytta, M., Torikka, H., Dickson, D. W., & Rinne, J. O. (1998). Cortical Lewy bodies and Alzheimer-type changes in patients with Parkinson's disease. *Acta Neuropathologica, 95,* 576–582.

McKeith, I. G., Dickson, D. W., Lowe, J., Emre, M., O'Brien, J. T., Feldman, H., et al. (2005). Diagnosis and management of dementia with Lewy bodies – Third report of the DLB consortium. *Neurology, 65,* 1863–1872.

McShane, R. H., Nagy, Z., Esiri, M. M., King, E., Joachim, C., Sullivan, N. et al. (2001). Anosmia in dementia is associated with Lewy bodies rather than Alzheimer's pathology. *Journal of Neurology, Neurosurgery & Psychiatry, 70,* 739–743.

Mesholam, R. I., Moberg, P. J., Mahr, R. N., & Doty, R. L. (1998). Olfaction in neurodegenerative disease: A meta-analysis of olfactory functioning in Alzheimer's and Parkinson's diseases. *Archives of Neurology, 55,* 84–90.

Narendra, D., Tanaka, A., Suen, D. F., & Youle, R. J. (2009). Parkin-induced mitophagy in the pathogenesis of Parkinson disease. *Autophagy, 5,* 706–708.

Olichney, J. M., Murphy, C., Hofstetter, C. R., Foster, K., Hansen, L. A., Thal, L. J., et al. (2005). Anosmia is very common in the Lewy body variant of Alzheimer's disease. *Journal of Neurology Neurosurgery & Psychiatry, 76,* 1342–1347.

Pearce, R. K., Hawkes, C. H., & Daniel, S. E. (1995). The anterior olfactory nucleus in Parkinson's disease. *Movement Disorders, 10,* 283–287.

Quinn, N. P., Rossor, M. N., & Marsden, C. D. (1987). Olfactory threshold in Parkinson's disease. *Journal of Neurology, Neurosurgery & Psychiatry, 50,* 88–89.

Ross, O. A., Spanaki, C., Griffith, A., Lin, C. H., Kachergus, J., Haugarvoll, K., et al. (2009). Haplotype analysis of Lrrk2 R1441H carriers with parkinsonism. *Parkinsonism and Related Disorders, 15,* 466–467.

Roth, J., Radil, T., Ruzicka, E., Jech, R., & Tichy, J. (1998). Apomorphine does not influence olfactory thresholds in Parkinson's disease. *Functional Neurology, 13,* 99–103.

Ruiz-Martinez, J., Gorostidi, A., Goyenechea, E., Alzualde, A., Poza, J. J., Rodriguez, F., Bergareche, A., Moreno, F., de Munain, A. L., & Masso, J. F. M. (2011). Olfactory deficits and cardiac 123I-MIBG in Parkinson's disease related to the LRRK2 R1441G and G2019S mutations. *Movement Disorders, 26,* 2026–31.

Sauders-Pullman, R., Stanley, K., Wang, C., San Luciano, M., Shanker, V., Hunt, A., Severt, L., Raymond, D., Ozelius, L. J., Lipton, R. B., & Bressman, S. B. (2011). Olfactory dysfunction in LRRK2 G2019S mutation carriers. *Neurology, 77,* 319–324.

Shah, M., Findley, L., Muhammed, N., & Hawkes, C. (2005). Olfaction is normal in essential tremor and can be used to distinguish it from Parkinson's disease. *Neurology, 64,* A261.

Shah, M., Muhammed, N., Findley, L. J., & Hawkes, C. H. (2008). Olfactory tests in the diagnosis of essential tremor. *Parkinsonism and Related Disorders, 14,* 563–568.

Shah, M., Muhammed, N., Findley, L. J., & Hawkes, C. H. (2008). Olfactory tests in the diagnosis of essential tremor. *Parkinsonism and Related Disorders, 14,* 563–568.

Shimada, H., Hirano, S., Shinotoh, H., Aotsuka, A., Sato, K., Tanaka, N., et al. (2009). Mapping of brain acetylcholinesterase alterations in Lewy body disease by PET. *Neurology, 73,* 273–278.

Siderowf, A., Newberg, A., Chou, K. L., Lloyd, M., Colcher, A., Hurtig, H. I., et al. (2005). [99mTc]TRODAT-1 SPECT imaging correlates with odor identification in early Parkinson disease. *Neurology, 64,* 1716–1720.

Silveira-Moriyama, L., Guedes, L. C., Kingsbury, A., Ayling, H., Shaw, K., Barbosa, E. R., et al. (2008). Hyposmia in G2019S LRRK2-related parkinsonism: Clinical and pathologic data. *Neurology, 71,* 1021–1026.

Silveira-Moriyama, L., Munhoz, R. P., de, J. C., Raskin, S., Rogaeva, E., de, C. A., et al. (2010). Olfactory heterogeneity in LRRK2 related Parkinsonism. *Movement Disorders, 25,* 2879–2883.

Tanner, C. M. (2008). Etiology: The role of environment and genetics. In S. A.Factor, & W. J. Weiner (Eds.), *Parkinson's Disease: Diagnosis and Clinical Management,* 2nd edition (pp. 387–405). New York: Demos.

Tanner, C. M., Ottman, R., Goldman, S. M., Ellenberg, J., Chan, P., Mayeux, R., et al. (1999). Parkinson disease in twins: an etiologic study. *JAMA, 281,* 341–346.

Tanner, C. M., Ross, G. W., Jewell, S. A., Hauser, R. A., Jankovic, J., Factor, S. A., et al. (2009). Occupation and risk of parkinsonism: a multicenter case-control study. *Archives of Neurology, 66,* 1106–1113.

Thal, D. R., Del, T. K., & Braak, H. (2004). Neurodegeneration in normal brain aging and disease 15. *Science of Aging, Knowledge Environment, 2004,* e26.

Thiruchelvam, M., Richfield, E. K., Goodman, B. M., Baggs, R. B., & Cory-Slechta, D. A. (2002). Developmental exposure to the pesticides paraquat and maneb and the Parkinson's disease phenotype. *Neurotoxicology, 23,* 621–633.

Tijero, B., Gomez-Esteban, J. C., Llorens, V., Lezcano, E., Gonzalez-Fernandez, M. C., de Pancorbo, M. M., et al. (2010). Cardiac sympathetic denervation precedes nigrostriatal loss in the E46K mutation of the alpha-synuclein gene (SNCA). *Clinical Autonomic Research, 20,* 267–269.

Tsuboi, Y., Wszolek, Z. K., Graff-Radford, N. R., Cookson, N., & Dickson, D. W. (2003). Tau pathology in the olfactory bulb correlates with Braak stage, Lewy body pathology and apolipoprotein epsilon4. *Neuropathology & Applied Neurobiology, 29,* 503–510.

Valldeoriola, F., Gaig, C., Muxi, A., Navales, I., Paredes, P., Lomena, F., et al. (2011). (123)I-MIBG cardiac uptake and smell identification in parkinsonian patients with LRRK2 mutations. *Journal of Neurology, 258,* 1126–1132.

Wenning, G. K., Shephard, B., Hawkes, C., Petruckevitch, A., Lees, A., & Quinn, N. (1995). Olfactory function in atypical parkinsonian syndromes. *Acta Neurologica Scandinavica, 91,* 247–250.

Wilson, R. S., Yu, L., Schneider, J. A., Arnold, S. E., Buchman, A. S., & Bennett, D. A. (2011). Lewy bodies and olfactory dysfunction in old age. *Chemical Senses, 36,* 367–373.

Yamada, M., Onodera, M., Mizuno, Y., & Mochizuki, H. (2004). Neurogenesis in olfactory bulb identified by retroviral labeling in normal and 1-methyl-4-phenyl-1,2,3,6-tetrahydropyridine-treated adult mice. *Neuroscience, 124*(1), 173–181.

Zijlmans, J. C. M., Thijssen, H. O. M., Vogels, O. J. M., Kremer, H. P. H., Poels, P. J. E., Schoonderwaldt, H. C., et al. (1995). MRI in patients with suspected vascular parkinsonism. *Neurology, 45,* 2183–2188.

CHAPTER 10

Remembering what the nose knows

Robert G. Mair
University of New Hampshire, Durham, USA

In 1973 Trygg Engen came up with the idea of studying olfaction by examining the effects of brain damage on odor perception. This work led to two fundamental insights that have stood the test of time. First, central brain lesions can disrupt higher functions like odor identification, naming, or recognition memory without affecting the ability to detect and discriminate differences between odors. Second, the ability to make sense of what we smell depends on brain regions beyond the olfactory bulb and its immediate projection areas in primary olfactory cortex. A lot has been learned about neural mechanisms that give rise to the sense of smell since these early neuropsychological studies were conducted. This chapter reviews this early work in light of more recent discoveries of olfactory neuroscience.

1. The neuropsychology of odor perception and memory – human studies

"We want to know what the nose knows" was Trygg Engen's gentle and repeated reminder that odors are not like colors or sounds: that beyond the realm of physiology "odor perception involves a more intimate relationship between the perceiver, the odor, and the event" (Engen, 1991: xii). Engen launched my graduate career with an almost apologetic promise that while the goals of my project examining the effects of brain lesions on odor identification might seem limited, he was quite certain that by the time I finished I would see much broader implications. He was of course correct. In this chapter I revisit this early work and consider its implications in the context of fundamental discoveries in chemosensory neuroscience over the last three decades.

In the early 1970's the neuropsychology of olfaction seemed an unlikely subject for study. Most treatments of olfaction emphasized primary sensory processes: sensitivity, intensity scaling, quality discrimination, and adaptation (Engen, 1971; Moncrieff, 1967; Mozell, 1972). There was little concern with cognitive aspects of olfaction, particularly functions impaired in conditions like aphasia, amnesia, and agnosia that tended to dominate the literature in neuropsychology. There was also uncertainty about how olfactory function is represented in the brain. Beyond projections from the olfactory bulb to primary olfactory cortex there was little known about higher order olfactory pathways and little thought given to how olfactory information might reach brain areas that mediate complex cognitive functions (MacLeod, 1971).

Trygg Engen's interest in neuropsychology was stimulated by a report that odor naming is impaired in aphasia to the same extent as naming in other sensory modalities (Goodglass, Barton, & Kaplan, 1968). He was intrigued by the possibility that careful studies of olfactory function in neurological patients might provide a unique window for understanding psychological processes that give rise to odor perception and to higher order processes like odor identification, naming, and memory. Engen's interests were less about neural systems or physiological mechanisms underlying olfaction. He wanted to understand the psychology of how we make sense of what we smell.

1.1 Cortical lesions (aphasia)

Our first study (Mair & Engen, 1976) was motivated by the implication from Goodglass et al. (1968) that odor naming is mediated by processes that occur after sensory processing is complete. Our approach was based on Engen's insight that odors can be identified in a matching task in two different ways: by specific (exact match) identification ("that is the smell of Ivory soap") or by less specific (similarity match) judgments based on general sensory properties ("it smells like soap, it must be the Ivory"). We carried out psychophysical studies with undergraduate students and developed a task matching odorants to visual targets (familiar containers) to compare these two means for distinguishing odors in a nonverbal task. We then tested groups of aphasic patients, non-aphasic brain-damaged controls, and non brain-damaged control patients at the Boston VA Medical Center. We also examined the ability of subjects to name and to describe sensory properties of the test odorants. For identification trials subjects sampled an odorant and selected from an array of 5 containers of common household products that were readily identified nonverbally. For exact match trials one of the containers matched the odorant precisely. For similarity trials one of the containers was associated with a highly similar odor. At the end of the session each odorant was presented for naming and description. We found that exact match odor identification is impaired in aphasic patients, who nevertheless retain the ability to sense odors and to match by similarity judgments (Figure 1). We confirmed that odor naming was also impaired in these patients and found that this deficit correlated

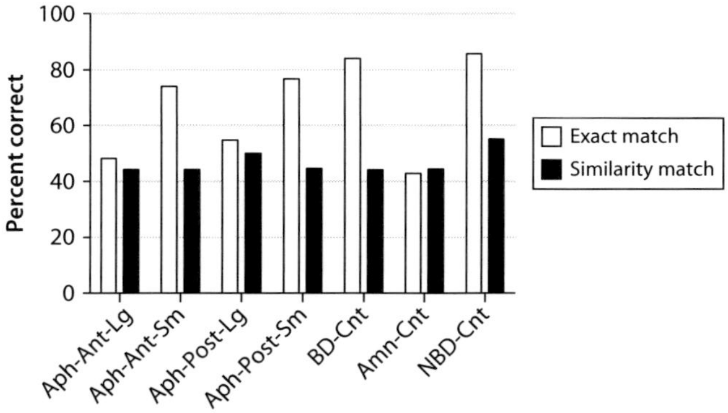

Figure 1. Effects of aphasia-producing lesions on exact-match and similarity-match odor identification tasks in Mair and Engen (1976). Results are shown for aphasics with large anterior lesions (Aph-Ant-Lg), small anterior lesions (Aph-Ant-Sm), large posterior lesions (Aph-Post-Lg), and small posterior lesions (Aph-Post-Sm). Results are shown separately for non-aphasic brain damaged controls with (Amn-Cnt) and without (BD-Cnt) signs of amnesia and for non-brain-damaged controls (NBD-Cnt) recruited from the inpatient population at the Boston VAMC. Aphasics with large cortical lesions were significantly impaired for the exact match task.

with odor identification in the exact match task ($r^2 = .46$, $p < .005$). Our results provided early evidence of a link between the ability to name and identify odors non-verbally that is distinct from primary sensory processes, like detection or perception of general odor qualities.

Odor identification deficits were only observed in aphasics with signs of extensive cortical pathology in either anterior (fronto-parietal) or posterior (temporo-parietal) language areas (these studies were carried out before CT-scans allowed more precise descriptions of lesions in living patients). Our results thus provided early evidence that higher-order olfactory functions, like naming or identification, depend on areas of cerebral cortex distinct from classic olfactory projection areas. These findings have since been confirmed by studies of olfactory function in patients with unilateral frontal or temporal removals (Eskenazi, Cain, Novelly, & Friend, 1983; Eskenazi, Cain, Novelly, & Mattson, 1986; Jones-Gotman & Zatorre, 1988; Jones-Gotman et al., 1997; Zatorre & Jones-Gotman, 1991) and by functional imaging evidence that pyriform and orbitofrontal cortices constitute a core olfactory area activated by passive smelling and that there are parallel, hierarchical systems involving frontal, temporal, and parietal cortices that are activated by tasks that require cognitive olfactory judgments (Dade, Zatorre, & Jones-Gotman, 2002; Savic, 2002; Zatorre, Jones-Gotman, Evans, & Meyer, 1992).

1.2 Korsakoff's syndrome (amnesia)

Interestingly, each of the three non-aphasic, brain damaged controls in Mair and Engen (1976) who performed as poorly as aphasics with large cortical lesions (Figure 1) exhibited mild signs of amnesia. We conducted a second study (Mair, Capra, McEntee, & Engen, 1980) to examine the impairment exhibited by these amnesic controls. By now a series of papers had come out from Nelson Butters' lab indicating that Korsakoff amnesics are impaired identifying odorants and have elevated thresholds for detecting odors, as extrapolated from magnitude estimation data (Jones, Butters, Moskowitz, & Montgomery, 1978; Jones, Moskowitz, Butters, & Glosser, 1975; Jones, Moskowitz, & Butters, 1975). To elucidate the precise nature of this impairment, we measured olfactory sensitivity directly, using a yes/no signal detection task to measure sensitivity and a short-term odor recognition memory task comparing similar and dissimilar odor pairs to examine odor discrimination and memory. Preliminary studies were conducted using ratio-scaling methods to create sets of similar and dissimilar odor pairs and to confirm that these differences were sufficient to affect recognition memory performance.

Korsakoff patients exhibited substantial and significant recognition memory impairment for both the similar and dissimilar odor pairs (Figure 2). There was no temporal decay in recognition memory for either Korsakoff or control subjects and thus no indication that the recognition deficit resulted from rapid decay of odor memory. The same Korsakoff patients were unimpaired detecting a weak odorant (2.05 mM n-butanol) in a signal detection task. Here subjects were given

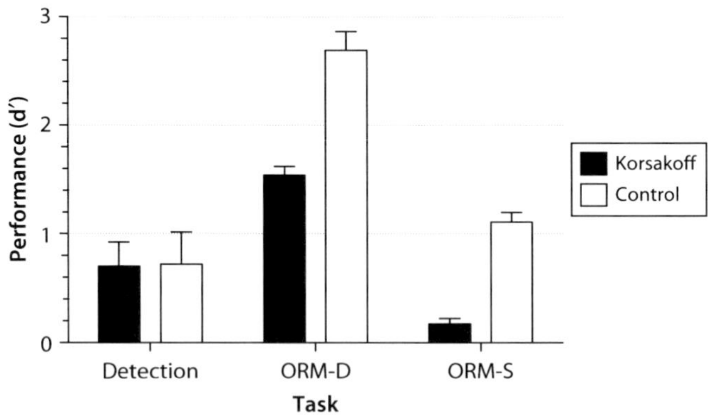

Figure 2. Effects of Korsakoff's disease on odor detection and on odor recognition memory for dissimilar (ORM-D) and similar (ORM-S) pairs of odors (Mair et al., 1980). Korsakoff patients were impaired for both recognition memory tasks, but performed normally for the detection task. Performances for all tasks are reported as d'. Error bars represent standard error of the mean.

10 practice trials and then asked to distinguish between the odorant and a blank in 60 test trials in which each stimulus was presented 30 times each in randomized order. There was 30 sec interval between stimulus presentations and feedback was given after each response. Our results indicate that Korsakoff's disease can impair the ability to discriminate between odorants presented for successive comparison, even when these odorants are quite dissimilar, while sparing odor detection and without affecting the temporal decay of odor recognition memory.

Korsakoff patients have been found to perform poorly on nonverbal measures of working memory involving minimal (immediate memory) delays (Cave & Squire, 1992). The patients exhibiting odor recognition memory deficits in our study also exhibited face recognition memory deficits in simultaneous and immediate matching conditions (Mair et al., 1980). It thus seems possible that the olfactory impairments we observed might have resulted from an inability to hold sensory information over the delays tested in the recognition memory task. Even at minimal delays, recognition memory tasks require subjects to remember information about the sample odor in working memory and to compare this internal representation with a second odorant presented on the test trial several seconds later. By contrast, detection tasks can be solved by rule based responding (reference memory): comparing a present odorant to fixed internal standards to decide whether it is the odor or the blank. There is little difference between a threshold-level odorant and the blank in a detection task. While detection presents a very difficult discrimination it does not require subjects to constantly update and remember information about sample odors that change from trial to trial.

We tested this possibility in two studies that examined the ability of Korsakoff patients to match odors to verbal descriptors in the Smell Identification Task (SIT). In the SIT subjects match odors to verbal labels that are present at the time that test stimuli are smelled (Mair et al., 1986). Working memory is required to hold sensory information while making the matching judgment, but these demands are minimized since labels are present when the odor is sampled and the odorant can be sampled repeatedly. All subjects in both studies were first screened to determine that they could reliably detect odorants (errorless performance for 8 detection trials) to confirm that primary sensory capacity was intact. In the first study (SIT-1 in Figure 3), 21 Korsakoff patients were found to exhibit substantial and significant SIT impairments compared to 21 age and gender matched controls drawn from a large computerized database. This impairment was remarkably reliable: only 2 Korsakoff subjects (9.5%) scored above 80% correct – a level achieved by 88% of subjects in a large normative study. These results showed a consistent impairment in odor identification in a task that allows immediate matching of odors to labels.

We conducted a second study to test the possibility that the SIT impairment resulted from non-olfactory demands of this task. To do this we compared SIT

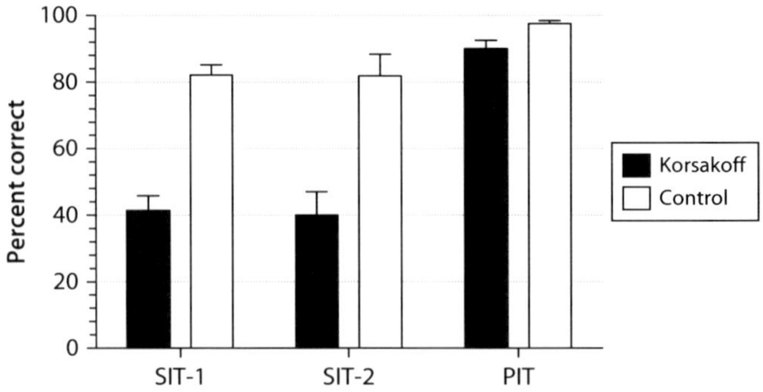

Figure 3. Performances of Korsakoff and Control subjects on Smell (SIT) and Picture (PIT) Identification Tasks (Mair et al., 1986). Korsakoff patients were consistently and significantly impaired for SIT in experiment 1 (SIT-1) and experiment 2 (SIT-2). Patients exhibiting SIT impairments in experiment 2 were unimpaired performing the PIT task.

with an equivalent Picture Identification Task (PIT) for 10 Korsakoff and 8 controls matched for age, gender, education, IQ, and drinking history (SIT-2, PIT in Figure 3). The results of this study showed a significant SIT impairment that was remarkably similar to the first study. These same patients were unimpaired for the PIT: there was no significant difference between the groups for the PIT. Korsakoff, but not control, subjects showed significant improvement for PIT compared to SIT performance. Thus patients who were unable to match smells to verbal labels on the SIT, were able to match equivalent labels to pictures under comparable conditions in the PIT. Taken together, our results indicate that Korsakoff patients are impaired making same-different judgments of odor pairs and in matching odors to verbal labels, even though they retain normal ability to detect odorants. Their impairments on these tasks are not easily explained as a memory impairment: Korsakoff patients showed no signs of temporal decay in the recognition memory task and were able to match pictures to labels in the PIT.

1.3 Summary – human studies

Overall, these early neuropsychological studies provide support for two important ideas. First, they helped to define syndromes of central olfactory impairment, affecting functions like naming, identification, and recognition memory while sparing primary sensory functions measured by detection tasks. These results provide direct evidence of olfactory agnosia: the loss of the ability to recognize or understand odorants with a preserved capacity to smell them.

Second, central olfactory impairments are associated with lesions damaging brain areas beyond primary olfactory cortex. The impairments observed in our study of aphasia (Mair & Engen, 1976) are consistent with other evidence implicating prefrontal and temporal areas of neocortex in higher olfactory function (Eskenazi et al., 1983, 1986; Jones-Gotman & Zatorre, 1988; Jones-Gotman et al., 1997; Zatorre et al., 1992; Savic, 2002). The consistent effects of Korsakoff's disease suggests an important role of systems involving midline areas of the diencephalon, particularly thalamus, that are consistently affected by this disease (Kopelman, Thomson, Guerrini, & Marshall, 2009; Mair, 1994). Other evidence indicates that central olfactory impairment is a consistent and early sign of Parkinson's and Alzheimer's diseases (Doty, 2009) and of schizophrenia (Tuertsky, Hahn, Borgmann-Winter, & Moberg, 2009). The diversity of brain systems affected by these neurological disorders suggests that olfaction may depend on large-scale distributed neural networks involving areas of neocortex not generally considered olfactory. While this possibility seems surprising in view of the modest role generally ascribed to olfaction in cognition, it is intriguing in view of recent evidence that expansion of olfactory function was an early driving force behind the evolution of the mammalian brain (Rowe, Macrini, & Luo, 2011). If olfaction played a key role in determining the organization of the mammalian brain, perhaps it should not be surprising that diverse parts of the brain continue to play a role in making sense of what we smell.

2. The neuropsychology of odor perception and memory – animal studies

Research at the University of New Hampshire has used a comparative neuropsychological approach, using experimental lesions in rats to elucidate the neurological basis of the olfactory impairments observed in Korsakoff patients. This work included dissertation projects by Sue Koger (1993) and Yue Ping Zhang (1996). Engen was a member of Zhang's dissertation committee in what turned out to be our last research collaboration.

We developed two related tasks to model measures found to be spared and impaired in human Korsakoff patients (Mair et al., 1980): simple discrimination and continuous olfactory delayed non-match to sample (cDNM). Both tasks are go/no go discriminations in which one odorant is presented on each trial (as in human detection and recognition memory tasks). In simple discrimination, one of two odorants (the S+ or the S–) is presented on each trial. Reinforcement (tap water) is delivered for responding (correctly) on trials when the S+ odorant is presented. Responding (incorrectly) on trials when the S– is presented is punished by a 5 s timeout. In the cDNM task responses are reinforced on trials when the odorant presented is different from the odorant presented on the previous trial (a correct nonmatching response). Responding on trials when the odorant is the same as on the preceding trial is punished with 5 s timeout (an incorrect matching response). Simple discrimination, like detection tasks in human studies, can be solved with reference memory by applying a simple rule: respond to the S+ but not the S–. By contrast cDNM, like recognition memory tasks in human studies, requires working memory to compare the particular odorants presented on successive trials. We use two task manipulations to examine possible causes of cDNM impairment: memory delays are varied between trials to examine the temporal decay of working memory and the size of the stimulus set is varied between sessions to examine effects of interference.

2.1 Lesion studies: Central thalamus vs. prefrontal cortex

Our first investigation compared the effects of electrolytic radiofrequency (RF) lesions of thalamus and prefrontal cortex (Koger & Mair, 1994). The thalamic lesions were aimed at lateral internal medullary lamina (L-IML), a location where RF lesions disrupt projections from the medial dorsal (MDn) and intralaminar (ILn) nuclei to prefrontal cortex. The prefrontal lesions were aimed at two different sites: cortex dorsal to the rhinal sulcus (RS) and along the medial wall (MW) of the cerebral hemispheres. RS includes orbital areas that receive olfactory and other sensory inputs and contains neurons that code olfactory information (Ongur & Price, 2000; Schoenbaum & Eichenbaum, 1995). MW includes areas that have been implicated in working memory, decision making, and other aspects of executive function (Bailey & Mair, 2007; Dalley, Cardinal, & Robbins, 2004).

The thalamic lesion produced a sharp decline in cDNM performance from presurgical levels that persisted throughout postsurgical training (Koger & Mair, 1994; Figure 4). Both MW and RS lesions produced initial impairments comparable to thalamic lesions, however these deficits proved evanescent- disappearing after the first block of 200 trials of post-surgical training. These results suggest that while MW and RS may affect cDNM performance, neither area alone plays a critical role in this task.

Analyses of task manipulations provided evidence that none of the lesions affected sensitivity to two manipulations known to increase demands on memory: the length of the memory delay or interference effects associated with reduced stimulus set size. Varying the length of the retention interval up to 40 s showed no sign of temporal decay for any group (Figure 4). The lack of temporal decay is consistent with results for human odor recognition memory (Engen, Kuisma, &

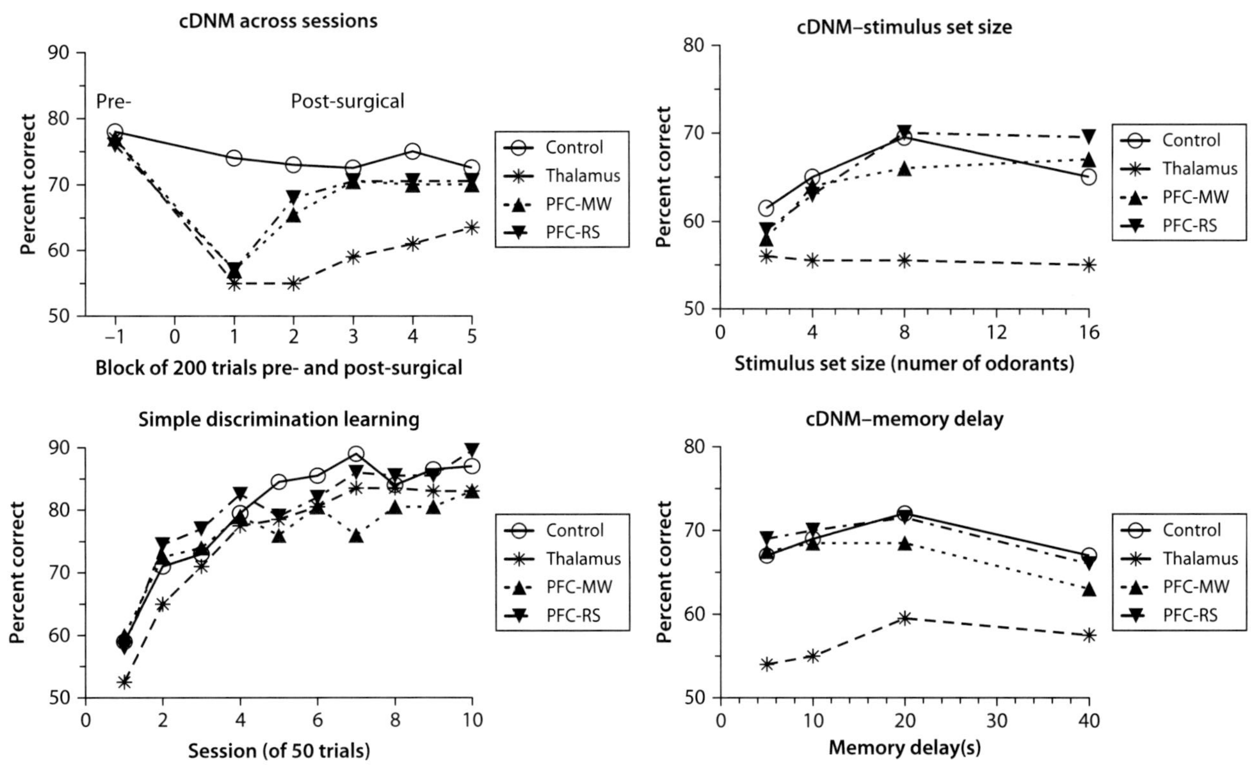

Figure 4. Effects of electrolytic RF lesions in the IML of thalamus and the medial wall (MW) and rhinal sulcus (RS) areas of prefrontal cortex (PFC) (Koger & Mair, 1994). The thalamic lesion produced persistent impairment while PFC lesions produced transient impairment of the continuous delayed non matching (cDNM). None of the lesions affected simple discrimination learning. Decreasing stimulus set size was associated with diminished accuracy while varying the length of the memory delay had no significant effect on cDNM. Neither of these task manipulations influenced effects of any of the lesions.

Eimas, 1973; Engen & Ross, 1973). The absence of an interaction between the effects of lesions and the length of the memory delay is consistent with evidence that Korsakoff's patients show no signs of temporal decay for short-term odor recognition memory (Mair et al., 1980). Reducing the size of the stimulus set below eight odorants produced the predicted decrease in performance for all groups except for the thalamic lesion group. Rats with thalamic lesions performed very poorly on this task so that the effects of manipulating stimulus set size were limited by a floor effect. Analyses of stimulus set size revealed decreased accuracy during later trials in sessions for small set sizes (2 or 4 odorants) that were not observed for larger sets (see Figure 3B in Koger & Mair, 1994). This result is consistent with the hypothesis that when set size is reduced beyond a critical value, interference effects should build up throughout cDNM training sessions as odors are presented repeatedly. None of the lesions affected this measure of interference.

The cDNM deficits observed with thalamic and (less permanently) cortical lesions could be explained by a number of non-olfactory factors associated with the go/no go task, for instance the ability to inhibit responding during S- trials or the ability to learn from reinforcement presented occasionally across a series of trials. To test this possibility rats were trained to perform two simple olfactory discriminations using the same go/no go procedures as the cDNM task. For the first discrimination eugenol was the S+ and amyl acetate the S–. For the second discrimination geraniol was the S+ and phenethyl alcohol the S–. Both discriminations were learned equivalently by controls and each of the three lesion groups. Results for the first discrimination problem are shown in Figure 4. The comparable performance of all groups learning these discriminations provide evidence that the thalamic and prefrontal lesions spared the ability to perform the go/no go procedures, the capacity to discriminate and to respond consistently to olfactory stimuli, and the ability to learn a rule-based discrimination across a series of daily training sessions.

The differential effects of thalamic L-IML lesions, impairing cDNM while sparing simple discrimination, is consistent with the pattern of central olfactory impairment observed in Korsakoff patients (Mair et al., 1980). Simple discrimination, like signal detection tasks in human studies, can be solved with reference memory: responding to stimuli that have consistent significance across trials. For discrimination learning, lesioned rats were able to learn and retain a consistent pairing of an odorant with reinforcement. Here thalamic and prefrontal lesions spared improvements within sessions and savings between daily sessions. By contrast cDNM, like odor recognition memory in human studies, requires that rats repeatedly store new information (the odor presented on a given trial) and to use this information to decide whether to respond on the subsequent trial.

2.2 Lesion studies: Thalamic vs. olfactory cortical lesions

Zhang et al. (1998) conducted two experiments to elucidate the neurological basis of the olfactory cDNM deficits observed by Koger and Mair (1994). Lesions were made with the excitotoxin NMDA to avoid damaging fibers of passage. The first experiment compared cortical lesions aimed at pyriform and entorhinal cortices and thalamic lesions aimed at the L-IML site, the mediodorsal nucleus (MDn) and the paracentral and centrolateral nuclei (PC-CL). Pyriform lesions produced substantial bilateral damage to primary olfactory cortex in all animals, including near-complete damage in three (of six) animals. Entorhinal cortex provides an important source of olfactory information to hippocampus. Entorhinal lesions produced substantial bilateral damage in all animals and near complete damage to this area in three (of seven). Our results (Figure 5) showed a persistent and significant cDNM deficit for rats with pyriform, but not entorhinal lesions (Zhang, Burk, Glode, & Mair, 1998).

The L-IML lesion was aimed at the same site where RF lesions were found to impair cDNM and spare discrimination learning in our earlier study (Koger & Mair, 1994). Comparing the effects of excitotoxic NMDA lesions of the same site provided a comparison between lesions damaging cell bodies while sparing fibers tracts (NMDA) to lesions damaging both (RF). The MDn lesion examined the effect of damaging specific projections to prefrontal cortex from thalamus. The PC-CL lesion examined the effect of damaging non-specific intralaminar nuclei involved directly in the L-IML lesion. We found a significant effect on cDNM for the L-IML lesion, but not for the MDn or PC-CL lesion. Thus while we replicated the impairment produced by the RF L-IML lesion in Koger and Mair (1994), neither of the two main areas affected by this lesion could account for this impairment when lesioned separately.

Manipulations of memory delay and stimulus set size produced effects similar to our earlier study: there was no effect of increasing memory delays (from 4 to 20 s) and reduced accuracy when stimulus set size was reduced below 8. None of these manipulations interacted with the effects of cortical or thalamic lesions. Thus none of the lesions appeared to increase sensitivity to the length of the memory delay or to interference effects between odorants. None of the cortical or thalamic lesions affected the acquisition of the simple discrimination between eugenol and amyl acetate (Figure 5). Thus the cDNM deficits produced by pyriform and L-IML lesions appear to be unrelated to the procedural demands of the go/no go task, the ability to discriminate between odorants, or the ability to learn and to respond consistently based on olfactory cues.

Zhang et al. conducted a second experiment to determine the neurological basis of cDNM deficits produced by L-IML lesions. There appeared to be two logical

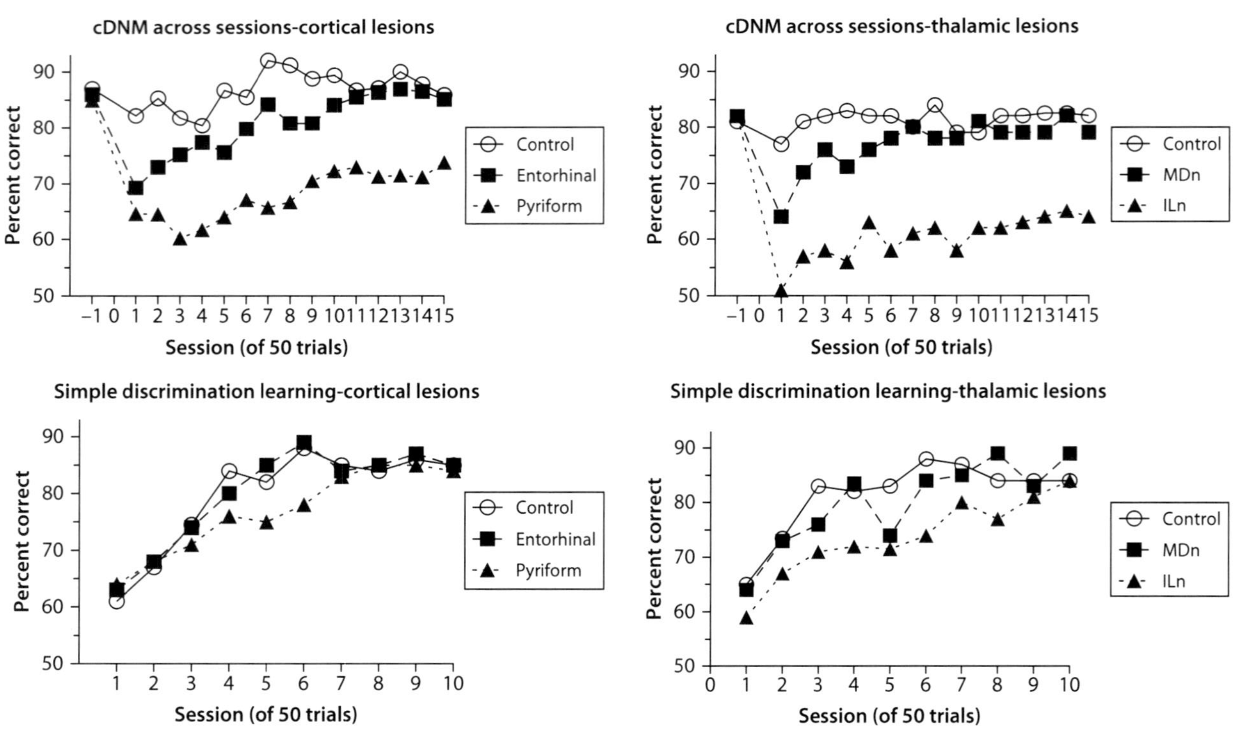

Figure 5. Effects of excitotoxic NMDA lesions on go/no go cDNM and simple discrimination tasks (Zhang et al., 1998). Lesions damaging pyriform cortex or intralaminar nuclei in thalamus produced significant cDNM impairments that persisted throughout post-surgical training. Lesions damaging entorhinal cortex or the mediodorsal thalamic nuclei did not. None of the lesions affected acquisition of simple discrimination task.

possibilities: the MDn and the intralaminar nuclei (ILn). The MDn are the main target of olfactory projections to thalamus and are generally presumed to be the source of olfactory impairment associated with thalamic lesions (Tham, Stevenson, & Miller, 2009). While the ILn is not known to receive specific olfactory inputs previous work in our lab has implicated the ILn as a source of pathology affecting memory in other sensory modalities (Bailey & Mair, 2005; Burk & Mair, 1998; Stevens & Mair, 1998). Here Zhang et al. compared the effects of more substantial MDn and intralaminar (ILn) lesions than in the first experiment. MDn lesions were made larger by increasing the volume of NMDA (0.6 vs. 0.4 µl of a 100 mM solution) used to lesion these nuclei. The ILn lesion included the same sites used to produce the PC-CL lesion along with more medial sites aimed at the centromedial nucleus. Our results demonstrated a substantial and persistent cDNM impairment for the ILn lesion, but not for the MDn lesion (Figure 5). Neither of the lesions affected the acquisition of the simple discrimination problem. This result was surprising given the lack of specific olfactory input to the ILn. In retrospect it seems consistent with evidence implicating "non-olfactory" areas in parietal, temporal, and frontal cortices as sources of central olfactory impairment.

The results of Zhang et al. (1998) revealed two key insights into the functional organization of central olfactory pathways. First, pyriform cortex lesions affect cDNM but spare discrimination learning. This is inconsistent with the functional properties of established areas of primary sensory cortex where lesions result in a loss of sensory capacity: scotomas with lesions in primary visual cortex, cortical deafness with primary auditory cortex lesions, and cortical sensory loss with somatosensory cortex lesions (Kim, 2007). The impairment observed with pyriform cortex lesions seems more consistent with effects of sensory association cortex lesions, an idea supported by anatomical evidence that the connections of pyriform cortex are organized more like association than primary sensory cortices (Johnson, Illig, Behan, & Haberly, 2000). Second, olfactory function was disrupted by ILn, but not MDn lesions. At first blush this is surprising. The MDn provides an indirect pathway linking pyriform cortex to prefrontal cortex, although olfaction is a minor source of input to MDn. The ILn do not receive direct olfactory projections. Rather they appear organized to control parallel, distributed neural pathways that connect frontal cortex with discrete areas of the basal ganglia and thalamus. The implication of the ILn in olfactory cDNM is consistent with the role that the ILn appear to play in delayed conditional discrimination tasks in other sensory modalities (Bailey & Mair, 2005; Burk & Mair, 1998; Stevens & Mair, 1998). These results provide further evidence that central olfactory function, our ability to make sense of what we smell, relies on brain systems well beyond areas considered specifically olfactory.

3. Remembering what the nose knows

Neuropsychology has provided important insights into central olfactory function: characterizing the role of central pathways in making sense of what we smell and the extent to which presumed non-olfactory parts of the brain are engaged in these processes. Both human and animal studies emphasize the important distinction between working and reference memory. Reference memory, or the ability to attach a relatively fixed or consistent meaning to an odor, appears to be impervious to the effects of central brain lesions. Thus Korsakoff patients can retain the ability to respond consistently to a threshold-level odorant in a detection task, while exhibiting substantial impairment in odor recognition memory or identification tasks that require working memory, the ability to hold incoming olfactory information over brief periods of time while performing cognitive operations. Likewise damage to temporal and frontal areas of cortex have been shown to spare odor detection while interfering with odor recognition memory, quality discrimination between successively presented odorants, matching odors to containers by visual or haptic cues, matching odors to verbal labels, and naming odorants (Eskenazi et al., 1983, 1986; Goodglass et al., 1968; Jones-Gotman & Zatorre, 1988; Jones-Gotman et al., 1997; Mair & Engen, 1976; Zatorre & Jones-Gotman, 1991). Similarly rats with extensive damage to pyriform cortex or central thalamus retain the ability to learn simple discriminations across a series of daily training sessions while exhibiting substantial cDNM deficits. These findings suggest that lesions damaging pyriform cortex or more central olfactory areas spare primary sensory capacity, at least to the extent required for reference memory in simple discrimination learning by animals or odor detection by human observers.

In rodents olfactory sensory neurons express one of approximately 1000 olfactory receptors and project to one of about 1800 glomeruli in the olfactory bulb (Mombaerts et al., 1996). Each receptor responds to multiple odorants and each odor is recognized by multiple receptors (Malnic, Hirono, Sato, & Buck, 1999). Sensory neurons expressing the same receptor are distributed randomly throughout the receptor epithelium and converge into one or two glomeruli (Mombaerts et al., 1996). Each glomerulus drives 50 to 100 mitral cells that send axons through the lateral olfactory tract to pyriform cortex and to other areas of olfactory cortex (anterior olfactory nucleus, olfactory tubercle, entorhinal cortex, amygdala). At the level of the olfactory bulb responses of glomeruli and the mitral cells they drive are determined by the common receptor of the sensory neurons from which they receive sensory input. Mitral cell activity reliably identifies odorants and exhibits plasticity in adjusting responses based on olfactory experience (Friedrich, 2006; Mandairon & Linster, 2009). More recently mitral cells have been shown to encode information about whether or not an odorant is associated with reinforcement (Doucette et al., 2011). The ability of animal and human subjects with central lesions to perform detection and discrimination tasks suggests that this early processing within the bulb is sufficient to support odor reference memory, presumably through projections from the bulb that bypass pyriform cortex.

Olfactory tasks affected by temporal or frontal lesions, unlike tasks spared by this damage, require working memory to remember a recently presented odorant while performing a cognitive operation (Eskenazi et al., 1983, 1986; Goodglass et al., 1968; Jones-Gotman & Zatorre, 1988; Jones-Gotman et al., 1997; Mair & Engen, 1976; Zatorre & Jones-Gotman, 1991). A similar pattern of impairment has been observed in Korsakoff patients (Mair et al., 1982, 1986). It is interesting in the context of these findings that lesions damaging central thalamus or pyriform cortex in the rat produce a similar dissociation: disrupting cDNM, a task requiring working memory, while sparing simple discrimination learning, a task that can be solved with reference memory. Central thalamus is consistently involved in Korsakoff's disease and pyriform cortex is potentially damaged by frontal or temporal lesions and is the main source of olfactory input to these areas.

The emphasis placed here on pyriform cortex in odor working memory is consistent with results from neuroimaging studies. Earlier work showed that pyriform activity is unchanged when odors are encoded, increased slightly during short-term odor recognition, and increased more substantially during long-term odor recognition (Dade et al., 2002). More recent evidence indicates that pyriform activity is increased and sustained throughout the delay period in a short-term odor recognition memory task (Zelano, Montag, Khan, & Sobel, 2009). Sustained delay-period activity is characteristic of neural responses in higher-order areas of prefrontal cortex of rats and monkeys performing visuospatial delayed conditional discrimination tasks (Fuster, 1989; Goldman-Rakic, 1996). Interestingly, this sustained activity was more pronounced in pyriform cortex for odors that were unnamable than nameable. By contrast, odors that were nameable produced more pronounced sustained activity in left prefrontal language areas. These results suggest that memory for odorants that are readily named (what Mair & Engen 1976 related to "exact match" identification) is mediated in part by higher associative areas involved in language functions.

The implication of pyriform cortex in odor working memory is consistent with anatomical evidence that pyriform cortex is organized more like association than primary sensory cortex. Although only two synapses removed from primary sensory neurons, pyriform pyramidal cells interact directly with prefrontal cortex, amygdala, and entorhinal and perirhinal cortices to mediate odor guided behavior (Johnson et al., 2000). Mitral cells provide diffuse innervation of pyriform cortex that has no apparent spatial bias and are indistinguishable between different projection neurons (Sosulski, Bloom, Cutforth, Axel, & Datta, 2011). By contrast projections to cortical amygdala are patchy, providing the possibility of spatial integration. This organization suggests a unique organization of the sense of smell in the brain in which primary sensory processing occurs at the level of second order sensory neurons and this information diverges at the level of third order neurons to interact with brain areas involved in distinct aspects of cognition and behavior. In many ways this organization seems more like the olfactory system of invertebrates like Drosophila, in which sensory neurons expressing a common receptor converge in spatially invariant glomeruli in the antennal lobe. The antennal lobe then projects to the mushroom body, a structure implicated in olfactory associative learning, and the lateral horn, which supports rudimentary odor guided behaviors (Fiala, 2007; Masse, Turner, & Jefferis, 2009)

While most vertebrate species have olfactory sensory epithelia that project to an olfactory bulb and then to central olfactory structures, it is unclear whether non-mammalian vertebrates have a structure homologous to pyriform cortex. Recent analyses of endocasts (internal casts of the cranial vault) suggest that early evolution of mammalian brains was driven by an increase in size of olfactory bulb and primary olfactory cortex (Rowe et al., 2011). This finding raises the interesting question of how the evolution of these areas might have benefitted our ancestors to the extent that it could become a driving force in our evolution. Rowe et al. note that the evolution of mammals corresponded with a substantial increase in the size of the olfactory receptor genome and argue that the expansion of olfactory bulbs and primary olfactory cortex reflected the development of a higher level of olfactory resolution, presumably related to the important role of olfaction, as compared to vision or audition in primitive mammals. Neuropsychological results suggest a more intriguing possibility. The development of pyriform cortex may have corresponded with the ability to hold odor information in working memory. The capacity to remember sensory information after it disappears from the senses is critical for complex behavioral functions like response-outcome learning, decision making, and attention control. Perhaps the early evolution of olfactory areas of the brain provided our ancient ancestors with a blueprint that eventually led to the capacity for working memory in other sensory modalities thus the cognitive capabilities that characterize modern mammalian species.

References

Bailey, K. R., & Mair, R. G. (2005). Lesions of specific and nonspecific thalamic nuclei affect prefrontal cortex-dependent aspects of spatial working memory. *Behavioral Neuroscience, 119*(2), 410–419.

Bailey, K. R., & Mair, R. G. (2007). Effects of frontal cortex lesions on action sequence learning in the rat. *European Journal of Neuroscience, 25*(9), 2905–2915.

Burk, J. A., & Mair, R. G. (1998). Thalamic amnesia reconsidered: Excitotoxic lesions of the intralaminar nuclei, but not the mediodorsal nucleus, disrupt place delayed matching-to-sample performance in rats (Rattus norvegicus). *Behavioral Neuroscience, 112*(1), 54–67.

Cave, C. B., & Squire, L. R. (1992). Intact verbal and nonverbal short-term memory following damage to the human hippocampus. *Hippocampus, 2*(2), 151–163.

Dade, L. A., Zatorre, R. J., & Jones-Gotman, M. (2002). Olfactory learning: Convergent findings from lesion and brain imaging studies in humans. *Brain, 125*(Pt 1), 86–101.

Dalley, J. W., Cardinal, R. N., & Robbins, T. W. (2004). Prefrontal executive and cognitive functions in rodents: neural and neurochemical substrates. *Neuroscience and Biobehavioral Reviews, 28*(7), 771–784.

Doucette, W., Gire, D. H., Whitesell, J., Carmean, V., Lucero, M. T., & Restrepo, D. (2011). Associative cortex features in the first olfactory brain relay station. *Neuron, 69*(6), 1176–1187. 10.1016/j.neuron.2011.02.024.

Engen, T. (1991). Preface in *Odor Sensation and Memory.* New York: Praeger. Pg. xii.

Engen, T. (1971). Olfactory psychphysics. In L. M. Beidler (Ed.), *Handbook of Sensory Physiology* (Vol. IV. Chemical Senses, pp. 216–244). Berlin: Springer Verlag.

Engen, T., Kuisma, J. E., & Eimas, P. D. (1973). Short-term memory of odors. *Journal of Experimental Psychology, 99*(2), 222–225.

Engen, T., & Ross, B. M. (1973). Long-term memory of odors with and without verbal descriptions. *Journal of Experimental Psychology, 100*(2), 221–227.

Eskenazi, B., Cain, W. S., Novelly, R. A., & Friend, K. B. (1983). Olfactory functioning in temporal lobectomy patients. *Neuropsychologia, 21*(4), 365–374.

Eskenazi, B., Cain, W. S., Novelly, R. A., & Mattson, R. (1986). Odor perception in temporal lobe epilepsy patients with and without temporal lobectomy. *Neuropsychologia, 24*(4), 553–562.

Fiala, A. (2007). Olfaction and olfactory learning in Drosophila: recent progress. *Current Opinion in Neurobiology, 17*(6), 720–726.

Friedrich, R. W. (2006). Mechanisms of odor discrimination: neurophysiological and behavioral approaches. *Trends in Neuroscience, 29*(1), 40–47.

Fuster, J. (1989). *The Prefrontal Cortex.* New York: Raven Press.

Goldman-Rakic, P. S. (1996). Regional and cellular fractionation of working memory. *Proceedings of the National Academy of Sciences USA, 93*(24), 13473–13480.

Goodglass, H., Barton, M. I., & Kaplan, E. F. (1968). Sensory modality and object-naming in aphasia. *Journal of Speech and Hearing Research, 11*, 488–496.

Johnson, D. M., Illig, K. R., Behan, M., & Haberly, L. B. (2000). New features of connectivity in piriform cortex visualized by intracellular injection of pyramidal cells suggest that "primary" olfactory cortex functions like "association" cortex in other sensory systems. *Journal of Neuroscience, 20*(18), 6974–6982.

Jones, B. P., Butters, N., Moskowitz, H. R., & Montgomery, K. (1978). Olfactory and gustatory capacities of alcoholic Korsakoff patients. *Neuropsychologia, 16*(3), 323–337.

Jones, B. P., Moskowitz, H. R., Butters, N., & Glosser, G. (1975). Psychophysical scaling of olfactory, visual, and auditory stimuli by alchoholic Korsakoff patients. *Neuropsychologia, 13*(4), 387–393.

Jones, B. P., Moskowitz, R. H., & Butters, N. (1975). Olfactory discrimination in alcoholic Korsakoff patients. *Neuropsychologia, 13*(2), 173–179.

Jones-Gotman, M., & Zatorre, R. J. (1988). Olfactory identification deficits in patients with focal cerebral excision. *Neuropsychologia, 26*(3), 387–400.
Jones-Gotman, M., Zatorre, R. J., Cendes, F., Olivier, A., Andermann, F., McMackin, D., & Wieser, H. G. (1997). Contribution of medial versus lateral temporal-lobe structures to human odour identification. *Brain, 120*(Pt 10), 1845–1856.
Kim, J. S. (2007). Patterns of sensory abnormality in cortical stroke: Evidence for a dichotomized sensory system. *Neurology, 68*(3), 174–180.
Koger, S. M., & Mair, R. G. (1994). Comparison of the effects of frontal cortical and thalamic lesions on measures of olfactory learning and memory in the rat. *Behavioral Neuroscience, 108*(6), 1088–1100.
Kopelman, M. D., Thomson, A. D., Guerrini, I., & Marshall, E. J. (2009). The Korsakoff syndrome: clinical aspects, psychology and treatment. *Alcohol and Alcoholism, 44*(2), 148–154.
MacLeod, P. (1971). Structure and function of higher olfactory centers. In L. M. Beidler (Ed.), *Handbook of Sensory Physiology* (Vol. IV. Chemical Senses, pp. 182–204). Berlin: Springer Verlag.
Mair, R. G. (1994). On the role of thalamic pathology in diencephalic amnesia. *Reviews in the Neurosciences, 5*(2), 105–140.
Mair, R. G., Capra, C., McEntee, W. J., & Engen, T. (1980). Odor discrimination and memory in Korsakoff's psychosis. *Journal of Experimental Psychology: Human Perception and Performance, 6*(3), 445–458.
Mair, R. G., Doty, R. L., Kelly, K. M., Wilson, C. S., Langlais, P. J., McEntee, W. J., & Vollmecke, T. A. (1986). Multimodal sensory discrimination deficits in Korsakoff's psychosis. *Neuropsychologia, 24*(6), 831–839.
Mair, R. G., & Engen, T. (1976). Some effects of aphasic lesions on odor perception. *Sensory Processes, 1*(1), 33–39.
Malnic, B., Hirono, J., Sato, T., & Buck, L. B. (1999). Combinatorial receptor codes for odors. *Cell, 96*(5), 713–723.
Mandairon, N., & Linster, C. (2009). Odor perception and olfactory bulb plasticity in adult mammals. *Journal of Neurophysiology, 101*(5), 2204–2209.
Masse, N. Y., Turner, G. C., & Jefferis, G. S. (2009). Olfactory information processing in Drosophila. *Current Biology, 19*(16), R700–713.
Mombaerts, P., Wang, F., Dulac, C., Chao, S. K., Nemes, A., Mendelsohn, M., & Axel, R. (1996). Visualizing an olfactory sensory map. *Cell, 87*(4), 675–686.
Moncrieff, R. W. (1967). *The Chemical Senses*. Londin: Leonard Hill Books.
Mozell, M. M. (1972). The Chemical Senses II. Olfaction. In K. J. L. a. R. L.A. (Ed.), *Experimental Psychology* (pp. 195–222). New York: Holt, Rinehard and Winston.
Ongur, D., & Price, J. L. (2000). The organization of networks within the orbital and medial prefrontal cortex of rats, monkeys and humans. *Cerebral Cortex, 10*(3), 206–219.
Otto, T., & Eichenbaum, H. (1992). Complementary roles of the orbital prefrontal cortex and the perirhinal-entorhinal cortices in an odor-guided delayed-nonmatching-to-sample task. *Behaioralv Neuroscience, 106*(5), 762–775.
Rowe, T. B., Macrini, T. E., & Luo, Z. X. (2011). Fossil evidence on origin of the mammalian brain. *Science, 332*(6032), 955–957.
Savic, I. (2002). Imaging of brain activation by odorants in humans. *Current Opinion in Neurobiology, 12*(4), 455–461.

Schoenbaum, G., & Eichenbaum, H. (1995). Information coding in the rodent prefrontal cortex. I. Single-neuron activity in orbitofrontal cortex compared with that in pyriform cortex. *Journal of Neurophysiology, 74*(2), 733–750.

Sosulski, D. L., Bloom, M. L., Cutforth, T., Axel, R., & Datta, S. R. (2011). Distinct representations of olfactory information in different cortical centres. *Nature, 472*(7342), 213–216.

Stevens, A. A., & Mair, R. G. (1998). Auditory conditional discrimination deficits without delays in rats with lesions of either frontal cortex or medial thalamus. *Psychobiology, 26*(3), 205–215.

Tham, W. W., Stevenson, R. J., & Miller, L. A. (2009). The functional role of the medio dorsal thalamic nucleus in olfaction. *Brain Research Reviews, 62*(1), 109–126.

Zatorre, R. J., & Jones-Gotman, M. (1991). Human olfactory discrimination after unilateral frontal or temporal lobectomy. *Brain, 114*(Pt 1A), 71–84.

Zatorre, R. J., Jones-Gotman, M., Evans, A. C., & Meyer, E. (1992). Functional localization and lateralization of human olfactory cortex. *Nature, 360*(6402), 339–340.

Zelano, C., Montag, J., Khan, R., & Sobel, N. (2009). A specialized odor memory buffer in primary olfactory cortex. *PLoS One, 4*(3), e4965.

Zhang, Y., Burk, J. A., Glode, B. M., & Mair, R. G. (1998). Effects of thalamic and olfactory cortical lesions on continuous olfactory delayed nonmatching-to-sample and olfactory discrimination in rats (Rattus norvegicus). *Behavioral Neuroscience, 112*(1), 39–53.

CHAPTER 11

Olfactory impairment in normal aging and Alzheimer's disease

Steven Nordin
Umeå University, Sweden

This chapter provides a broad review of early and recent findings on the neuropsychology of olfactory impairment in normal aging and Alzheimer's disease (AD) in a growing population of elderly. Neural and anatomical changes in these populations underlying olfactory impairment are described, and effects on behavioral measures of olfactory function are presented. These functions span from sensory to cognitive in character, and include detection sensitivity, discrimination, recognition memory, familiarity, identification and semantic networks. Furthermore, the role of awareness of olfactory impairment is discussed with respect to normal aging and AD. Other aspects provided in this chapter include olfactory testing for early signs of AD, and consequences of olfactory impairment on quality of life, food intake and health. Finally, future research needs are identified and discussed.

Introduction

The proportion of elderly in the general population is growing fast in the western part of the world, and an associated problem is malnutrition. Studies show that 3–10% of non-institutionalized elderly and 25–60% of institutionalized elderly suffer from malnutrition (Swedish National Food Administration, 1998; Vellas et al., 2001). Consequences of malnutrition are extensive, and age-related changes in chemosensory function, in particular olfactory function, are likely to contribute to these problems. Age-related changes in this sensory function can also contribute to poor quality of life and poor safety (Hummel & Nordin, 2005). These consequences can be referred to both normal aging and several forms of age-related degenerative disorders such as Alzheimer's disease (AD).

AD is a neurological disorder, characterized by progressive memory loss, and the most common cause of dementia. About 5% of the population above the age of 65 years suffer from dementia, and AD accounts for 50 to 60% of the dementing diseases. Sensitive tools for early diagnosis of AD become increasingly important as the efficacy of medication is improved to decelerate the dementia progression. In this respect olfaction is of particular interest due to its very early and pronounced brain neuropathology and functional pathology in AD.

The objectives of this chapter are to review and discuss (i) the neurohistological, neurophysiological, and functional changes in normal aging, and conditions and exposures that may contribute to these changes, and (ii) the neurohistological, neurophysiological, and functional changes in AD, the roles of Apolipoprotein E (ApoE) and awareness of olfactory impairment, and the potential of olfactory testing for early diagnosis of AD.

1. Normal aging

1.1 Neurohistological and neurophysiological changes

Normal aging is accompanied by an increase in degeneration of olfactory receptor neurons, and a diminished ability to replace these neurons with new olfactory neurons. Instead, they are replaced with respiratory epithelium with an apparently equal distribution throughout the mucosa (Rawson, Gomez, Cowart, & Restrepo, 1998). At a slightly higher level of the olfactory system, age-associated atrophy of mitral cells and glomeruli is common (Bahtnagar, Kennedy, Baron, & Greenberg, 1987). Although considerably more present in AD, neuritic plaques (NP) and neurofibrillary tangles (NFT) are often found in the normal aging brain in higher-order olfactory regions, including the olfactory tract, entorhinal cortex, hippocampus, and amygdala (Price, Davis, Morris, & White, 1991).

Age-related changes in neurophysiological processing of odorous substances have also been reported. Recordings of electrophysiological cortical activity by means of olfactory event-related potentials (ERP) suggest that neuronal allocation is diminished and that processing speed is decreased in normal aging for both relatively sensory and cognitive components of the ERP (Murphy, Nordin, de Wijk, Cain, & Polich 1994; Hummel, Barz, Pauli, & Kobal, 1998; Morgan, Geisler, Covington, Polich, & Murphy, 1999; Murphy, Morgan, Geisler, Wetter, Covington, Madowitz, Nordin, & Polich, 2000). Neuroimaging suggests age-related decrease in activity in the enthorinal cortex, piriform cortex, amygdala and periamygloid cortex, hippocampus and parahippocampal gyrus, orbitofrontal cortex and insula when conducting tasks requiring low-cognitive olfactory functioning (Cerf-Ducastel & Murphy, 2003; Wang, Eslinger, Smith, & Yang, 2005).

1.2 Functional changes

1.2.1 *Sensory functions*

An important age-related olfactory impairment of sensory character lies in the ability to detect odorous substances. A large number of studies have quite consistently demonstrated orthonasal loss in this function, for both food and non-food odorants (Schiffman, Moss, & Erickson, 1976; Stevens & Cain, 1987; Stevens, Cain, & Weinstein, 1987; Cain & Gent, 1991; Hummel, Sekinger, Wolf, Pauli, & Kobal, 1997). The age-related impairment in detectability seems to cover a wide range of odorous substances across studies (Schiffman et al., 1976), and Stevens and colleagues (1989) have demonstrated stronger odor adaptation and slower recovery. However, loss in odor sensitivity does not appear to be inevitable to the aging individual. Odor detection thresholds were carefully determined for so-called successfully-aged elderly who were extensively screened for medical and cognitive factors known to affect olfaction. Interestingly, their thresholds were found to be very similar to those for young adults (Nordin, Almkvist, & Berglund, in press).

Compromised ability to discriminate between odor qualities (Schiffman & Leffingwell, 1981; Stevens & Lawless, 1981; Hummel et al., 1997; Kaneda, Maeshima, Goto, Kobayakawa, Ayabe-Kanamura, & Saito, 2000), and diminished perceived odor intensity (Stevens & Cain, 1985, 1987; Wysocki & Gilbert, 1989) has also been reported for the old population. When odorants are perceived as very weak the discrimination ability will, naturally, be compromised. However, with this exception, a study of neuroimaging (Savic, Gulyas, Larsson, & Roland, 2000) suggests largely parallel processing of odor intensity and quality discrimination. Further support for relative parallel processing is provided by psychophysical data (Cain, de Wijk, Nordin, & Nordin, 2008). In this study young, middle-aged and elderly adults were given an odor discrimination task with both perceptually homogeneous and heterogeneous odorous substances. The stimuli were either at intensity-matched concentrations far above detection threshold, but fixed across participants, or at levels adjusted to neutralize differences in sensitivity and, thus, assumed to be similar in perceived intensity across participants. Performance in quality discrimination was found to decline progressively with age, not only for the stimuli fixed in concentration across participants, but also for the stimuli adjusted to neutralize for differences in sensitivity (Figure 1). This may be explained by broader tuning in receptor cells for old than for young participants (Rawson et al., 1998).

Effects of aging have to some extent also been investigated regarding retronasal olfaction. In accordance with orthonasal stimulation, impairment has been found with respect to detection and quality discrimination (Cain, Reid, & Stevens

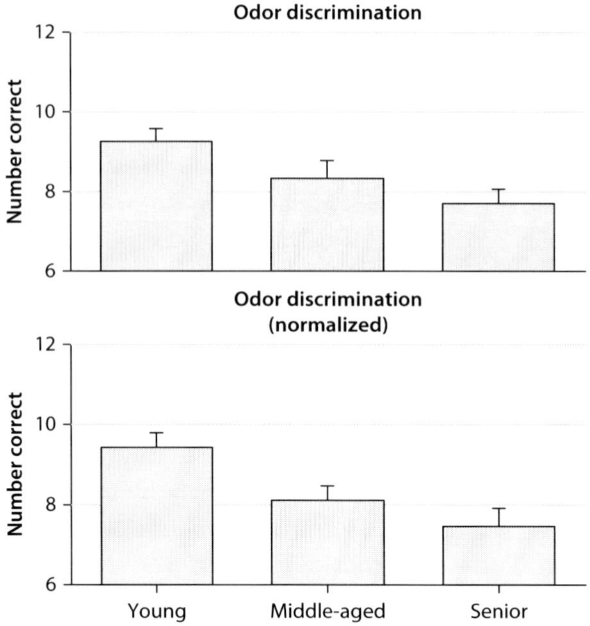

Figure 1. Upper diagram: Performance in odor quality discrimination for stimuli at fixed concentrations. Lower diagram: Performance in odor quality discrimination for stimuli normalized in concentration by participant. Error bars indicate +1 standard error. [With kind permission from Springer Science: Chemosensory Perception, Independence of odor quality and absolute sensitivity in a study of aging, vol. 1, 2008, p. 28, Cain, de Wijk, Nordin, & Nordin.]

1990; Stevens & Cain, 1993; Duffy, Cain, & Ferris, 1999), and perceived intensity (Stevens & Cain, 1986).

1.2.2 *Cognitive functions*

Performance on tests of olfactory cognition is related to performance on non-olfactory cognition. For example, performance in non-olfactory executive functioning, semantic memory and recall predicts odor identification performance (Hedner, Larsson, Arnold, Zucco, & Hummel, 2010; Wehling, Nordin, Espeseth, Reinvang, & Lundervold, 2010). In line with results from studies of sensory function, elderly show impairment in the ability to identify odors (Schemper, Voss, & Cain, 1981; Doty, Shaman, Applebaum, Giberson, Siksorski, & Rosenberg, 1984; Murphy, 1985; Stevens & Cain, 1987; Wysocki & Gilbert, 1989; Larsson & Bäckman, 1993; Cain, Stevens, Nickou, Giles, Johnston, & Garcia-Medina, 1995; Duffy, Backstrand, & Ferris, 1995; Hummel et al., 1997; Larsson, Nilsson, Olofsson, & Nordin, 2004). Results from population-based studies show that prevalence rates for olfactory impairment, by means of tests of cued odor identification, increase from 11–24% in middle-aged individuals to 37–70% at the age of 70 years (Murphy, Schubert, Cruickshanks, Klein, Klein, & Nondahl, 2002; Brämerson, Johansson, Ek, Nordin, & Bende 2004).

The ability to recognize previously experienced odors is predominantly a matter of episodic memory, but also influenced by semantic knowledge of the odorous item (Larsson, 1997). As for odor identification, odor recognition memory is typically compromised in normal aging (Murphy, Cain, Gilmore, & Skinner, 1991; Larsson & Bäckman, 1997). Gilbert and associates (2006) assessed source memory by asking young and old adults to indicate whether an odor from a study phase had been presented by a male or female. Although they found no significant age-related decline in recognition memory for odorous items, they reported lower performance on source memory in the elderly compared to the young. Studies of age-related effects on odor-recognition memory have traditionally used a relatively long interval (e.g., 20 min) between the study and test phase. Zucco (2011) recently developed and applied a test to assess short-term recognition memory for odors. With a 10-s interval between the study and test phase he demonstrated that also short-term recognition memory for odors is compromised in normal aging.

Another cognitive aspect is the ability of verbal recall of previously presented odors. In using an olfactory test analogous to the California Verbal Learning Test, and by adding a test of odor identification, Murphy and collaborators (1997) compared performance on recall, identification and recognition memory with respect to age-related decline, and investigated predictors of odor recall. They found that recall and recognition were significantly affected by aging and that the odor-recall decline cannot simply be referred to poor identification. The authors also reported very similar age-related effect sizes for the three tasks, and that the combined ability to encode, store, and retrieve odors predicts overall recall performance better than do identification and recognition.

1.3 Medical conditions, drugs and environmental exposures

Studies with animal models imply that age-related neurohistological, neurophysiological and functional pathology is present even when the impact of poor health conditions and environmental exposures on olfaction is carefully controlled for (Enwere, Shingo, Gregg, Fujikawa, Ohta, & Weiss, 2004). Nevertheless, these conditions and exposures can, in addition to aging per se, contribute significantly to olfactory impairment in the elderly population. The most common etiologies of olfactory impairment among patients seeking medical attention at ENT clinics include upper respiratory infection, chronic sinusitis, nasal polyposis, allergic rhinitis, and head trauma (Nordin & Brämerson, 2008). In addition, Murphy, Doty and Duncan (2003) have listed a large number of medical conditions that may have an adverse impact on olfactory functioning: endocrine and metabolic diseases, infections, medical interventions, tumors, and neurological and psychiatric diseases. Importantly, old persons are likely to be more susceptible than young persons to the same causes of olfactory impairment (Genter & Ali, 1988). Certain causes are also more specific for aging (Seiberling & Conley, 2004).

Adults over 65 years of age take in average 2.9 to 3.7 medications, and institutionalized elderly take about twice as many (Finkelstein & Schiffman, 1999). It has been suggested that more than 250 drugs may alter chemosensory sensations. These include antihistamines, lipid-lowering drugs, antimicrobial medications, antineoplastic medications, asthma medications, antihypertensives, muscle relaxants and antidepressants (Schiffman, 1991). Nutritional deficits, such as reduced levels of zinc and vitamin A and B, are known to affect smell sensitivity and hedonics (Schiffman, 1997). Moreover, poor oral health in elderly, such as tooth loss, may affect retronasal smell perception (Seiberling & Conley, 2004).

The unprotected position of the olfactory receptor neurons makes them subject to direct exposure to potentially toxic environmental compounds. Accordingly, there are a large number of compounds and processes with documented potential toxic impact on the olfactory sense. In a review of such compounds and processes, Hastings and Miller (2003) include metallic compounds, metallurgical processes, non-metallic inorganic compounds, organic compounds, dusts, and various manufacturing processes.

2. Alzheimer's disease

2.1 Neurohistological and neurophysiological changes

When referring to AD, one usually refers to the diagnosis probable AD, which is established by clinical examination, whereas definite AD is diagnosed reliably only by autopsy. In this chapter probable AD will be referred to as AD when not further specified. The neuropathology of AD is characterized by NP, NFT, and cell loss. It has been suggested that regions of importance for olfactory processing are among those predominantly affected, and may even be the site of initial involvement in the disease (Braak & Braak, 1997). NP and NFT have been found in the anterior olfactory nucleus (Price et al., 1991), and NFT (Attems, Lintner, & Jellinger, 2005) and axonal loss (Davies, Brooks, & Lewis, 1993) have been found in the olfactory bulb. Signs of pathology in the olfactory epithelium have been less consistent.

The entorhinal cortex and periamygdaloid nucleus, which are parts of the olfactory cortex in the temporal lobe, are especially damaged. Work of Price et al. (1991) confirms the presence of NFT in areas that mediate olfactory function, particularly the anterior olfactory nucleus, entorhinal cortex and amygdala in AD patients with very mild dementia. Reyes, Deems and Suarez (1993) reported NP and especially NFT in the entorhinal and prepiriform cortices and in the periamygdaloid nucleus, and concluded that there is greater evidence of pathology in olfactory cortex than in the hippocampus. According to Hyman (1997), the most severe lesions are found in the entorhinal and perirhinal cortices, CA1/subicular area of hippocampus, amygdala and the association cortices. Braak and Braak (1992) have argued that the lesions in the entorhinal and transentorhinal areas effectively disconnect the hippocampus from the isocortex, preventing the transfer of information essential to memory function. Positron emission tomography and ERP studies of AD show low metabolic activity in response to odorants in the anterior portion of the medial- and sub-frontal temporal lobes (Buchsbaum et al., 1991; Kareken et al., 2001) as well as increased cortical processing time (Morgan & Murphy, 2002).

2.2 Functional changes

2.2.1 *Sensory functions*

Loss in absolute detection sensitivity in AD has been demonstrated for a large number of substances (Knupfer & Spiegel, 1986; Doty, Reyes, & Gregor, 1987; Rezek, 1987; Murphy, Gilmore, Seery, Salmon, & Lasker, 1990; Serby, Larson, & Kalkstein, 1991; Lehrner, Brücke, Dal-Bianco, Gatterer, & Kryspin-Exner, 1997; Nordin, Almkvist, Berglund, & Wahlund, 1997). The poor performance in detection is most likely due to neurological rather than rhinological status (Feldman, Murphy, Davidson, Jalowayski, & Galindo de Jaime, 1991), and to a sensory loss rather than a lack of comprehension of the detection task per se (Nordin, Monsch, & Murphy, 1995). Loss in detection sensitivity is further supported by its association with degree of dementia (Murphy et al., 1990; Nordin et al., 1997) and by an association between a fast progression in dementia and a fast decline in sensitivity (Murphy, Nordin, & Jinich, 1999). Odor threshold has, interestingly, also been found to correlate with family history of AD (Schiffman, Clark, & Warwick, 1990). Impairment in odor quality discrimination has been shown in AD when applying same-or-different (Koss, 1986) and match-to-sample paradigms (Kesslak et al., 1988; Buchsbaum et al., 1991; Kesslak, Nalcioglu, & Cotman, 1991).

2.2.2 *Cognitive functions*

Deficit in recognition memory for odors has been documented in mild to moderate cases of AD for both short-term (Knupfer & Spiegel, 1986) and long-term (Moberg, Pearlson, Speedie, Lipsey, Strauss, & Folstein, 1987; Lehrner et al., 1997) memory. Odor-recognition memory appears to be affected earlier in the disease progression than both visual-recognition memory and odor sensitivity (Murphy et al., 1999). Further support for an early loss in this function is provided by data from persons with questionable AD (Nordin & Murphy, 1996). Persons with this diagnosis show impairment in cognitive functions, similar to patients with AD, but, in contrast, they do not yet show changes in everyday functioning as reported by significant others (Bondi, Monsch, Galasko, Butters, Salmon, & Delis, 1994).

Familiarity with a certain object or event can be considered as a form of remote memory; memories for stimuli encoded at some unspecified previous time. It has been reported that AD patients rate odorants, but not faces or symbols, as significantly less familiar than do controls, indicative of deficit in remote odor memory. Follow-up testing, one year later, again, showed this result (Niccoli-Waller, Harvey, Nordin, & Murphy, 1999).

Difficulties in odor identification by naming have been demonstrated in AD when assessed by an odor match-to-name paradigm (Knupfer & Spiegel, 1986) and without available response alternatives (Rezek, 1987; Bacon Moore, Paulsen, & Murphy, 1999). However, the most commonly used procedure to study odor identification in AD is by means of response alternatives provided for each test odor. Results from these studies consistently show considerable loss in the ability to name odors with the aid of written words available (Warner, Peabody, Flattery, & Tinklenberg, 1986; Doty et al., 1987, 1991; Kesslak et al., 1991; Serby et al., 1991; Moberg et al., 1997). The decline in identification can most likely not be referred to difficulties in lexical functioning since poor odor identification also has been demonstrated when using pictures as response alternatives, in both probable and questionable AD (Morgan, Nordin, & Murphy, 1995). This conclusion is supported by a study by Larsson and associates (1999) who compared performance on odor identification when providing support as either written response alternatives or as real-life objects as alternatives, and when providing no support.

Performance in odor identification in AD has been shown to correlate with non-olfactory cognitive performance (Makowska, Kloszewska, Grabowska, Szatkowska, & Rymarczyk, 2011), hippocampal volume (Murphy, Jernigan, & Fennema-Notestine, 2003) and NFT density in hippocampus and enthorinal cortex (Wilson, Arnold, Schneider, Tang, & Bennett, 2007), and to be strongly associated with 5-year incidence of cognitive impairment (Schubert, Carmichael, Murphy, Klein, Klein, Cruickshanks, 2008). Support for a genetic link between olfactory impairment and AD has also been provided by data suggesting that performance on odor identification is impaired in first-degree relatives to patients with AD (Serby, Mohan, Aryan, Williams, Mohs, & Davis, 1996; Handley, Morrison, Miles, & Bayer, 2006). Mild cognitive impairment (MCI) is a major risk factor for developing AD, and poor performance on odor identification in MCI has been shown to predict dementia conversion (Devanand et al., 2000; Djordjevic, Jones-Gotman, De Sousa, & Chertkow, 2007; Fusetti et al., 2010). Laakso and associates (2009) have further shown that identification and delayed recall for odors is more impaired than that for pictures in MCI.

An investigation was conducted to determine the correct classification rate (CCR; ability to correctly classify subjects as healthy or diseased) of tests of odor identification for persons with questionable AD and controls. The results showed a CCR of 83% for the 6-item San Diego Odor Identification Test (~5 min), and a CCR of 100% for the 40-item University of Pennsylvania Odor Identification Test (15–20 min; Morgan et al., 1995).

Closely related to identification is the ability of verbal fluency. Bacon Moore and associates (1999) investigated this domain by asking AD patients to generate, within 60 s, as many names as possible of odor items (i) with respect to a certain category, (ii) beginning with a certain letter, and (iii) that came to mind when presented with two odorants. The results demonstrated impairment in AD in all three measures, and support the claims that the poor odor identification in AD is better explained by cognitive than sensory impairment, and that difficulties generating item names to odorants may reflect a breakdown in the semantic network for odors. Indeed, the claim for such a breakdown in AD has been supported by applying multidimensional scaling, showing deterioration of the structure of semantic knowledge for odors but not for colors in AD (Razani, Chan, Nordin, & Murphy, 2010).

2.3 The roles of the ApoE-ε4 allele and unawareness of olfactory impairment

2.3.1 *The ApoE-ε4 allele*

Another approach to the investigation of early progression of AD is by genotyping of ApoE. The ApoE gene is expressed in the central nervous system, including the olfactory bulb and epithelium. This protein is found in increased amounts in patients with AD, specifically in NP and NFT (Namba, Tomonaga, Kawasaki, Otomo, & Ikeda, 1991), suggesting that ApoE is a genetic risk factor for AD (Corder et al., 1993). Pairs of three primary alleles, ε2, ε3, and ε4, determine an individual's ApoE status. About 25% of the general population has at least one ε4 allele, yet 80% of patients with familial AD and 64% of patients with the sporadic form of the disease have one or both of the ε4 alleles (Sing & Davignon, 1985). Individuals who are homozygous for the ApoE-ε4 allele are at higher risk for developing AD than are those who are heterozygous. Tsuboi and associates (2003) have demonstrated a correlation between ApoE-ε4 allele genotype and presence of NFT in the olfactory bulb in a gene dose-dependent manner. Studies also suggest that healthy elderly who are ApoE-ε4 positive, compared to ApoE-ε4 negative, show longer cortical processing time in response to odorants (Wetter & Murphy, 2001) and perform poorer on odor identification (Murphy, Baco, Bondi, & Salmon, 1998).

Hyposmia and, in particular, anosmia has been shown to increase the risk of subsequent cognitive decline, and that there is a close-to 5-fold risk of subsequent cognitive decline among anosmics with at least one ApoE-ε4 allele (Graves et al., 1999). Further support for the role of odor identification and ApoE is provided by a recent study showing main effects of both poor odor identification and ApoE-ε4 allele on three cognitive domains after the age of 65 years: verbal, spatial, and speed (Finkel, Reynolds, Larsson, Gatz, & Pedersen, 2011). Calhoun-Haney and Murphy (2005) compared ApoE-ε4 positive and negative elderly, and for the positive group reported a decline over time in odor identification, but not in odor threshold, picture identification or score on the Dementia Rating Scale. The authors concluded that decline in odor identification occurs before decline in other measures of dementia in persons at risk for AD because of their ApoE allele status.

Results from a large, population-based sample from northern Sweden followed longitudinally, taking part in the Betula Study, show that ApoE-ε4 is associated with impairment in odor identification even when vocabulary ability, global cognitive status and various health variables have been partialled out. The results further suggest that the ApoE gene plays a role in olfactory functioning that is independent of dementia conversion within 5 years (Olofsson, Nordin, Wiens, Hedner, Nilsson, & Larsson, 2010). As shown in Figure 2, further results from the Betula Study suggest that poor odor identification in combination with old age and ApoE-ε4 predicts large prospective global cognitive decline (Olofsson, Rönnlund, Nordin, Nyberg, Nilsson, & Larsson, 2009).

2.3.2 *Unawareness of olfactory impairment*

It is in general rather common not to be aware of an olfactory impairment. Although self-report of impairment increases with age (Nordin, Brämerson, & Bende, 2004), it seems not to increase with age to the same extent as the actual impairment (Murphy et al., 2002). Using an odor detection task, more than 70% of normal elderly who reported a normal sense of smell had in fact an olfactory

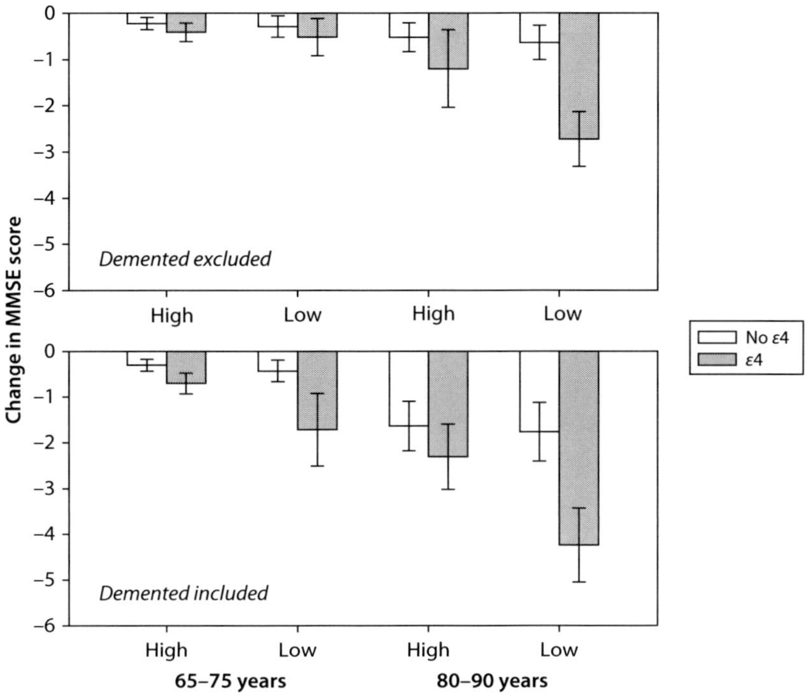

Figure 2. Change in MMSE scores (mean ± standard error) from baseline to follow-up as a function of odor identification ability (high, low) and presence of ApoE-ε4 in participants aged 65–75 and 80–90 years. [With kind permission from Springer Science: Behavior Genetics, Odor identification deficit as a predictor of five-year global cognitive change: Interactive effects with age and ApoE-ε4, vol. 39, 2009, p. 501, Olofsson, Rönnlund, Nordin, Nyberg, Nilsson, & Larsson.]

impairment (Nordin et al., 1995), and unawareness is even more common in AD (Doty et al., 1987; Nordin et al., 1995). It has been proposed that unawareness of deficits of functional abilities may be predictive of dementia in MCI (Tabert et al., 2002). In accordance with this, Devanand and associates (2000) reported that the combination of poor odor identification and unawareness of olfactory deficit predicted conversion from MCI to AD after two years with a sensitivity of 72% and a specificity of 64%. Further support for the role of unawareness in cognitive decline is provided by data from a recent study showing that unawareness of olfactory impairment is associated with poor performance in verbal learning, memory and attention/processing speed (Wehling, Nordin, Espeseth, Reinvang, & Lundervold, 2011).

2.4 Assessment of odor identification for early diagnosis of Alzheimer's disease

The need for sensitive tools for early diagnosis of AD increases as the efficacy of medication is improved to decelerate the dementia progression caused by AD. This calls for search of tools also within non-traditional domains of dementia assessment. There are several reasons for proposing that clinical evaluation of olfaction in general, and of odor identification in particular, is a strong candidate in this respect. As reviewed above, these are: (1) the neurohistological and physiological pathology that characterize AD is initiated and most severe in the regions that process olfactory information, (2) impairment is found in olfactory function of both sensory and cognitive nature, but appears to be more consistent and pronounced in the cognitive domain, in particular for identification, (3) performance in odor identification correlates with neurophysiological and histological pathology in AD, (4) impairment in olfactory cognition, and in particular in identification, seems earlier than various non-olfactory functions to be affected in the dementia progression, (5) important risk factors for AD are associated with olfactory impairment, and in particular with odor identification, (6) test of odor identification requires limited time, is non-invasive, inexpensive, and has a high CCR. The addition of ApoE-typing and assessment of awareness of a possible impairment in olfactory function may improve the diagnostic accuracy.

When including test of odor identification as a supplement to traditional assessment tools for clinical investigation of AD it is important to bear in mind that many conditions can affect olfactory functioning, and that it therefore is important to be able to exclude alternative causes of olfactory impairment. Since olfactory impairment is present in several forms of neurodegenerative diseases that can cause dementia (Kovács, 2004) there is at present not sufficient documentation to suggest procedures for olfactory testing to differentially diagnose between these conditions. Future research may provide such information.

3. Conclusion

The fast growing proportion of elderly and the consequences of olfactory impairment for malnutrition, safety and poor quality of life in general makes age-related olfactory impairment a societal concern. Studies consistently show neurohistological and physiological changes in the olfactory nervous system in the normal elderly population. As a result, functional olfactory impairment of both sensory and cognition character is typically found among elderly, and includes detection sensitivity, discrimination, perceived intensity, recognition memory, identification and recall. Although aging per se is most likely to cause olfactory impairment, a vast range of medical health conditions, drugs and environmental exposures contribute to this impairment.

Sensitive tools for early diagnosis of AD become increasingly important, and the very early and pronounced brain neuropathology in olfactory regions in AD places tests of olfactory function in the spotlight. This pathology includes the presence of NP and NFT, which are the neuropathological hallmarks of AD. Documented olfactory functional impairment in AD includes detection sensitivity, discrimination, recognition memory, familiarity, and identification. Impairment in olfactory cognitive domains seems to be more pronounced than impairment in olfactory sensory domains, and more pronounced than in corresponding functions in other sensory modalities. Several risk factors for AD have been shown to be associated with olfactory impairment: family history of AD, questionable AD, MCI, the ApoE-ε4 allele, and unawareness of the impairment. There is reason to believe that clinical evaluation of odor identification is an appropriate supplement to traditional assessment tools for early diagnosis of AD.

References

Attems, J., Lintner, F., & Jellinger, K. A. (2005). Olfactory involvement in aging and Alzheimer's disease: An autopsy study. *Journal of Alzheimers Disease, 7,* 149–157.

Bacon Moore A., Paulsen J. S., & Murphy, C. (1999). A test of odor fluency in patients with Alzheimer's and Huntington's disease. *Journal of Clinical and Experimental Neuropsychology, 21,* 341–351.

Bahtnagar, K. P., Kennedy, R. C., Baron, G., & Greenberg, R. A. (1987). Number of mitral cells and the bulb volume in the aging human olfactory bulb: A quantitative morphological study. *Annatomical Records, 218,* 73–87.

Bondi, M. W., Monsch, A. U., Galasko, D., Butters, N., Salmon, D. P., & Delis, D. C. (1994). Preclinical cognitive markers of dementia of the Alzheimer type. *Neuropsychology, 8,* 374–384.

Braak, H., & Braak, E. (1992). The human entorhinal cortex: Normal morphology and lamina-specific pathology in various diseases. *Neuroscience Research, 15,* 6–31.

Braak, H., & Braak, E. (1997). Frequency of stages of Alzheimer-related lesions in different age categories. *Neurobiology of Aging, 18,* 351–357.

Brämerson, A., Johansson, L., Ek, L., Nordin, S., & Bende, M. (2004). Prevalence of olfactory dysfunction: The Skövde population-based study. *Laryngoscope, 114,* 733–737.

Buchsbaum, M. S., Kesslak, J. P., Lynch, G., Chui, H., Wu, J., Sicotte, N., & Cotman, C. W. (1991). Temporal and hippocampal metabolic rate during an olfactory memory task assessed by positron emission tomography in patients with dementia of the Alzheimer type and controls. Preliminary studies. *Archives of General Psychiatry, 48,* 840–847.

Cain, W. S., de Wijk, R. A., Nordin, S., & Nordin, M. (2008). Independence of odor quality and absolute sensitivity in a study of aging. *Chemosensory Perception, 1,* 24–33.

Cain, W. S., & Gent, J. F. (1991). Olfactory sensitivity: Reliability, generality, and association with aging. *Journal of Experimental Psychology: Human Perception & Performance, 17,* 382–391.
Cain, W. S., Reid, F., & Stevens, J. C. (1990). Missing ingredients: Aging and the discrimination of flavor. *Journal of Nutrition for the Elderly, 9,* 3–15.
Cain, W. S., Stevens, J. C., Nickou, C. M., Giles, A., Johnston, I., & Garcia-Medina, M. R. (1995). Life-span development of odor identification, learning, and olfactory sensitivity. *Perception, 24,* 1457–1472.
Calhoun-Haney, R., & Murphy, C. (2005). Apolipoprotein epsilon 4 is associated with more rapid decline in odor identification than in odor threshold or dementia rating scale scores. *Brain and Cognition, 58,* 178–182.
Cerf-Ducastel, B., & Murphy, C. (2003). FMRI brain activation in response to odors is reduced in primary olfactory areas of elderly subjects. *Brain Research, 986,* 39–53.
Corder, E. H., Saunders, A. M., Strittmatter, W. J., Schmechel, D. E., Gaskell, P. C., Small, G. W., & Pericak-Vance, M. A. (1993). Gene dose of apolipoprotein E type 4 allele and the risk of Alzheimer's disease in late onset families. *Science, 261,* 921–923.
Davies, D. C., Brooks, J. W., & Lewis, D. A. (1993). Axonal loss from the olfactory tracts in Alzheimer's disease. *Neurobiology of Aging, 14,* 353–357.
Devanand, D. P., Michaels-Marston, K. S., Liu, X., Pelton, G. H., Padilla, M., Marder, K., & Mayeux, R. (2000). Olfactory deficits in patients with mild cognitive impairment predict Alzheimer's disease at follow-up. *American Journal of Psychiatry, 157,* 1399–1405.
Djordjevic, J., Jones-Gotman, M., De Sousa, K., & Chertkow, H. (2007). Olfaction in patients with mild cognitive impairment and Alzheimer's disease. *Neurobiology of Aging, 29,* 693–706.
Doty, R. L., Perl, D. P., Steele, J. C., Chen, K. M., Pierce, J. D., Reyes, P., & Kurland, L. T. (1991). Olfactory dysfunction in three neurodegenerative diseases. *Geriatrics, 46 (Suppl. 1),* 47–51.
Doty, R. L., Reyes, P. F., & Gregor, T. (1987). Presence of both identification and detection deficits in Alzheimer's disease. *Brain Research Bulletin, 18,* 597–600.
Doty, R. L., Shamam, P., Applebaum, S. L., Giberson, R., Siksorski, L., Rosenberg, L. (1984). Smell identification ability: Changes with age. *Science, 226,* 1441–1443.
Duffy, V. B., Backstrand, J., & Ferris, A. (1995). Olfactory dysfunction and related nutritional risk in free-living, elderly women. *Journal of the American Dietary Assocication, 95,* 879–884.
Duffy, V., Cain, W. S., & Ferris, A. (1999). Measurement of sensitivity to olfactory flavor: Application in a study of aging and dentures. *Chemicl Senses, 24,* 671–677.
Enwere, E., Shingo, T., Gregg, C., Fujikawa, H., Ohta, S., & Weiss, S. (2004). Aging results in reduced epidermal growth factor receptor signaling, diminished olfactory neurogenesis, and deficits in fine olfactory discrimination. *Journal of Neuroscience, 24,* 8354–8365.
Feldman, J. I., Murphy, C., Davidson, T. M., Jalowayski, A. A., & Galindo de Jaime, G. (1991). The rhinologic evaluation of Alzheimer's disease. *Laryngoscope, 101,* 1198–1202.
Finkel, D., Reynolds, C. A., Larsson, M., Gatz, M., & Pedersen, N. L. (2011). Both odor identification and ApoE-ε4 contribute to normative cognitive aging. *Psychology and Aging.* Advance online publication. doi: 10.1037/a0023371.
Finkelstein, J. A., & Schiffman, S. S. (1999). Workshop on taste and smell in the elderly: An overview. *Physiology & Behavior, 66,* 173–176.

Fusetti, M., Fioretti, A. B., Silvagni, F., Simaskou, M., Sucapane, P., Necozione, S., & Eibenstein, A. (2010). Smell and preclinical Alzheimer disease: Study of 29 patients with amnesic mild cognitive impairment. *Journal of Otolaryngology-Head and Neck Surgery, 39,* 175–181.

Genter, M. B., & Ali, S. F. (1988). Age-related susceptibility to 3,3'-iminodipropionitrile-induced olfactory mucosal damage. *Neurobiology of Aging, 19,* 569–574.

Gilbert, P. E., Pirogovsky, E., Ferdon, S., & Murphy, C. (2006). The effects of normal aging on source memory for odors. *Journal of Gerontology: Psychological Sciences, 61B,* 58–60.

Graves, A. B., Bowen, J. D., Rajaram, L., McCormick, W. C., McCurry, S. M., Schellenberg, G. D., & Larson, E. B. (1999). Impaired olfaction as a marker for cognitive decline: Interaction with apolipoprotein E epsilon4 status. *Neurology, 53,* 1480–1487.

Handley, O. J., Morrison, C. M., Miles, C., & Bayer, A. J. (2006). ApoE gene and familial risk of Alzheimer's disease as predictors of odour identification in older adults. *Neurobiology of Aging, 27,* 1425–1430.

Hastings, L., & Miller, M. L. (2003). Influence of environmental toxicants on olfactory function. In Doty, R. L (Ed.), *Handbook of olfaction and gustation* (pp. 575–592). New York: Marcel Dekker.

Hedner, M., Larsson, M., Arnold, N., Zucco, G. M., & Hummel, T. (2010). Cognitive factors in odor detection, odor discrimination, and odor identification tasks. *Journal of Clinical and Experimental Neuropsychology, 32,* 1062–1067.

Hummel, T., Barz, S., Pauli, E., & Kobal, G. (1998). Chemosensory event-related potentials change as a function of age. *Electroencephalography and Clinical Neurophysiology, 108,* 208–217.

Hummel, T., & Nordin, S. (2005). Olfactory disorders and their consequences for quality of life: A review. *Acta Oto-Laryngologica, 125,* 116–121.

Hummel, T., Sekinger, B., Wolf, S. R., Pauli, E., & Kobal, G. (1997). 'Sniffin' sticks': Olfactory performance assessed by the combined testing of odor identification, odor discrimination and olfactory threshold. *Chemical Senses, 22,* 39–52.

Hyman, B. T. (1997). The neuropathological diagnosis of Alzheimer's disease: Clinical-pathological studies. *Neurobiology of Aging, 18,* S27–32. *Chemical Senses,* 22, 39–52.

Kaneda, H., Maeshima, K., Goto, N., Kobayakawa, T., Ayabe-Kanamura, S., & Saito, S. (2000). Decline in taste and odor discrimination abilities with age, and relationship between gustation and olfaction. *Chemical Senses, 25,* 331–337.

Kareken, D. A., Doty, R. L., Moberg, P. J., Mosnik, D., Chen, S. H., Farlow, M. R., & Hutchins, G. D. (2001). Olfactory-evoked regional cerebral blood flow in Alzheimer's disease. *Neuropsychology, 15,* 18–29.

Kesslak, J. P., Cotman, C. W., Chui, H. C., van den Noort, S., Fang, H., Pfeffer, R., & Lynch, G. (1988). Olfactory tests as possible probes for detecting and monitoring Alzheimer's disease. *Neurobiology of Aging, 9,* 399–403.

Kesslak, J. P., Nalcioglu, O., & Cotman, C. W. (1991). Quantification of magnetic resonance scans for hippocampal and parahippocampal atrophy in Alzheimer's disease. *Neurology, 41,* 51–54.

Knupfer, L., & Spiegel, R. (1986). Differences in olfactory test performance between normal aged, Alzheimer and vascular type dementia individuals. *International Journal of Geriatric Psychiatry, 1,* 3–14.

Koss, E. (1986). Olfactory dysfunction in Alzheimer's disease. *Developmental Neuropsychology, 2,* 89–99.

Kovács, T. (2004). Mechanisms of olfactory dysfunction in aging and neurodegenerative disorders. *Ageing Research Reviews, 3,* 215–232.

Laakso, M. P., Tervo, S., Hänninen, T., Vanhanen, M., Hallikainen, M., & Soininen, H. (2009). Olfactory identification in non-demented elderly population and in mild cognitive impairment: A comparison of performance in clinical odor identification versus Boston Naming Test. *Journal of Neural Transmission, 116,* 891–895.

Larson, M. (1997). Semantic factors in episodic recognition of common odors in early and late adulthood: A review. *Chemical Senses, 22,* 623–633.

Larsson, M., & Bäckman, L. (1993). Semantic activation and episodic odor recognition in young and older adults. *Psychology and Aging, 8,* 582–588.

Larsson, M., & Bäckman, L. (1997). Age-related differences in episodic odour recognition: The role of access to specific odour names. *Memory, 5,* 361–378.

Larsson, M., Nilsson, L.-G., Olofsson, J., & Nordin, S. (2004). Demographic and cognitive predictors of cued odor identification: Evidence from a population-based study. *Chemical Senses,* 29, 547–554.

Larsson, M., Semb, H., Winblad, B., Amberl, K., Wahlund, L.-O., & Bäckman, L. (1999). Odor identification in normal aging and early Alzheimer's disease: Effects of retrieval support. *Neuropsychology, 13,* 47–53.

Lehrner, J. P., Brücke, T., Dal-Bianco, P., Gatterer, G., & Kryspin-Exner, I. (1997). Olfactory functions in Parkinson's disease and Alzheimer's disease. *Chemical Senses, 22,* 105–110.

Makowska, I., Kloszewska, I., Grabowska, A., Szatkowska, I., & Rymarczyk, K. (2011). Olfactory deficits in normal aging and Alzheimer's disease in the polish elderly population. *Archives of Clinical Neuropsychology, 26,* 270–279.

Moberg, P. J., Doty, R. L., Mahr, R. N., Mesholam, R. I., Arnold, S. E., Turetsky, B. I., & Gur, R. E. (1997). Olfactory identification in elderly schizophrenia and Alzheimer's disease. *Neurobiology of Aging, 18,* 163–167.

Moberg, P. J., Pearlson, G. D., Speedie, L. J., Lipsey, J. R., Strauss, M. E., & Folstein, S. E. (1987). Olfactory recognition: Differential impairments in early and late Huntington's and Alzheimer's disease. *Journal of Clinical and Experimental Neuropsychology, 9,* 650–664.

Morgan, C. D., Geisler, M. W., Covington, J. W., Polich, M., & Murphy, C. (1999). Olfactory P3 in young and older adults. *Psychophysiology, 36,* 281–287.

Morgan, C. D., & Murphy, C. (2002). Olfactory event-related potentials in Alzheimer's disease. *Journal of the International Neuropsychological Society, 8,* 753–763.

Morgan, C. D., Nordin, S., & Murphy, C. (1995). Odor identification as an early marker for Alzheimer's disease: Impact of lexical functioning and detection sensitivity. *Journal of Clinical and Experimental Neuropsychology, 17,* 793–803.

Murphy, C. (1985). Cognitive and chemosensory influences on age-related changes in the ability to identify blended foods. *Journal of Gerontology, 40,* 47–52.

Murphy, C., Baco, A. W., Bondi, M. W., & Salmon, D. P. (1998). Apolipoprotein status is associated with odor identification deficits in nondemented older persons. *Annals of the New York Academy of Sciences, 855,* 744–750.

Murphy, C., Cain, W. S., Gilmore, M. M., & Skinner, R. B. (1991). Sensory and semantic factors in recognition memory for odors and graphic stimuli: elderly versus young persons. *American Journal of Psychology, 104,* 161–192.

Murphy, C., Doty, R. L., & Duncan, H. J. (2003). Clinical disorders of olfaction. In R. L. Doty (Ed.), *Handbook of olfaction and gustation* (pp. 461–478). New York: Marcel Dekker.

Murphy, C., Gilmore, M. M., Seery, C. S., Salmon, D. P., & Lasker, B. R. (1990). Olfactory thresholds are associated with degree of dementia in Alzheimer's disease. *Neurobiology of Aging, 11,* 465–469.
Murphy, C., Jernigan, T. L., & Fennema-Notestine, C. (2003). Left hippocampal volume loss in Alzheimer's disease is reflected in performance on odor identification: A structural MRI study. *Journal of the International Neuropsychologicl Society, 9,* 459–471.
Murphy, C., Morgan, C. D., Geisler, M. W., Wetter, S., Covington, J. W., Madowitz, M. D., & Polich, J. M. (2000). Olfactory event-related potentials and aging: Normative data. *International Journal of Psychophysiology, 36,* 133–145.
Murphy, C., Nordin, S., & Acosta, L. (1997). Odor learning, recall, and recognition memory in young and elderly adults. *Neuropsychology, 11,* 126–137.
Murphy, C., Nordin, S., & Jinich, S. (1999). Very early decline in recognition memory for odors in Alzheimer's disease. *Aging, Neuropsychology, & Cognition, 6,* 229–240.
Murphy, C., Nordin, S., de Wijk, R. A., Cain, W. S., & Polich, J. (1994). Olfactory evoked potentials: Assessment of young and elderly, and comparison to psychophysical threshold. *Chemical Senses, 19,* 47–56.
Murphy, C., Schubert, C. R., Cruickshanks, K. J., Klein, B. E., Klein, R., & Nondahl, D. M. (2002). Prevalence of olfactory impairment in older adults. *Journal of the American Medical Association, 288,* 2307–2312.
Namba, Y., Tomonaga, M., Kawasaki, H., Otomo, E., & Ikeda, K. (1991). Apolipoprotein E immunoreactivity in cerebral amyloid deposits and neurofibrillary tangles in Alzheimer's disease. *Brain Research, 541,* 163–166.
Niccoli-Waller, C. A., Harvey, J., Nordin, S., & Murphy, C. (1999). Remote odor memory in Alzheimer's disease: Deficits as measured by familiarity. *Journal of Adult Development, 6,* 131–136.
Nordin, S., Almkvist, O., & Berglund, B. Odor detectability is not impaired in successfully-aged elderly. *Chemosensory Perception.* DOI. 10.1007/s12078-011-9102-8.
Nordin, S., Almkvist, O., Berglund, B., & Wahlund, L.-O. (1997). Olfactory dysfunction for pyridine and dementia progression in Alzheimer's disease. *Archives of Neurology, 54,* 993–998.
Nordin, S., & Brämerson, A. (2008). Complaints of olfactory disorders: Epidemiology, assessment and clinical implications. *Current Opinion in Allergy and Clinical Immunology, 8,* 10–15.
Nordin, S., Brämerson, A., & Bende, M. (2004). Prevalence of self-reported poor odor-detection sensitivity: The Skövde population-based study. *Acta Oto-Laryngologica, 124,* 1171–1173.
Nordin, S., Monsch, A. U., & Murphy, C. (1995). Unawareness of smell loss in normal aging and Alzheimer's disease: Discrepancy between self-reported and diagnosed smell sensitivity. *Journal of Gerontology: Psychological Sciences, 50B,* P187–P192.
Nordin, S., & Murphy, C. (1996). Impaired sensory and cognitive olfactory function in questionable Alzheimer's disease. *Neuropsychology, 10,* 113–119.
Olofsson, J. K., Rönnlund, M., Nordin, S., Nyberg, L., Nilsson, L.-G., & Larsson, M. (2009). Odor identification deficit as a predictor of five-year global cognitive change: Interactive effects with age and ApoE-ε4. *Behavior Genetics, 39,* 496–503.
Olofsson, J. K., Nordin, S., Wiens, S., Hedner, M., Nilsson, L.-G., & Larsson, M. (2010). Odor identification impairment in carriers of ApoE-ε4 is independent of clinical dementia. *Neurobiology of Aging, 31,* 567–577.

Peabody, C. A., & Tinklenberg, J. R. (1985). Olfactory deficits and primary degenerative dementia. *American Journal of Psychiatry, 142,* 524–525.
Price, J. L., Davis, P. B., Morris, J. C., & White, D. L. (1991). The distribution of tangles, plaques, and related immunohistochemical markers in healthy aging and Alzheimer's disease. *Neurobiology of Aging, 12,* 295–312.
Rawson, N. E., Gomez, G., Cowart, B., & Restrepo, D. (1998). The use of olfactory receptor neurons (ORNs) from biopsies to study changes in aging and neurodegenerative disease. *Annals of the New York Academy of Sciences, 855,* 701–707.
Razani, J. L., Chan, A., Nordin, S., & Murphy, C. (2010). Semantic networks for odors and colors in Alzheimer's disease. *Neuropsychology, 24,* 291–299.
Reyes, P. F., Deems, D. A., & Suarez, M. G. (1993). Olfactory-related changes in Alzheimer's disease: A quantitative neuropathologic study. *Brain Research Bulletin, 32,* 1–5.
Rezek, D. L. (1987). Olfactory deficits as a neurologic sign in dementia of the Alzheimer type. *Archives of Neurology, 44,* 1030–1032.
Savic, I., Gulyas, B., Larsson, M., & Roland, P. (2000). Olfactory functions are mediated by parallel and hierarchical processing. *Neuron, 26,* 735–745.
Schemper, T., Voss, S., & Cain, W. S. (1981). Odor identification in young and elderly persons: Sensory and cognitive limitations. *Journal of Gerontology, 36,* 446–452.
Schiffman, S. S. (1991). Drugs influencing taste and smell perception. In T. V. Getchell, L. M. Bartoshuk, R. L. Doty, & J. B. Snow, Jr. (Eds.), *Smell and Taste in Health and Disease* (pp. 845–50). New York: Raven Press.
Schiffman, S. S. (1997). Taste and smell losses in normal aging and disease. *Journal of the American Medical Association, 278,* 1357–1362.
Schiffman, S. S., Clark, C. M., & Warwick, Z. S. (1990). Gustatory and olfactory dysfunction in dementia: Not specific to Alzheimer's disease. *Neurobiology of Aging, 11,* 597–600.
Schiffman, S. S., & Leffingwell, J. C. (1981). Perception of odors of simple pyrazines by young and elderly subjects: A multidimensional analysis. *Pharmacology, Biochemistry and Behavior, 14,* 787–798.
Schiffman, S. S., Moss, J., & Erickson, R. P. (1976). Thresholds of food odors in the elderly. *Experimental Aging Research, 2,* 389–398.
Schubert, C. R., Carmichael, L. L., Murphy, C., Klein, B. E., Klein, R., & Cruickshanks, K. J. (2008). Olfaction and the 5-year incidence of cognitive impairment in an epidemiological study of older adults. *Journal of the American Geriatric Society, 56,* 1517–1521.
Seiberling, K. A., & Conley, D. B. (2004). Aging and olfactory and taste function. *Otolaryngological Clinics of North America, 37,* 1209–1228.
Serby, M., Larson, P., & Kalkstein, D. (1991). The nature and course of olfactory deficits in Alzheimer's disease. *American Journal of Psychiatry, 148,* 357–360.
Serby, M., Mohan, C., Aryan, M., Williams, L., Mohs, R. C., & Davis, K. L. (1996). Olfactory identification deficits in relatives of Alzheimer's disease patients. *Biological Psychiatry, 39,* 375–377.
Sing, C. F., & Davignon, J. (1985). Role of the apolipoprotein E polymorphism in determining normal plasma lipid and lipoprotein variation. *American Journal of Human Genetics, 37,* 268–285.
Stevens, D. A., & Lawless, H. T. (1981). Age-related changes in flavor perception. *Appetite, 2,* 127–136.
Stevens, J. C., & Cain, W. S. (1985). Age-related deficiency in the perceived strength of six odorants. *Chemical Senses, 10,* 517–529.

Stevens, J. C., & Cain, W. S. (1986). Smelling via the mouth: Effect of aging. *Perception & Psychophysics, 40,* 142–146.
Stevens, J. C., & Cain, W. S. (1987). Old-age deficits in the sense of smell as gauged by thresholds, magnitude matching, and odor identification. *Psychology and Aging, 2,* 36–42.
Stevens, J. C., & Cain, W. S. (1993). Changes in taste and flavor in aging. *Critical Reviews in Food Science and Nutrition, 33,* 27–37.
Stevens, J. C., Cain, W. S., Schiet, F. T., & Oatley, M. W. (1989). Olfactory adaptation and recovery in old age. *Perception, 18,* 265–276.
Stevens, J. C., Cain, W. S., & Weinstein, D. E. (1987). Aging impairs the ability to detect gas odor. *Fire Technology, 23,* 198–204.
Swedish National Food Administration (1998). *Mat och kostbehandling för äldre: Problem och möjligheter.* Uppsala: Livsmedelsverket.
Tabert, M. H., Albert, S. M., Borukhova-Milov, L., Camacho, Y., Pelton, G., Liu, X., & Devanand, D. P. (2002). Functional deficits in patients with mild cognitive impairment: Prediction of AD. *Neurology, 58,* 758–764.
Tsuboi, Y., Wszolek, Z. K., Graff-Radford, N. R., Cookson, N., & Dickson, D. W. (2003). Tau pathology in the olfactory bulb correlates with Braak stage, Lewy body pathology and apolipoprotein epsilon4. *Neuropathology and Applied Neurobiology, 29,* 503–510.
Vellas, B., Lauque, S., Andrieu, S., Nourhashemi, F., Rolland, Y., Baumgartner, R., & Garry, P. (2001). Nutrition assessment in the elderly. *Current Opinion in Clinical Nutrition & Metabolic Care, 4,* 5–8.
Wang, J., Eslinger, P. J., Smith, M. B., & Yang, Q. X. (2005). Functional magnetic resonance imaging study of human olfaction and normal aging', *Journal of Gerontol: Biological Sciences and Medical Sciences, 60,* 510–514.
Warner, M. D., Peabody, C. A., Flattery, J. J., & Tinklenberg, J. R. (1986). Olfactory deficits and Alzheimer's disease. *Biological Psychiatry, 21,* 116–118.
Wehling, E. I., Nordin, S., Espeseth, T., Reinvang, I., & Lundervold, A. J. (2010). Familiarity, cued and free odor identification and their association with cognitive functioning in middle aged and older adults. *Neuropsychology, Development, and Cognition: Section B, Aging, Neuropsychology and Cognition, 17,* 205–219.
Wehling, E. I., Nordin, S., Espeseth, T., Reinvang, I., & Lundervold, A. J. (2011). Unawareness of olfactory dysfunction and its association with cognitive functioning in middle aged and old adults. *Archives of Clinical Neuropsychology, 26,* 260–269.
Wetter, S., & Murphy, C. (2001). Apolipoprotein E epsilon4 positive individuals demonstrate delayed olfactory event-related potentials. *Neurobiology of Aging, 22,* 439–447.
Wilson, R. S., Arnold, S. E., Schneider, J. A., Tang, Y., & Bennett, D. A. (2007). The relationship between cerebral Alzheimer's disease pathology and odour identification in old age. *Journal of Neurology Neurosurgery and Psychiatry, 78,* 30–35.
Wysocki, C. J., & Gilbert, A. N. (1989). National Geographic Smell Survey: Effects of age are heterogenous. *Annals of the New York Academy of Sciences, 561,* 12–28.
Zucco, G. M. (2011). Olfactory performance assessed via a new odour recognition test: Reliability and normative data. *Journal of Cognitive Pychology, 23,* 1–7.

PART IV

Odor hedonic perception and development

CHAPTER 12

The psychophysics of olfaction in the human newborn

Habituation and cross-adaptation

Lewis P. Lipsitt and Carolyn Rovee-Collier
Brown University, Providence, USA / Rutgers University, Piscataway, USA

Until the early 1960s, little was known about the human newborn's ability to detect odors. At that time, the modern study of newborn olfaction was introduced by Trygg Engen in collaboration with Lewis Lipsitt, colleagues in the Psychology Department of Brown University. In two seminal studies, they demonstrated that newborns could not only detect pure olfactory stimuli, as opposed to odorous trigeminal irritants, but also could habituate to them over repeated presentations. These studies introduced numerous now-standard control procedures and compound stimuli to the habituation literature.

In another series of experiments on newborn olfactory psychophysics, infants directly scaled the subjective intensity of a $\log_2$ dilution series of concentrations of five members of an homologous series of aliphatic alcohols, which differ only in the number of carbons in their chemical chains. Newborn's data were described by power functions that were inversely related to chain length. These functions permitted the role of chain length in newborn sensory adaptation to be examined. Two alcohol pairs, each composed of different chain lengths, were formed from concentrations that were equated for magnitude of responding within pairs. Members of each pair served as both the adapting stimulus and the test stimulus after two adapting intervals. Newborns exhibited either selective cross-adaptation or selective cross-facilitation, depending on the physical properties of odorants with different chain lengths, their order of presentation, and the adapting interval within each odor pair.

This body of work exemplifies Engen's innovations in the science of olfaction with the human newborn. The systematic studies of Engen and his colleagues were the first to document the newborn's highly developed and acute sense of smell and the significant effect of olfactory experience on newborn learning.

Introduction

In the early 1960s, little was known about the human newborn's sense of smell. Summarizing the state of knowledge, Peiper (1963: 97) wrote, "many newborns respond to odors by turning toward them but more often with defensive movements. It has not been proved that they can find the mother's nipple with the aid of the olfactory sense. At seven to nine months conditioned reflexes to certain odors can be obtained." Even in the early 1970s, a popular introductory text reiterated earlier statements that the newborn's sense of smell was rudimentary at best (Morgan & King, 1972). Yet, nothing could have been further from the truth. The picture had changed dramatically when Trygg Engen and Lewis Lipsitt, colleagues at Brown University, initiated the modern study of newborn olfaction a decade earlier.

Lipsitt had recently founded the Newborn Sensory and Conditioning Laboratory at the Providence Lying-In Hospital,[1] where their research took place. The laboratory was equipped with a stabilimeter bed and associated activity integrator, a cardiotachometer, and a multi-channel polygraph that printed a continuous record of newborn behaviors and stimulus presentations on a moving time base (see Figure 1; Lipsitt & DeLucia, 1960). A rabbit-sized pneumobelt that was strapped around the infant's abdomen (see Figure 2) and associated with a pressure transducer provided the concurrent polygraphic record of breathing. Before testing, infants were swaddled to minimize spontaneous activity, and the sensitivity of the stabilimeter was weight-calibrated for each infant. This equipment enabled researchers to undertake the first systematic studies of sensory (absolute) thresholds, habituation, and conditioning.

Over the course of their collaboration, Engen and Lipsitt introduced innovative and objective methodology that has since become standard in studies of habituation in all modalities with newborns and older infants. First, they obtained permanent, automated and time-based records of concurrent multiple responses (e.g., breathing and stabilimeter oscillations, cardiac responses, leg withdrawal), their integrated (cumulative) signals, and stimulus presentations. Second, they controlled for changes in unlearned activity and random startles by alternating sham stimulus presentations with presentations of the odorant. Third, they controlled for infant state by requiring closed eyes, regular and steady respiration, and minimum stabilimeter activity prior to the initiation of each trial. Fourth, they introduced posttest trials with a different odorant. Fifth, they used independent judges to evaluate infants' behavioral records and reported interobserver

1. Now called The Hospital for Mothers and Infants.

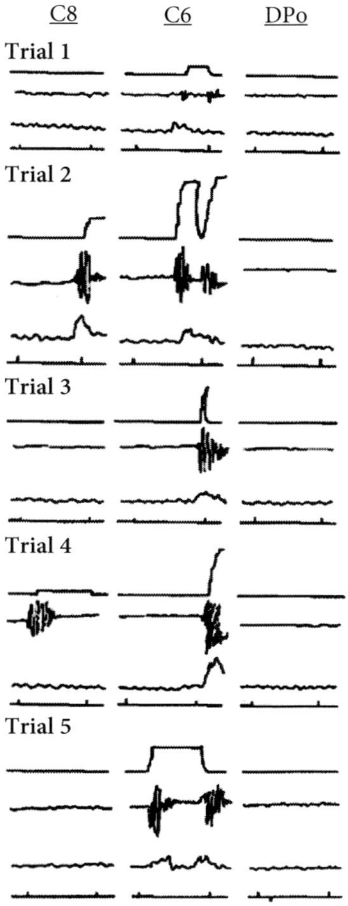

Figure 1. Sample record from a Grass S5 Polygraph, run at a speed of 5 mm/s, showing a newborn's response over five trials to a 25%-C6 (hexanol) stimulus that was presented 30 s after a 100%-C8 (octanol) stimulus. The third stimulus on each trial is diethyl phthlate (DP_0), the odorless diluent used as a control stimulus. The top line on each trial is the integrated stabilimeter activity of line 2; line 3 shows respiration; and line 4 shows the onset and offset of each 10-s stimulus. Intertrial intervals are not shown.
(From Rovee, 1972.)

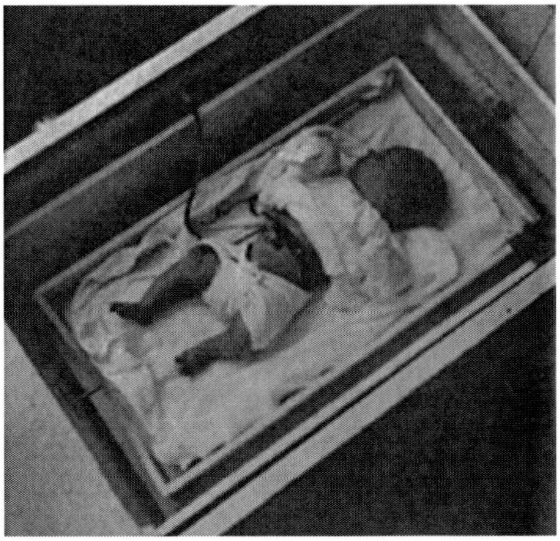

Figure 2. A newborn with a pneumobelt in the stabilimeter bed.

reliabilities. Sixth, they habituated infants to a compound stimulus and afterwards tested infants with each component. And seventh, they matched newborns' magnitude of responding to different odorants and then tested newborns for equivalent habituation and posttest responding to them.

1. Developmental changes in newborn olfactory thresholds

Newborn sensory thresholds are usually high at birth but decline at different rates over the first 4 days of life. Early researchers had concluded that olfactory thresholds do not change over the first 10 postnatal days, but they had used odorants that were either ungraded in intensity or graded in only four steps. Lipsitt, Engen, and Kaye (1963) reexamined this conclusion with 10 infants who were repeatedly tested on four successive days with an ascending $\log_2$ dilution series of seven concentrations of tincture asafoetida and a control stimulus (the odorless diluent, diethyl phthalate). Stimuli were presented for 10 s at approximately 1-min intervals, with the weakest intensity presented first. If an infant did not respond to the weakest concentration, then the next strongest concentration was presented and so forth until the infant responded to a concentration for the first time. This concentration defined that infant's absolute threshold on that day. They found that the absolute threshold for asafoetida, like absolute thresholds in other sensory modalities (e.g., Lipsitt & Levy, 1959), decreased dramatically over the first few days of life and was essentially stable by 30 h postpartum. As a result, an age criterion of >30 h was established for participation in studies of human newborn olfaction.

The fact that olfactory absolute thresholds are among the first to stabilize after birth reflects an evolutionarily selected, survival-related benefit of an acute sense of smell that develops in utero (Mennella, Johnson, & Beauchamp, 1995). Its functional significance includes maternal recognition (Porter, Balogh, & Makin, 1988; Sullivan, Taborsky-Barba, Mendoza, Itano, Leon, Cotman, Payne, & Lott, 1991), the facilitation of learning and long-term memory (Rubin, Carroll, & Fagen, 1998; Schroers, Prigot, & Fagen, 2007; Wigal, Kucharski, & Spear, 1984), and the establishment of long-term conditioned preferences and aversions (Delaunay-El Allam, Soussignan, Patris, Marlier, & Schaal, 2010; Schaal, Marlier, & Soussignan, 1998; Sevelinges, Mouly, Raineki, Moriceau, Forest, & Sullivan, 2011; Shionoya, Moriceau, Lunday, Miner, Roth, & Sullivan, 2006).

2. Newborn habituation to odorants

Habituation is widely considered to be the most primitive form of learning. It is a nonassociative, stimulus-specific decrement in elicited responding over repeated presentations of a stimulus that is inconsequential (i.e., nonpredictive). Because the decline in responding is specific to the eliciting stimulus, the greater the difference between infants' magnitude of responding on posthabituation tests with both a physically different stimulus and the original habituating stimulus, the greater is the subjective dissimilarity between them. Engen and Lipsitt exploited the habituation/posttest paradigm to assess the newborn's sense of smell.

Their general procedure was the same in all studies unless otherwise indicated. Trials lasted 10 s and were 60 s apart. Stimuli were presented on a Q-tip held 5 mm below the nostrils. A control trial with a clean (dry) Q-tip preceded each experimental trial with a liquid odor-saturated Q-tip for 20 trials total. The control trial provided a running baseline for judging responding on the succeeding experimental trial. In experiments that infant exhibited a response decrement, they received a posttest with all or part of the original odorant.

In an initial study, Engen, Lipsitt, and Kaye (1963) asked whether newborn infants could (1) respond to pure olfactory stimuli, and (2) habituate to a pure olfactory stimulus over the course of repeated presentations. Disher (1934) had originally studied newborn olfactory habituation, but she did not distinguish between stimulation of the olfactory nerve and the trigeminal nerve. Although both nerves respond simultaneously to olfactory stimuli, only the olfactory nerve mediates the sense of smell; the trigeminal nerve mediates the tactile sensation that accompanies the inhalation of odorous irritants. In Disher's study, infants might have exhibited trigeminal nerve response to higher stimulus intensities, olfactory nerve response to lower stimulus intensities, or olfactory nerve response only.

In Experiment 1, therefore, infants received a series of habituation trials with each of two odors – acetic acid (an irritant) and phenylethyl alcohol (a pure odor), counterbalanced for order. They responded to both odorants but failed to habituate to either. Nonetheless, Experiment 1 answered their first question: *Newborns can respond to a pure odor*. In Experiment 2, a parallel experiment, infants' responses to both asafoetida (a pure odor) and anise (an irritant) decreased significantly over trials. Infants also exhibited an order effect, responding less to both odorants during the second habituation series than during the first series (i.e., a cross-adaptation effect). After completing both habituation series, each order group renewed responding during a posttest with the odorant they had received in the first series. Whether the response decrement was due to habituation (learning in the central nervous system) or to sensory adaptation (decreased receptor sensitivity at the periphery), however, was still unclear. The odorants had differed in subjective intensity, physical and chemical properties, trigeminal contribution, receptor sites, and degree of cross-adaptation.

In another attempt to determine whether newborn infants could habituate to repeated presentations of a pure olfactory stimulus, Engen and Lipsitt (1965, Experiment 1) habituated newborns to a compound stimulus, a 50%–50% asafoetida-anise mixture (pure odors), and then gave them a posttest with one of its components – either 50% asafoetida or 50% anise – plus 50% diethyl phthalate (an odorless diluent). They reasoned that if sensory adaptation rather than habituation accounts for the response decrement, then newborns will not respond to a component of the compound during the posttest. Infants' responding recovered significantly to asafoetida but not to anise, indicating that they perceived 50% anise – but not 50% asafoetida – as highly similar to the mixture. Adults psychophysically scaled the similarity between each test stimulus and the mixture and also reported that anise was subjectively more similar to the original compound than asafoetida. Because a component of the original mixture had reinstated responding during the posttest, Engen and Lipsitt tentatively concluded that sensory adaptation was not responsible for the response decrement. Their earlier finding that newborns' responding to anise was reduced by prior exposure to asafoetida, however, was troubling.

In Experiment 2, Engen and Lipsitt (1965) habituated independent groups of newborns to a compound consisting of two components that were equated for subjective intensity – 33.3% amyl acetate and 16.7% heptanal; diethyl phthalate was the diluent. During the posttest, infants responded more to the 33.3% amyl acetate component (see Figure 3a) than to the 16.7% heptanal component (see Figure 3b), indicating that they perceived heptanal as more similar to the mixture than amyl acetate. Adults who psychophysically scaled the similarity between each posttest stimulus and the mixture also judged heptanal as more similar to the mixture than amyl acetate, cross-validating the infant data. This experiment answered their second question: *Newborn infants are capable of olfactory habituation.*

3. Direct psychophysical scaling of stimulus intensity in newborns

In place of the laborious and time-consuming traditional psychophysical methods, Engen had introduced a new, direct psychophysical scaling method (Engen, Cain, & Rovee, 1968; Stevens, 1955) to the study of newborn olfaction. By this method, the infant's magnitude of response (e.g., stabilimeter activity) to an odorant was measured on a ratio scale, and the simple average of infants' responses to

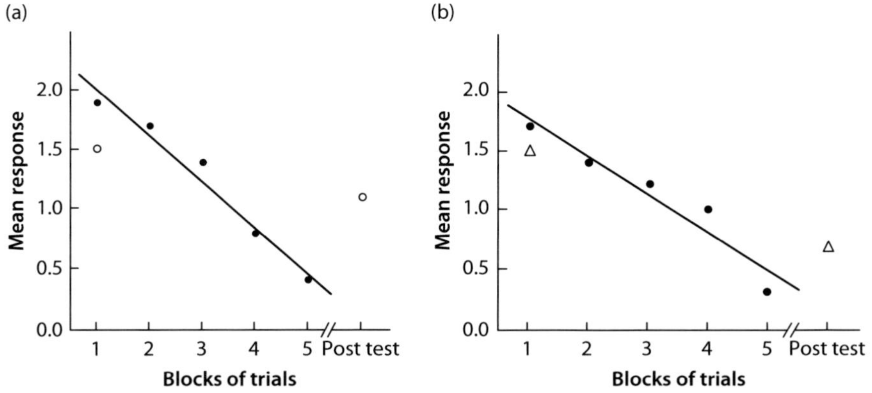

Figure 3. Left panel, Figure 3a: Average number of responses to a mixture of 16.7% heptanal and 33% amyl acetate over five blocks of four trials each (*filled circles*) and a posttest with 33% amyl acetate (*open circle*). An independent group provided the basal response to amyl acetate (*open circle, trial block 1*). Right panel, Figure 3b: Average number of responses to a mixture of 16.7% heptanal and 33% amyl acetate over five blocks of four trials each (*filled circles*) and a posttest with 16.7% heptanal (*open triangle*). An independent group provided the basal response to amyl acetate (*open triangle, trial block 1*). (From Engen & Lipsitt, 1965.)

a given odorant in a $\log_2$ series of concentrations was fitted by the method of least squares. Adults' numerical magnitude estimates of the same concentration series that were obtained with this method yielded a power function.

Engen (1963) tested newborns with an homologous series of aliphatic alcohols (organic compounds that differ chemically only in the number of carbon atoms that are linked in an open chain). The physical properties and subjective quality of aliphatic alcohols change gradually with increasing chain length, but their chemical and structural properties change minimally. Shorter chain alcohols, for example, are water-soluble with high vapor pressures, whereas longer chain alcohols are lipid-soluble with low vapor pressures. Also, naive adults judged shorter chain alcohols to be "cooler" and "fresher" and longer chain alcohols to be "oilier" and "heavier" (Engen, 1964).

Following the procedure of Beck, Kruger, and Calabresi (1954), Engen (1963) prepared a geometric dilution series by pipette with an odorless diluent (benzyl benzolate or diethyl phthalate), such that the strongest stimulus contained 100% of a particular odorant, the next strongest contained 50% of the odorant and 50% of the diluent, the next contained 25% of the odorant and 75% of the diluent, and so forth. Kruger, Feldzamen, and Miles (1955) had reported that adult psychophysical judgments of the intensities of an homologous series of aliphatic alcohols were correlated with alcohol chain length. He found that prior exposure to 100%

C3 and C7 (adapting stimuli) produced cross-adaptation effects on the threshold intensities of C4, C5, and C6 (test stimuli). Prior exposure to 100% C3 and C11 (adapting stimuli) produced the same effects on C4, C7, and C10 (test stimuli). Unlike Kruger et al. (1955), however, he found no relation between cross-adaptation effects and the degree of physical similarity in the stimuli as defined by chain length.

Previously, the relation between alcohol chain length and intensity was based solely on either the detection threshold concentration or a suprathreshold concentration of each alcohol. Conjecturing that detection threshold data and suprathreshold data might not be related, however, Engen (1965) measured newborns' respiratory responses to both 100% concentrations (Experiment 1) and detection threshold concentrations (Experiment 2) of a series of aliphatic alcohols ranging from C2 to C10 (except C9); diethyl phthalate (odorless) was the diluent. He found that shorter chain alcohols evoked larger behavioral responses (Experiment 1), but longer chain alcohols had lower absolute thresholds (Experiment 2). Engen had to turn to adults, however, to obtain magnitude estimates of the intensity of intermediate concentrations. Using C3 and C8 to represent short and long chain-length alcohols, respectively, he found that the slope of the psychophysical functions decreased with increasing chain length.

Engen concluded, "It is clear that one cannot generalize about the intensity of a whole range of effective stimulus concentrations with information about only thresholds or other selected values. This kind of prediction must be made on the basis of the appropriate psychophysical scale, which in the present case, ought to be determined for the infant" (Engen, 1965: 115–116).

4. Newborn psychophysical scaling of stimulus intensity

Although Engen (1965) had found that newborn infants could systematically order the intensity of aliphatic alcohols by chain length at 100% and threshold concentrations, his response measure – respiratory change versus no change – had not permitted the determination of the magnitude of infants' response to intermediate concentrations. Researchers in the Lipsitt lab had previously found that summated stabilimeter activity, calibrated for infant weight, correlated highly with respiratory change (Levin & Kaye, 1964) and permitted valid within-subjects comparisons (Rovee & Levin, 1966). Using the summated stabilimeter activity measure, therefore, Rovee (1969) determined newborns' magnitude of response to a $\log_2$ dilution series of five concentrations (100%, 50%, 25%, 12.5%, and 6.25%) of each of five aliphatic alcohols (C3, C5, C6, C8, and C10). Each newborn was tested with a single dilution series of only one chain length.

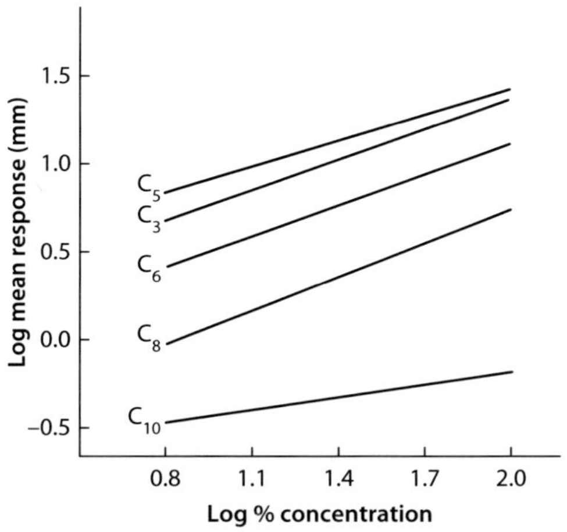

Figure 4. Power functions, corrected for absolute threshold, describing the relation between response magnitude and concentration for an homologous series of five aliphatic alcohols. (From Rovee, 1969.)

As before (Engen, 1965), Rovee found that the magnitude of response varied inversely with the number of carbons in the alcohol chain. In addition, she found that the relation between the physical and subjective intensity of the dilution series for each chain length was described by a power function, but the slopes of infants were steeper than those of adults (Engen, 1965) for corresponding alcohols. When adjusted for threshold differences, the five newborn power functions did not overlap (see Figure 4).

These data suggested that different concentrations (log % concentration; see Figure 4, *x-axis*) of different chain-length alcohols could be matched or equated for subjective intensity (log magnitude of response; see Figure 4, *y-axis*).

5. Cross-adaptation and cross-facilitation effects

When Engen (1963) had successively exposed adults to two members of an homologous series of aliphatic alcohols that differed only in chain length, the first alcohol that was exposed (the adapting stimulus) had affected both the detection threshold and suprathreshold intensities of the second alcohol (the adapting stimulus). Although he had initially hypothesized that the magnitude of cross-adaptation effects would be related to the physical similarity of the chain lengths of the adapting and test odorants, his hypothesis was not supported. Engen concluded, “Perhaps it would be more fruitful to consider cross-adaptation on the basis of scaled psychological similarity” (Engen, 1963: 102).

Rovee (1972) addressed this possibility. Her primary research question was whether newborn infants would adapt *nonselectively* or *selectively* to alternating presentations of two concentrations of different aliphatic alcohols that were matched for log magnitude of response. If infants adapted *nonselectively*, then repeated alternating presentations of the different members of each matched pair would yield the same adaptation function as successive presentations of the same member, irrespective of carbon chain length. If newborns adapted *selectively*, however, then repeated alternating presentations of the members of each matched pair would yield different cross-adaptation functions, perhaps related to alcohol chain length.

Based on the data in Figure 4, one pair of concentrations from two different shorter chain-length alcohols and a second pair from two different longer chain-length alcohols were equated for subjective intensity (log response magnitude). Twenty infants received a matched pair of 25% C3 and 100% C6, and 20 infants received a matched pair of 25% C6 and 100% C8. *Within pairs*, each odorant served as the adapting stimulus for half of the infants and as the test stimulus for the other half. Further, half of the infants in each condition received an adapting interval of 30 s; the other half received an adapting interval of 60 s.

These adapting intervals were selected to reflect differences in the recovery state of the olfactory epithelium. Electrophysiological records of the potential change in the frog olfactory organ (Ottoson, 1956) revealed complete neural recovery after 60 s but only partial neural recovery after 30 s. Engen's (1964, 1965) finding that newborns exhibited behavioral cross-adaptation after a 60-s adapting interval, however, indicated that the behavioral effects of stimulation outlast the underlying neural event. In order to capture behavioral effects during the transient neural event, a 30-s adapting interval was included.

Relatively uncontaminated normal adaptation functions were obtained for each odorant when it served as the adapting stimulus, succeeding itself at 100-s intervals. The normal functions differed from the selective cross-adaptation effects within each matched pair. Within the C3-C6 pair, normal responding to 25% C3 did not diminish over trials. When 100% C6 was the adapting stimulus, responding to 25% C3 was not affected after 30 s on trial 1, but it was depressed thereafter; after 60 s, responding to 25% C3 was depressed on all trials. Conversely, normal responding to 100% C6 decreased by trial 2, but the response depression was advanced to trial 1 after the 25% C3 adapting stimulus (see Figure 5).

In contrast, when 100% C8 was the adapting stimulus, responding to 25% C6 was *enhanced after both 30 s and 60 s* relative to its normal adaptation function. When 25% C6 was the adapting stimulus, cross-facilitation of responding to 100% C8 was not seen after either interval (see Figure 5). Thus, despite being

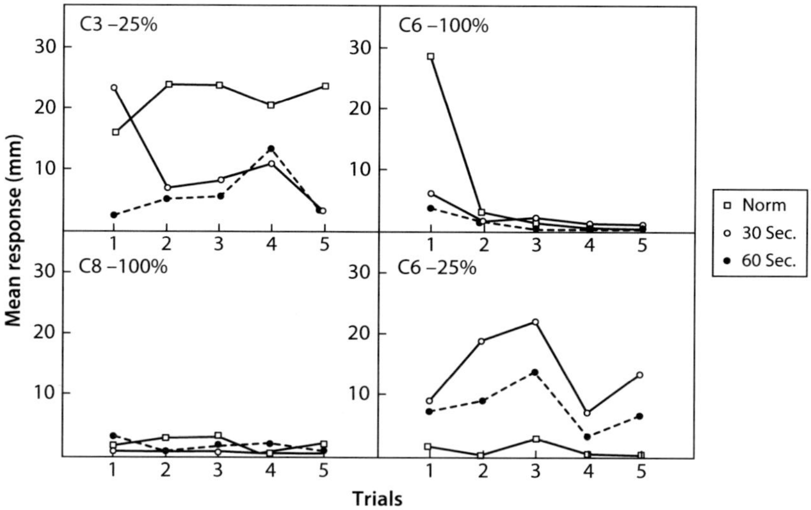

Figure 5. Normal adaptation functions (*open circles*) and cross-adaptation effects on the magnitude of response to each test concentration within each matched alcohol pair after a 30-s or 60-s adapting interval. (From Rovee, 1972.)

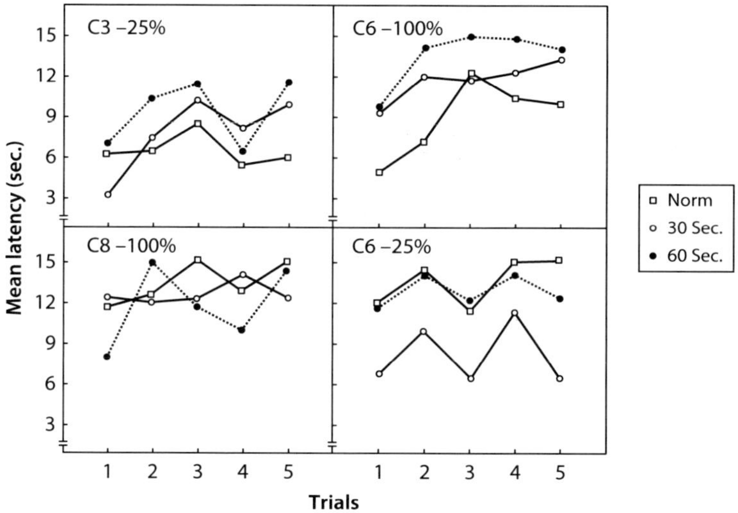

Figure 6. Normal adaptation functions (*open circles*) and cross-adaptation effects on the latency to respond to each test concentration within each matched alcohol pair after a 30-s or 60-s adapting interval. (From Rovee, 1972.)

matched for response magnitude, both the original adaptation and cross-adaptation functions of members within the pair of shorter chain-length alcohols differed, as did the original adaptation and cross-facilitation functions of members within the pair of longer chain-length alcohols.

Selective cross-adaptation and cross-facilitation effects were also obtained for response latencies (see Figure 6). Within the C3-C6 pair, response latencies over trials were longer than the normal response latencies, but response latencies within members of the C6-C8 pair did not differ from the normal response latencies.

Although the adapting interval played a major role in the results within each alcohol pair, it likely did so for different reasons. The cross-adaptation effects within the C3-C6 pair reflected the state of the olfactory epithelium after adapting intervals of 30 s and 60 s. The enhancement of responding to 25% C6 by 100% C8 at the 30-s adapting interval, however, probably reflected trigeminal involvement due to the summation of 25% C6 and residual 100% C8 (both oily stimuli) after the briefer delay. Alternatively, if the 100% C8 odorant did not persist for 30 s, then perhaps it increased the sensitivity of the trigeminal receptors, allowing the weaker C6 to stimulate them. The reverse effect of C6 on C8 would not be expected: A weak C6 is unlikely to persist throughout a 30-s adapting interval, much less throughout a 60-s adapting interval. The asymmetry between pairs of odorants in a cross-adaptation paradigm has also been documented with adults (Cain, 1970; Engen, Cain, & Rovee, 1968).

6. The adaptive role of olfactory experience for newborn mammals

Engen and his colleagues demonstrated more broadly that at birth, the human infant's sense of smell is already highly developed and exquisitely tuned. This work laid the foundation for the groundswell of ensuing research on its prenatal origins and adaptive postnatal significance.

Prenatal origins in amniotic fluid

Teicher and Blass (1977) determined that post-parturient rat dams spread amniotic fluid over their nipples, which attracts her newborn pups to attach and suckle. Attachment was disrupted by washing her nipples (Pederson & Blass, 1982; Pedersen, Steward, Greer, & Sheperd, 1983). Development of a surgical procedure that allowed the introduction of novel chemosensory stimuli directly into rats' amnion (Smotherman, 1982; Smotherman & Robinson, 1985; Stickrod, Kimble, & Smotherman, 1982) led to the first experimental demonstration of prenatal taste/odor conditioning, which was manifested 16 days after birth. The rat model of in utero learning was later found to generalize across all mammalian species, including humans.

7. Summary

Engen and his colleagues were the first to systematically document the newborn's acute sense of smell and the newborn's capacity for olfactory habituation and discrimination. These studies, conducted more than a half-century ago, are foundational for the understanding of newborn learning.

References

Beck, L. H., Kruger, L., & Calabresi, P. (1954). Observations on olfactory intensity. I. Training procedures, methods, and data for two aliphatic homologous series. *Annals of the New York Academy of Sciences, 58*, 225–238.

Cain, W. S. (1970). Odor intensity after self-adaptation and cross-adaptation. *Perception & Psychophysics, 7*, 271–275.

Disher, D. R. (1934). The reactions of newborn infants to chemical stimuli administered nasally. In F. C. Dockeray (Ed.), *Studies of Infant Behavior* (pp. 1–52). Columbus, OH: Ohio State University Press.

Davis, L. B., & Porter, R. H. (1991). Persistent effects of early odor exposure in human neonates. *Chemical Senses, 16*, 169–174.

Delaunay-El Allam, M., Soussignan, R., Patris, B., Marlier, L., & Schaal, B. (2010). Long-lasting memory for an odor acquired at the mother's breast. *Developmental Science.* doi: 10.1111/j.14677687.2009.00941.x

Engen, T. (1963). Cross-adaptation to the aliphatic alcohols. *American Journal of Psychology, 76*, 96–102.

Engen, T. (1964). Psychophysical scaling of odor intensity and quality. *Annals of the New York Academy of Sciences, 116*, 504–516.

Engen, T. (1965). Psychophysical analysis of the odor intensity of homologous alcohols. *Journal of Experimental Psychology, 70*, 611–616.

Engen, T., Cain, W. S., & Rovee, C. K. (1968). Direct scaling of olfaction in the newborn infant and the adult human observer. In N. Tanyolic (Ed.), *Theories of Odors and Odor Measurement* (pp. 271–294). London: Technivision.

Engen, T., & Lipsitt, L. P. (1965). Decrement and recovery of responses to olfactory stimuli in the human neonate. *Journal of Comparative and Physiological Psychology, 59*, 312–316.

Engen, T., Lipsitt, L. P., & Kaye, H. (1963). Olfactory responses and adaptation in the human neonate. *Journal of Comparative and Physiological Psychology, 59*, 73–77.

Kruger, L., Feldzamen, A. N., & Miles, W. R. (1955). Comparative olfactory intensities of the aliphatic alcohols in man. *American Journal of Psychology, 68*, 386–395.

Levin, G. R., & Kaye, H. (1964). Non-nutritive sucking by human neonates. *Child Development, 35*, 749–758.

Lipsitt, L. P., & De Lucia, C. A. (1960). An apparatus for the measurement of specific response and general activity of the human neonate. *American Journal of Psychology, 7*, 630–632.

Lipsitt, L. P., & Levy, N. (1959). Electrotactual threshold in the neonate. *Child Development, 30*, 547–554.

Mennella, J. A., Johnson, A., & Beauchamp, G. K. (1995). Garlic ingestion by pregnant women alters the odor of amniotic fluid. *Chemical Senses, 20*, 207–209.

Morgan, C. T., & King, R. A. (1972). *Introduction to Psychology*. New York: McGraw-Hill.

Ottoson, D. (1956). Analysis of the electrical activity of the olfactory epithelium. *Acta Physiologica Scandinavic, 35* (Suppl. 122), 1–83.

Pedersen, P. E., & Blass, E. M. (1982). Prenatal and postnatal determinants of the first suckling episode in albino rats. *Developmental Psychobiology, 15*, 349–355.

Pedersen, P. E., Steward, W. B., Greer, C. A., & Shepherd, G. M. (1983). Evidence for olfactory function in utero. *Science, 221*, 478–480.

Peiper, A. (1963). *Cerebral Function in Infancy and Childhood*. New York: Consultants Bureau.

Porter, R. H., Balogh, R. D., & Makin, J. W. (1988). Olfactory Influences on mother-infant interactions. In C. Rovee-Collier, & L. P. Lipsitt (Eds.), *Advances in Infancy Research 8* (pp. 39–68). Norwood, NJ: Ablex.

Rovee, C. K. (1969). Psychophysical scaling of olfactory response to the aliphatic alcohols in human neonates. *Journal of Experimental Child Psychology, 7*, 245–254.

Rovee, C. K. (1972). Olfactory cross-adaptation and facilitation in human neonates. *Journal of Experimental Child Psychology, 13*, 368–381.

Rovee, C. K., & Levin, G. R. (1966). Oral 'pacification' and arousal in the human newborn. *Journal of Experimental Child Psychology, 3*, 1–17.

Schaal, B., Marlier, L., & Soussignan, R. (1998). Olfactory function in the human fetus: Evidence from selective neonatal responsiveness to the odor of amniotic fluid. *Behavioral Neuroscience, 112*, 1438–1449.

Schroers, M., Prigot, J., & Fagen, J. W. (2007). The effect of a salient odor context on memory retrieval in young infants. *Infant Behavior & Development, 30*, 685–689.

Sevelinges, Y., Mouly, A.-M., Raineki, C., Moriceau, S., Forest, C., & Sullivan, R. M. (2011). Adult depression-like behavior, amygdala and olfactory cortex functions are restored by an odor previously paired with shock during infants' sensitive period for attachment learning. *Developmental Cognitive Neuroscience, 1*, 77–87.

Shionoya, K., Moriceau, S., Lunday, L., Miner, C., Roth, T. L., & Sullivan, R. M. (2006). Developmental switch in neural circuitry underlying odor-malaise learning. *Learning & Memory, 13*, 801–808.

Smotherman, W. P. (1982). Odor aversion learning by the rat fetus. *Physiology and Behavior, 29*, 769–771.

Smotherman, W. P., & Robinson, S. R. (1985). The rat fetus in its environment: Behavioral adjustment to novel, familiar aversive and conditioned stimuli presented in utero. *Behavioral Neuroscience, 99*, 521–530.

Stevens, S. S. (1955). On the averaging of data. *Science, 121*, 113–116.

Stickrod, G., Kimble, D. P., & Smotherman, W. P. (1982). In utero taste/odor aversion conditioning in the rat. *Physiology & Behavior, 28*, 5–7.

Sullivan, R. M., Taborsky-Barba, S., Mendoza, R., Itano, A., Leon, M., Cotman, C., Payne, T., & Lott, I. (1991). Olfactory classical conditioning in neonates. *Pediatrics, 87*, 511–518.

Teicher, M. H., & Blass, E. M. 1977. First suckling response of the newborn albino rat: The roles of olfaction and amniotic fluid. *Science, 198*, 635–636.

Wigal, T., Kucharski, D., & Spear, N. E. (1984). Familiar contextual odors promote discrimination learning in preweanling but not in older rats. *Developmental Psychobiology, 17*, 555–570.

CHAPTER 13

Emerging chemosensory preferences

Another playground for the innate-acquired dichotomy in human cognition

Benoist Schaal
CNRS-Université de Bourgogne, Dijon, France

As in many other domains of perception and cognition, two opposite conceptions about the causes of emergence of odor preferences were raised by empirical research in newborn infants, one defending the blank slate view while the other endorsed the innate view. Although these conceptions were expressed some decades ago, they remain still alive in the domain of chemosensory research. The present chapter aims to weigh the evidence for both of these views on human newborns' hedonic propensities towards odors in the light of current knowledge. It will address first how far we can consider that neonatal responses to odors depend exclusively on learning. Secondly, it will address how far we can consider that infantile responses to odors, or at least some of them, are "innate". This will lead us to attempt clarifying the meaning of the concept of innateness and how it is applied in theorizing chemosensory abilities. Finally, we will consider whether the concept of innateness can stand the considerable plasticity of the olfactory system that appears to be generalized at all levels of its functioning.

Introduction

Since the dawn of philosophy, biology and psychology, scholars have eagerly sought to understand the factors that shape structures and functioning in living organisms. This quest about primordial determinants of the phenotype has powered an endless and passionate controversy on whether a given physical trait or a given psychological process are innate or acquired. This debate comes and goes. It has been peaking first when psychobiologists criticized the radical nativist positions raised by classical ethologists to explain imprinting or the sensory mechanisms releasing fixed action patterns in various animals (Lehrman, 1953). It continues to blossom in the writings of contemporary developmentalists (e.g.,

Moore, 2001; Karmiloff-Smith, 2001; Lickliter & Honeycutt, 2003; Lewkowicz, 2011) who object cognitive and evolutionary psychologists that invoke innate processes and core knowledge to explain complex processes of human cognition, such as face recognition, language development, numerical abilities, theory of mind, or moral propensities (e.g., Spelke & Newport, 1998; Pinker, 2002).

The debate between nativists and empiricists goes on very actively in the cognitive sciences (e.g., Rose & Rose, 2001; Spencer et al., 2009), although different perspectives are proposed by developmental scientists to shift from origin-oriented approaches to process-oriented approaches. Issues articulated in terms of underlying processes promote indeed investigations on how given traits do emerge from the co-operation of biological and experiential mechanisms along development rather than attempting to parcel out how much genes and how much experience contribute to their progress (e.g., Gottlieb, 2007; Lickliter & Honeycutt, 2003). This essay is a preliminary elaboration towards a process-oriented approach to the development of chemosensation, especially to the emerging abilities to differentiate chemostimuli in terms of hedonic appraisal in neonatal organisms. Currently, these emerging perceptual abilities are generally described or explained in using concepts that fully endorse the innate-acquired dichotomy.

1. Two views on the causes of emerging olfactory preferences

To begin with, one may highlight two views of preference development in olfaction that prevailed during the last decades, and which are still widely around. The first is linked with Trygg Engen. In close collaboration with Lewis P. Lipsitt and their students, Engen has pioneered the experimental investigation of the earliest manifestations of human olfaction. They considered neonates as ideal psychophysical sensors, assuming that they were relatively uncontaminated by previous experience. Based on the data of his time, Engen espoused a radical empiricist view about the ways odor and flavor preferences develop, supposing an exclusive role for conditioning and direct odor exposure effects. He assumed that "odor preferences are absent at birth and acquired with age" (Engen, 1982, p. 130), and that "… odour hedonics … is a matter of learning by association in a Pavlovian sense and through modeling by the modes of one's culture" (Engen, 1988, p. 80). Engen repeatedly dismissed the possibility that the newborn brain could support associative or experiential links between incoming odor stimuli and response systems mediating approach or withdrawal tendencies.

The second view of emerging chemosensory preferences came up during the same period. Describing the behavioral responses of newborn infants to food-related odorants before any postnatal food exposure, Jacob Steiner concluded that "odors are properly evaluated by the human neonate in the very first hours of its extra-uterine life according to their hedonic value" (Steiner, 1974, p. 232). Further, he assumed that the "hedonics of taste and smell have been found to be reflected by fixed and innate, most probably inherited, stereotyped motion features" (Perl et al., 1992, p. 779). Steiner was comforted in that position by exposing neonates devoid of cortical structures to hedonically-contrasted odor and taste stimuli, and noting that they responded in a way similar to healthy neonates (Steiner, 1974).

The present chapter aims to weigh the evidence for both of these views on human newborns' hedonic propensities towards odors in the light of current knowledge. It will address first how far neonatal responses to odors depend exclusively on learning, and secondly, how far infantile responses to odors are "innate". This will lead to attempt clarifying the meaning of the concept of innateness and how it is applied in theorizing chemosensory abilities. Finally, we will consider whether the concept of innateness can stand the considerable plasticity of the olfactory system.

2. Back to the facts: Do human neonates express odor preferences at all?

The first point to settle is whether human neonates express actions that can be considered to reflect an ability to process odor stimuli as pleasant or unpleasant. Generally, when facing an odorous object or context, the initial reactions of more grown-up humans pertain to their hedonic appreciation, which can be expressed through an array of response possibilities – ranging from autonomic to behavioral and declarative responses. Typically, a pleasant odorant elicits increased stimulus sampling and positive responses at all levels reflecting attention and investigative proneness, and then open appreciation. Conversely, after initial awareness, an unpleasant odorant tends to mobilize responses that discontinue stimulus intake and introduce distance between the subject and the source.

The debate on whether neonates do feel the "pleasure of sensations" has been ongoing over the centuries before the 20th, and the demonstration of their ability to express preferences for odors has certainly contributed to solve it not so long ago, in the second half of the previous century. Our understanding of odor preferences was indeed initially flawed. First were issues pertaining to the nature of the stimuli used to probe early responsiveness: these were then often chosen without ecological consideration, and they were then as diverse as petroleum, wild bed-bugs, putrefied meat, oil of mustard, sheep brine, amber oil, tar, aqua foetida anti-hysterica, ammonia, etc. (e.g., Kroner, 1882; Preyer, 1884). Accordingly, the triggered reactions were often extreme and interpreted to be caused by startling novelty or irritation rather than only by "pure" olfactory impact. Second went the issues about procedures to evaluate what reactions meant attraction and liking in neonatal organisms. Engen and Lipsitt were seminal in improving objective quantitative methods to investigate early human responses to odorants (e.g., Engen, 1974, 1986; Rovee-Collier, 1969, 1972). A further decisive leap was made when ethologically-primed experimenters came into the scene in devising age-adapted methods. One line of such research was inaugurated by Stirnimann (1936) and systematized by Steiner, looking for oro-facial responses elicited by arbitrary chemosensory stimulations. Another line of research was inaugurated by psychologists Ripin et al. (1930), and sharpened by Macfarlane (1975) and Russell (1976), who investigated neonatal orientation and oral responses to odors associated with the mother's body.

2.1 Passive preference responses

Stirnimann (1936), and then Steiner have shown that newborns mobilize distinct response patterns in the face and mouth when exposed to odorants that evoke (un)pleasantness for adults. The hedonic nature of the infant's orofacial responses was inferred in two ways. First, single photographs taken at the apex of the odor-induced expression were shown to adult judges who had to rate the hedonic nature of the stimulus infants had under the nose. These judges categorized the responses of neonates either into "acceptance, like, enjoyment, satisfaction" or "aversion, dislike, disgust" (Steiner, 1979, p. 274), resembling the responses elicited by sweet or bitter taste, respectively. Interjudge-agreement was greater for responses elicited by unpleasant odors than for responses to pleasant odors, suggesting either that unpleasant odors elicit less ambiguous facial responses in infants, or that adults are more sensitive to negative facial motions.

A second approach used a more quantitative and dynamic way to code infant responses. The responses were videotaped and then decoded using Ekman and Friesen's Facial Action Coding System. Soussignan et al. (1997) presented newborns with 12 stimulations [4 unfamiliar biological odors (amniotic fluid, human milk and 2 formula milks bearing very different odors for adults) and the corresponding intensities of vanillin and of butyric acid], as well as odorless controls. The pure odorants were of very low intensities, matching those of biological secretions. The spatiotemporal configurations of facial responses were differentiable according to the odor: the "unpleasant" butyric acid provoked more negatively-valenced responses, while the "pleasant" vanilla did not generate any more positive facial response than butyric acid. As both odorants were equalized in intensity and trigeminal impact, the negative response was based on the infants' aptitude to differentiate odors in qualitative and/or hedonic terms, thus confirming in part Steiner's results. However, the hedonic polarity of facial responses elicited by biological odorants remained equivocal. Among these, only the (unfamiliar) amniotic fluid evoked more negative facial responses than the control stimulus. Unfamiliar formula milk, although olfactorily perceived as highly unpleasant by adults, was not associated with negative responses in infants. Therefore, the hedonic facets that infants and adults detect in odorants cannot be completely superimposed. Nevertheless, considering that the morphological analogy of facial expressions between newborns and adults may be taken to imply analogy in underlying emotional states, Steiner (1979) postulated that certain olfactory stimulations are more acceptable than others for newborns. According to him, the perceptual attributes used by infants to differentiate odors is their hedonic connotation, suggesting that hedonic value may be inherent to odorants.

2.2 Active preference responses

The keenness of neonatal hedonic processing is more strongly corroborated by experiments involving an active choice behavior between two paired odorants. When infants are confronted with two attractive odors, they are capable of "showing" which one they prefer by the relative duration of head-turning. Macfarlane (1975) used this index to infer differential processing of odors. He videotaped supine-laying newborns while exposed to two odor pads hanging bilaterally. In a test opposing odors of their mother's breast and of another nursing female's breast, 2-day-olds treated both stimulations equivalently, while 6-day-olds oriented more frankly to the mother's breast odor. Thus, as early as the end of the first week of life, infants can distinguish very low-intensity odors and designate which one they prefer approaching. In subsequent experiments, 15 day-old bottle-fed infants displayed greater attraction to the breast odor of an unfamiliar nursing mother than to the breast odor of a non-nursing woman (Porter et al., 1991), suggesting that the lactating breast emits an odor attractant, which is also present in milk itself. Breast-fed newborns turn indeed more insistently to the of odor human milk from an unrelated woman rather than to an iso-intense unfamiliar odor (Marlier & Schaal, 2005). Thus, human milk may convey an attractive odor factor.

Once oriented to a preferred odor stimulus, infants further display responses indicative of their motivation to approach and to seize it orally. Such oral response has been considered as an additional index of early preferential behavior. Thus, 4 day-old breastfed infants express appetitive oral responses to the odor of any human milk as opposed to the odor of formula milk; their bottle-fed counterparts also evince this preference for human milk odor, even more than toward their own formula milk (Marlier & Schaal, 2005). The secretion from the areolar glands, which become active during lactation (Schaal et al., 2009; Doucet et al., 2011), is even stronger in eliciting oral responses (Doucet et al., 2009). Again, these responses do not depend on previous exposure, as 3 day-old bottle-fed infants react to them as strongly as breastfed infants. Finally, milk odor is already effective in infants born prematurely (Bingham et al., 2003).

Finally, odors motivate infants to move closer to an odor source placed at distance (10–15 cm). Right after delivery, newborns laid supine on the mother's chest can indeed reach a breast by crawling. When placed on the chest of their post-parturient mother, given the choice between a deodorised (washed) breast and an unwashed breast, newborns (aged 5–13 minutes) moved more often towards the former than the latter (Varendi et al., 1994). Although the infants' choice response was not clear in these conditions, they exhibited sustained crawling towards the breast. A subsequent experiment presenting an odorant at distance indicated that newborns could reach a target carrying the odor of mother's breast (Varendi & Porter, 2001).

Taken together, the findings summarized above concur to establish that human newborn infants, as well as the other mammalian infants studied so far (Rosenblatt, 1983; Alberts, 1984; Schaal et al., 2009) do express clear-cut odor preferences. The questions coming next pertain to the developmental processes that initiate preferential responses so early.

3. How far do neonatal responses to chemostimuli depend on learning?

The notion of learning subsumes cognitive mechanisms such as habituation, familiarization, conditioning, training, trial-and-error integration, imitation, and any other way to acquire expertise. In principle, such cognitive effects depend on the direct exposure to the learned stimulus, or on some kind of generalization to resembling stimuli. Thus, we will ask firstly, when chemosensation begins, and secondly when chemosensory learning sets on in the human infant.

3.1 Early competence in human chemosensation: Structures

Chemoreception is mediated by dissociable subsystems located in the nasal and oral cavities (as well as in the digestive tract). Human nasal chemoreception is mediated by the main olfactory and the trigeminal subsystems. Oral chemoreception is supported by specific gustatory pathways beginning in taste buds, and by tacto-chemical pathways actualized in trigeminal innervation. From a functional point of view, the olfactory system is considered to detect innumerable low-concentration compounds carried in the incoming airflow, whereas the trigeminal system senses higher concentrations of irritants. These nasal subsystems are, however, functionally overlapping and synergistic. The gustatory subsystem mediates elemental taste modalities (*viz.*, sweet, sour, salted, bitter, savory or umami, and perhaps fat) and their combinations, while the oral trigeminal subsystem mediates various aspects of mouthfeel (*viz.* burning, cooling, astringency, stickiness). Finally, despite separate receptors and neural pathways, olfactory, gustatory and trigeminal sensations result in the emergence of unitary percepts.

3.2 Early competence in human chemosensation: Functioning

a. Neural processing abilities

How do all these intricate chemoreceptive competences develop functionally, and when? They are all characterized by their ontogenetic precocity in mammalian embryos. In humans, the olfactory system shows morphologically adult-like olfactory sensory neurons (OSNs) by the 11th gestational (G) week (reviewed in Schaal et al., 1995). When they reach functional maturity, OSNs express a specific olfactory marker protein (OMP) that is correlated with their selective reactivity and connectivity with mitral cells in the olfactory bulb. In the human fetus, the OMP is not detected before G week 28. Between weeks 28 and 32, it is expressed in the cell bodies and dendrites of OSN, and, by week 32, in the entire OSN. Finally, by week 35, the OMP is detected in the olfactory bulb (Chuah & Zheng, 1987), indicating connections with the next neurons in the functional chain towards higher brain structures. The pattern of neural maturation indexed by the OMP is linked with odor-related responsiveness in premature infants. Before the G week 28, they are almost unresponsive to mint odor, whereas they react more and more consistently after 29 weeks. By 33–36 weeks their responsiveness to mint odor has the same frequency than those of term-born newborns (Sarnat, 1978). Finally, trigeminal fibers do innervate the nasal and oral mucosae by G week 4, and react to touch stimulation by G week 7. By that age, they may thus also sense chemical stimuli in the womb. Thus, to sum up, it is clear that human nasal peripheral chemoreceptive systems undergo an anatomical development that is compatible with sensory function from late gestation.

b. From neural detection to perception

The onset of olfactory functioning was deciphered by converging approaches in animal models and in human infants. In mammalian animal models, olfaction operates from the fetal period. At the behavioral level, the strongest evidence for prenatal olfactory function comes from studies on rats, mice and sheep tested *in utero* (Smotherman & Robinson, 1987; Schaal & Orgeur, 1992). All these animal fetuses exhibit acute motor and autonomic responses when exposed to an odorant. But such fetal responses are different with qualitatively and/or quantitatively distinct odorants, indicating discriminative abilities *in utero* (e.g., Smotherman & Robinson, 1987).

As testing olfaction in human fetuses is difficult *in utero*, investigation has been privileged *ex utero*, either in infants born pre term or in term-born newborns being re-exposed to prenatal stimuli. As mentioned above for mint odor, premature infants have the ability to detect and discriminate odorants (Sarnat, 1978; Schaal et al., 2004). Pihet et al. (1996) further found that nonanoic acid or eucalyptol, selected for their low and high trigeminal potency, respectively (but matched for subjective intensity), released differential responses in premature infants (born within 30–37 G week). Although, 76 and 80% of the infants responded to nonanoic acid and eucalyptol, respectively, both stimuli were discriminated in terms of time-course of response. Still another study (Goubet et al., 2002) relied on habituation in prematures (G age: 31.8 weeks) and in term newborns. After habituating infants to odorant A, the delivery of a novel odorant prompted an increase in negative facial responses in both groups of infants, supporting similar odor discrimination abilities in premature and term-born infants. Finally, a study consisted in exposing preterm infants (G age: 33.7 weeks) to disinfectant or detergent odors carrying strong trigeminal potency while measuring variations of cerebral blood flow by near infra-red spectroscopy (Bartocci et al., 2001). After such stimulation, decrease in oxygenated hemoglobin was noted over the parietal region, supposed to correlate with the detection of the irritants. Thus, as early as the last gestational trimester, can the human brain discriminate between odorants, and between odorants and irritants.

c. From discrimination to learning

Habituation reflects short-term memory supporting the decrease in novelty-related response, and hence the decrease in attention allocated to the stimulus. One study compared habituation of behavioral and autonomic responses to odors in prematures (G age: 31.8 weeks) and in term-born infants (Goubet et al., 2002). Repeated exposure to either vanillin or anethole resulted in reduced facial responsiveness in term, but not in preterm, newborns (but, in the latter, heart rate habituated, suggesting that motor and autonomic reactivity do not habituate at the same rate as a function of development). The ability to familiarize to odorants may facilitate their engagement in learning. Irzhanskaia and Felberbaum (1954) assessed this possibility in exposing preterm infants to mint odor during bottle-feeding for 2 weeks. They then evaluated the strength and stability of a conditioned response wherein the mint was paired with an air-puff into the eyes to elicit an eye-blink; after several such pairing trials, the eye-blink response was elicited with mint alone. When applied to 3 groups of infants differing in gestational age at birth (6.5, 7 and 8 G months) and in postnatal age at testing (1.5, 2 and 2.5 months), fewer associative episodes were needed to elicit the conditioned eye-blink with the familiar mint odor than with a novel odor; and fewer pairings were necessary to establish the eye-blink in the preterm infants born closer to term and oldest at the moment of testing. These results suggest that odorants can be engaged very early as conditioned stimuli.

d. Prenatal learning feeds postnatal preferences

As mentioned above, normal newborn infants show hedonic differentiation of odorants by displaying facial actions indicative of acceptance/rejection for different odorants. So far, only few studies investigated the very initial manifestation of such stimulus-response links. One experiment in preterm infants showed already contrastive responses to odorants chosen for their opposed hedonic value for term newborns and adults – vanilla and butyric acid – both being delivered at very low intensity assumed to be devoid of trigeminal effect (i.e., matched with the odor of milk) (Marlier et al., 2001). Respiratory responses induced by both odorants were in opposite directions, vanilla odor eliciting acceleration and butyric acid odor eliciting deceleration of respiratory rate. In adults, such odor-induced variations in respiratory rate are indicative of the affective value of the inhaled stimulus, a negative smell being followed by reduced, and a positive smell by increased, inspiratory intake. This interpretation in affective terms is corroborated by the facial responses of these infants (Soussignan & Schaal, 2004). When videotapes from the above olfactory tests were decoded using the Facial Action Coding System, it came out that facial actions released by vanilla odor included more appetitive oral responses than the reactions following butyric acid odor. Conversely, butyric acid odor more often than vanilla odor induced facial indicators of disgust (Gaugler et al., 2003). These results suggest that, as early as 31 G weeks, preterm infants can display affective responses to odorants. Thus, differentiated olfacto-facial responses to hedonically-contrasted odorants recorded in neonates born at normal term may derive from pre-existing stimulus-response relations.

This was directly assessed in experiments where neonates were re-exposed to stimuli they could only encounter in the womb. For example, neonates born to mothers who consumed anise-flavored foods during the last days of pregnancy exhibited all facial signs of acceptance when re-exposed to a dilute anise stimulus 4 days after birth. In contrast, infants born to mothers who never consumed anise had oro-facial expressions indicative of rejection (Schaal et al., 2000). Similar results obtained with other flavors suggest that fetal exposure effects on neonatal responsiveness are not odor-specific (Hepper, 1995; Faas et al., 2000; Mennella et al., 2001). Thus, from late-gestation the fetal brain can detect, encode, and memorize available odor information, and the neonatal brain can detect and decode the same odors, and behaves selectively towards them.

e. Learning goes on in the neonate

The brain continues to encode the surrounding world after birth, presumably with increasing rate in the more complex, multimodal postnatal environment (Schaal & Durand, 2012). Feats of neonatal learning are now well established with all kinds of paradigms (non-associative, operant and classical conditioning, imitation, etc.) in all domains of adaptive necessities (self-regulatory, social, food) (e.g., Field, 2007; Lécuyer, 2005). Some examples will illustrate the rapid dynamics of olfactory learning, as well as its extreme malleability. The mammalian nursing situation is the prototypical context of neonatal learning, as it amalgamates into a same context learnable stimuli (all sense modalities being exposed), actions that are instrumental to learning (rooting and sucking), and all kinds of reinforcing processes to consolidate learning (brain reward processes, sucking-based operant processes, milk-borne bioactive agents, metabolic factors). Any initially-neutral odorant associated with breastfeeding is thus easily turned into an attractive stimulus in non-human neonates (Alberts, 1987; Blass, 1990; Schaal, 2005), as well as in the human newborn (Schleidt & Genzel, 1990; Delaunay-El Allam et al., 2010).

The proneness of newborn infants to learn odors is so boundless that it operates even outside the hyper-reinforcing context of nursing. Thus, for example, the arousal state induced by a 5-minute massage in newborns aged less than 24 hours is enough to change a neutral odor into a cue that elicits positive head orientation (Sullivan et al., 1991). Finally, even the mere fact of exposing newborn to an odorant in their crib for several hours without planned exposure to reinforcing agents (but not without the possibility that such reinforcers operate) is enough to reverse an arbitrary neutral odorant into an attractive stimulus (Davis & Porter, 1991). Such mere exposure effects of odors can be engaged very rapidly after birth, as only 30 minutes of exposure right after birth suffice to psychologically shift an odorant from novelty to familiarity (Romantshik et al., 2007).

Thus, odor learning goes on very actively in the neonatal brain, and the more accessible and controllable conditions of such learning in the just-born infant may suggest underlying phenomena that do already operate in the late-gestation fetus. Any stimulus-response link observed in the newborn can result from fetal acquisition or even from prior postnatal acquisition that has taken place during labor or immediately after birth, right before the neonatal test.

3.3 Early chemosensory experience and the emergence of preferences

To fully understand the effects of the environment on the development of olfactory function, we need to go beyond the traditional, mentalistic acception of learning through conditioning to encompass the wider notion of experience (Michel & Moore, 1995). In the epigenesis conceptualization proposed by Schneirla (1965) and others (Gottlieb, Lickliter), the term "experience" incorporates the various stimulative aspects to which organisms are subject during prenatal and postnatal life, including aspects that originate from within or outside the organism. Gottlieb (1976) categorized experiential effects in terms of maintenance, facilitation, or induction, as a function of their endpoints in the development of a given structure or functional ability. Maintenance experience favors the preservation of a structure or ability. Facilitative experience ensures its appearance at a given time in boosting or inhibiting the rate of its development/maturation. Finally, inductive experience is mandatory to the developmental achievement of the structure or ability.

Although these classes of plasticity processes are questionable, they are useful to describe what is going on in the development of olfaction, and especially which effects early experience can have at different levels of emergent hedonic responsiveness. The maintenance effect of experience is exemplified in the consequences of suppressing any input to the olfactory mucosa by obstructing the nasal pathways (in rodents). Blocking one nare leads to a considerable reduction in the volume of the ipsilateral olfactory bulb, and in an atypical structuration of its constitutive neural layers (Brunjes & Frazier, 1986). Facilitating experience is exemplified in the consequences of manipulating the odor environment during development. For example, enriching an organism's atmosphere with a given odorant canalizes the structural and functional development of the olfactory bulb in the young rat, resulting in an odor-specific spatial degeneration pattern of mitral cells (Döving & Pinching, 1973). The young rats' sensitivity to the exposure odor appeared preserved, but they explored it more than other odorants in a relative preference test (Laing & Panhuber, 1978). Finally, the induction effect of the odor environment was shown in a study that exposed adult androstenone-anosmic mice to androstenone for several weeks (Wang et al., 1993). These mice became more sensitive to that odorant. A similar inductive effect of odor experience was described in fetuses (e.g., Youngentob et al., 2007).

The level(s) at which the odor environment impinges on the chain of events leading to selective odor responsiveness is poorly understood. In various animal models studied so far, one or several levels have been concerned with experiential induction of olfactory function, from the OSN (gene expression during OSN development, olfactory receptor expression; e.g., Youngentob & Kent, 1995; Yee & Wisocki, 2001) to the first relay (glomeruli) in the olfactory bulb (e.g., Todrank et al., 2010), and from detection performance to higher-brain levels involved in perception, cognition, motivation and action (e.g., Jones et al., 2008).

3.4 Consequences for the conceptions on emerging chemosensory preferences

The data summarized above clearly demonstrate that, in our own species as well as in other mammals, the brain supports preferential responsiveness for given odor agents from the neonatal stage onwards, and that such preferences can reach back into earlier experience (in the fetus or the neonate). Thus, on the one hand, Engen was not accurate in denying human infants the possibility of expressing odor preferences from birth onwards. However, his more general statement that "the meaning of an odor is determined by its environmental associations" (Engen, 1991, p. 117) has been largely verified from the earliest developmental stages examined so far in humans, the fetus and prematurely born infant. When one considers "environmental associations" and learning as one type among many of the effects of experience, Engen's view appears even more accurate. The odor ecology has indeed a strong formative bearing on the structural and functional properties of the olfactory system from the very periphery to higher levels of perceptual processing. On the other hand, Steiner has been accurate in claiming that "odors are properly evaluated by the human neonate in the very first hours of its extra-uterine life". But was he right in affirming that such abilities are "reflected by fixed and innate, most probably inherited, stereotyped motion features"?

4. How far are neonatal responses to odors innate?

4.1 Innateness-related concepts in the chemosensory literature

Steiner interpreted the responses of human newborns to given odorants (and tastants) in the background of classical ethology. He explained his behavioral observations in infants, as well as older children and adults, as resulting from fixed motor patterns that are unlocked by key stimuli (Steiner, 1977). In the black-box between stimulus and response, he conceived the "decision making" process to rely on a cerebral "hedonical monitor system" (1979, p. 261). Accordingly, "the "decision" between accepting hedonically positive and rejecting hedonically negative food-related stimuli relies on an innate faculty of the nervous system"

(1979, p. 261). In sum, in Steiner's view the stimulus-response loop involved in olfacto-facial (and gusto-facial) responses are highly specific as regards both eliciting stimuli and response patterns, and they depend on integrative brain mechanisms which mature very early and do not obligatorily rely on cortical processes. Finally, "the innate motor coordinations … are *programmed* in terms of activation sequence and in terms of features of their appearance" (1979, p. 258), indicating that all the facets of their manifestation may depend on the expression of genetic factors. In other words, "Since the responses are elicitable at the date of birth their innate, possibly inherited character is demonstrated" (1979, p. 278). Thus, one may sum up Steiner's view as supporting that the neonatal stimulus-response loops triggered by food-related odorants (and tastants) are "innate" in the strongest sense of genetic inheritance at the level of sensory reception, brain integration, action decision and motor execution.

But the conception of hedonic responsiveness to chemosensory stimuli as supported by innate or inborn processes goes far beyond Steiner's work, arching all over the literature on early chemosensory preference development (including my own work), and even more generally over the domain of early development of perception and action, and underlying brain processes. For example, certain stimulus-behavior loops in neonatal organisms were termed "prefunctional" because they appear ready-to-use in advance of normal functional demand and before any fine-tuning by first trials could occur (Hogan 1998). Others (e.g., Bolhuis 1996; Horn, 2004; Schaal et al., 2009) rather use the term "predisposition" or "preexisting bias" for such processes, which have the advantage that no prescience of function is needed. These terms denote that neonates are biased to respond in particular ways to particular classes of stimuli. In the context of emerging visual preferences, for example, Johnson and Morton (1991) hypothesized that the neonatal brain may contain structural information about conspecifics that directs visual attention and regulates orientation responses toward faces. This "predisposed device" implies that a specific stimulus-response association does not obligatorily result from previous sensory exposure to specific stimuli, although it does not exclude that non-specific processes are involved.

In a final example, Alberts and Decsy (1990) used the concept of "predetermination" to describe the relatively time-fixed appearance of species-typical behavioral traits in developing organisms. Such predetermined phenotypic attributes follow the criteria of chronotypy (fixed timing in terms of chronology or maturation of the phenotypic trait), sender-receiver coupling (when cues presented by the sender are stable, species-typical, and heritable), and heritability of phenotype (proportion of the phenotypic variance due to genetic variance). According to Alberts and Decsy (1990, p. 571), "in the course of ontogenesis, mammals sequentially manifest attributes that are developmentally orderly, stereotyped, and which, if they were present congenitally, might be suspected of being "innate".

4.2 What does "innate" mean?

Acceptions of the term "innate" are multiple in the biological or psychological explanations of behavior. Vernacular and dictionary uses of the word pertain to such traits as: present at birth (the literal meaning of *innatus*), not learned, distributed pan-specifically, correlated with genetic difference, resulting of evolutionary adaptation, unchanging through development, or endogenously determined (Bateson, 1991). Mameli and Bateson (2006) have attempted to identify conditions for a trait to be defined as innate. They identified 26 conditions, many of them being considered irrelevant, problematic or redundant, and finally only 8 properties, called innateness properties, were retained. These proposed properties imply that a trait is innate if and only if: (1) it reliably appears in a particular and well-defined stage of the life cycle; (2) all environmental conditions capable of producing an alternative trait are evolutionarily abnormal; (3) it is not produced by an experiential mechanism adapted to map environmental conditions onto different phenotypes; (4) it is a source of developmental constraint in the design of an adaptive feature; (5) it is not developmentally canalized by the environment; (6) it is not post-developmentally canalized by the environment; (7) it is a species-typical trait; (8) it is a standard Darwinian adaptation. These different proposals "capture some aspect of the vernacular distinction between innate and non-innate traits" (Mameli & Bateson, 2006, p. 179), but they are not equivalent and the question of their correlation has never been addressed systematically (p. 178). At the end, Mameli and Bateson propose that a trait might be considered innate if it can be characterized by a large number of the above innateness properties.

4.3 Consequences for the conceptions on emerging olfactory preferences

The validity of these properties to empirically assess the innateness of a given attribute needs to be tested one by one, or, more demandingly, as a cluster of definitional criteria. In an attempt to characterize the above criteria operationally, criterion 1 would mean that the trait is developmentally regular, e.g., that it occurs at birth or at puberty; criterion 2, that the emergence of the trait is independent from prior environmental influence; criterion 3, that the normal emergence of the trait is unaffected by adaptive plasticity and learning; criterion 4, that the trait does constrain the normal development of other traits (e.g., facilitate learning); criterion 5, that the trait's emergence is robust in the context of normal environmental variation; criterion 6, that once emerged the trait cannot be altered by normal environmental variation; criterion 7, that the trait is expressed in all members of a species/genus; and finally, criterion 8 would mean that the trait is maintained and spread in the population due to selection for it. When this grid of criteria is applied onto the olfacto-facial response as investigated by Steiner, only some of the above criteria appear to be confirmed: it is present right at birth (1), emerges in many different developmental contexts (5), and seems to be expressed pan-specifically (7). The remaining criteria seem more difficult (or impossible?) to verify. Indeed, the olfacto-facial response cannot be considered: as developing in independence from environmental influence (3), or to hinder normal development if not present (2); further, it is highly alterable in morphology and timing during development (6), does not seem constraining on the development of other traits (4), and its selective advantage in the population is not established (8). As the three innateness properties (1, 5, 7) that best apply onto the olfacto-facial response do not exclude the intervention of previous experience, this behavior should not be characterized as innate, if one follows the above logic.

Then, one may ask whether elements or partial processes supporting the olfacto-facial response loop are, in the terms of the above criteria, more "innate" than others. The functional chain supporting the olfacto-facial response can be decomposed in terms of receptor processes involved in the detection of odorants, brain processes that assign psychological salience and reward value to the stimuli, and motor processes that direct the responses of the organism. At all these levels gene-environment interactions occur during development to specify the olfactory periphery, to induce the expression of olfactory receptors in OSN and target their axons into bulbar glomeruli, to connect second-order and x-order neurons with limbic and neocortical structures involved in discrimination, recognition, memory, on the one hand, and in assigning reward and meaning (familiarity), on the other hand, and finally, to tie together with areas involved in motor programming and output. Experiential mechanisms are certainly involved at all these levels of processing (see above).

Steiner (1979) made strong case of the fact that newborns afflicted with severe brain malformations (anencephaly, hydranencephaly) displayed expressive facial responses similar to those of normal infants in response to tastants and odorants. For tastants, he stated that "discrimination between "pleasant" and "aversive" does not result from cognitive mental processes based on learning but reflects a biological existentional approach to the acceptable and repulsive taste" (p. 262). Likewise, the olfacto-facial response is considered not to be "a result of previous acquisition of experience but rather is part of a complex innate mechanism" (p. 263). However, even in the absence of cortical structures, a range of low-level cognitive capacities are operative. For example, experimentally decerebrated animals show habituation or classical conditioning at the same rate than intact animals (e.g., Ronca et al., 1985). Moreover, human infants lacking cerebral hemispheres show evidence of associative learning (e.g., Berntson et al., 1983). Thus, in the newborn, and probably already in the advanced fetus, cerebral structures in the lower mesencephalic, diencephalic or medullary levels can sustain integrative processes (Joseph, 2000). Accordingly, structures in the lower neuraxis should also support the acquisition of functional links between odor/taste stimuli, reward systems and non-random facial responses. The question is now when and through which mechanisms these processes are becoming functionally coordinated during early development. As such functional links are already well stabilized in the newly born infant, they should emerge in previous stages of development, namely in the late gestation fetus, when chemosensation is mature enough to sense odorants and tastants, when facial activity goes on, and when reward systems are becoming organized.

5. How do odorants become pleasant/unpleasant odors in newborn organisms?

The hedonic value of a stimulus, as attested by approach/rejection tendencies, results from the interaction between the stimulus and the sensing organism. This interaction takes place in different conditions, at different time points, after different prerequisites. First, it presides to the here-and-now decision to act toward a stimulus sensed for the first time, because it resonates enjoyably/aversively to the brain without reference to any prior knowledge of it. Second, it may reflect prior organism-stimulus interactions and, more broadly, rely on individual-specific encoding of beneficial/harmful consequences. Third, an organism-stimulus interaction may be assimilated after generation in a repertoire of sensory processes which adaptive value may have led individuals reacting positively (or negatively) to a given stimulus to survive more, thrive better, and reproduce optimally; such advantage may have led to some kind of hard-wiring of the sensory apparel, most probably in terms of receptor expression at the most peripheral levels or in terms of an evolutionarily determined threshold or bias for responding at higher levels of integration. All these different levels of organism-stimulus interaction encoding – instantaneous, experience-based, and inherited – are nested in the explanation of response direction of an organism encountering an odor stimulus for the first time after birth. They are uneasy to dissociate empirically in an adult organism. These multiple ways to encode the environment may not be much easier to disentangle in immature organisms, which have already a long past, but one may reasonably consider that the number and range of first-time exposures is then reduced, or at least controllable.

Here, we will address whether odorants bear intrinsic affective value, i.e. before they have had any direct interaction with the chemoreceptors of a receiving neonatal organism. In other words, is there an unconditional hedonic polarity in the olfactory domain? Conversely, we will address whether the initial stage of perception in neonates is exclusively shaped by the environment created by a knowledgeable nurturing female. Alternatively, we will consider whether the initial impact of experience depends on a species-specific level of brain development that restricts the range/nature of sensory inputs at given periods of the organism's interaction with the environment, so that any stimulus can be registered as either pleasant or unpleasant more by virtue of the developmental stage of the recipient organism than by the nature of the incoming stimuli.

5.1 Can unprecedentedly perceived odorants be intrinsically pleasant/ unpleasant?

Steiner's experiments indicate that artificial flavors chosen because of their unambiguous appreciation (butter, banana, vanilla, chocolate, and strawberry) or rejection (shrimp flavor, rotten egg) by adults elicit neonatal responses which hedonic direction tends to be congruous with the adults' judgments. Does this mean that newborns react to hedonic information that is carried within the stimuli themselves? Steiner's olfacto-facial response tests were run some time after birth so that one cannot exclude that the newborns' differential responses were, at least in part, a consequence of earlier-encoded chemosensory experience. Further, the food-related odor qualities presented to the newborns might have been encountered in the womb or in milk, so that the tested infants could be responding to more or less familiar stimuli. Alternatively, the stimuli selected to represent the negative side of the hedonic scale have been delivered in high intensity that may have elicited aversive responses based on trigeminal sensation (for example, the rotten egg and fishy odorants were reported to be "concentrated" and judged as "offensive" by adults; Steiner, 1979, p. 274). This might explain why infants' responses to both "offensive" odorants were the least ambiguous in terms of adults' judgments of elicited facial reactions. Thus, there may indeed be alternative explanations than an exclusively hedonic discrimination to the diffentiated facial responses of neonatal response to the odorants.

However, recent evidence suggests that one may, to a certain extent, foretell the hedonic valence of odor molecules from their structural traits. Khan et al. (2007) probed indeed the relationship between a wide range of molecular properties in a set of 52 monomolecular compounds and the description of these compounds by adults. The mathematical model derived from the physicochemical properties of these compounds predicted fairly well their (un)pleasantness. When this model was applied to a novel set of odorants, their pleasantness could also be well predicted by adults from different cultures. Based on this and other studies (Haddad et al., 2010; Mandairon et al., 2009), it may be concluded that odor preferences are not only in the nose of the beholder, but may derive at least in part from structural properties that are inherent to the odor compounds. However, these results do not systematically consider the intensity of the stimuli, which is strongly linked with hedonic judgments (see below). Also, in biological mixtures, odor compounds are rarely encountered in pure form and one is eager to understand how the above physicochemical model of odor hedonics of individual odorants will predict the hedonic value of mixtures.

Another stimulus-bound influence on perception relates to intensity. In adults, there is indeed a link between odor intensity (i.e., odor concentration) and the hedonic quality of a stimulus (Doty, 1975), and between intensity and the qualitative account of the odor source (Gross-Isseroff & Lancet, 1988). So far, very rare odor intensity studies have been conducted in early development, but they pertain more to psychophysical responses without hedonic specification (Engen et al., 1968; Rovee, 1969). But, refering to Schneirla's (1965) principle of intensity-based dichotomous responses in neonates and to the fact that during prenatal development the brain is sheltered from intense chemostimuli (Schaal, 2005), it may be predicted that low-intensity odor stimuli may elicit more positive responses in neonates than high-intensity odorants.

A further inherent property of odorants is their trigeminal impact, i.e. their confounded effect on olfaction and nasal somatosensation (cf. Section 3). In a classical study, Doty et al. (1978) showed indeed that among a series of undiluted pure odorants most elicited a nasal sensation in adults lacking a functional olfactory nerve. Some odorants of this same series had been earlier used in experiments with newborn (Engen et al., 1963, 1968; Rovee, 1969). In general, volatile compounds that bear irritation release aversive reactions, although these negative responses can be inverted by exposure effects. In 3 day-old newborns, it was clearly shown that a strong trigeminal stimulus (ammonia) reliably elicits head-turning away from the stimulus (Rieser et al., 1976).

Finally, another line of evidence that odorants may have potential intrinsic hedonic value is reflected in their impact on autonomic and central nervous systems' reactivity. For example, some pure odorants have the capacity to alter psychophysiological indices (e.g., Alaoui-Ismaili et al., 1997) in a coordinated pattern correlated with discrete affective states (e.g., Christie & Friedman, 2004). Odor stimuli may thus be able to instantaneously switch on affective states in the recipient organism. However, such studies have almost exclusively been conducted in adults, with generally little consideration of previous exposure to the investigated odorants (but see Haller et al., 2000; Poncelet et al., 2010).

To sum up, odorant stimuli might carry inherent sensory features that may in part constrain the direction of neonatal responses at the very first postnatal exposure. Odorant-based properties through which early responses are hedonically oriented appear thus multiple and interdependent, and attributable to lower as well as higher sensory-perceptual processes.

5.2 Exposure effects as moderators of novelty avoidance?

As noted earlier, the inherent hedonic meaning of odorants can be easily redirected by exposure effects. So easily and inconspicuously that one may be inclined to think that odor preferences do exclusively depend of prior exposure. The newborn brain is indeed extremely "absorptive" of all kinds of sensory influences mediated by non-associative and associative ways of organism-environment interactions.

Habituation experiments applying regularly-paced odor stimuli over short periods of time indicate that neonates react to exposure reiteration of a stimulus by lessening responsiveness (Engen et al., 1963; Engen & Lipsitt, 1965). This response decrement is taken as an index of information extraction and short term encoding, and interpreted as loss in the arousing properties, disengagement of interest, or decreased reactivity to novelty. Continuous exposure to an odor in newborn infants can further lead to more stable familiarization, i.e. change in the affective meaning of an initially neutral stimulus. This process was shown in the newborn (Davis & Porter, 1991). Then, reinforcing episodes of nurturance strongly affect the hedonic appraisal of odors. For example, 1 day-old infants exposed to gentle massage (for 10 30-second periods) in presence of a lemon odor, exhibit positive head turning responses to that odor given separately the following day (Sullivan et al., 1991). Thus, either the stage of arousal elicited by the tactile interaction or the positive feelings associated with it are transferred onto the odor stimulus in only 3 minutes.

Finally, several studies focused on nursing as a potent context to influence the acquisition of odor hedonics. This process has been well analyzed in animal newborns where tactile stimuli received from nursing females and events linked with milk intake are causal in assigning positive value to any arbitrarily associated odor stimulus (e.g., Rosenblatt 1983; Alberts 1987; Blass 1990). Although similar data are currently scarce in the human newborn, converging evidence exists to suggest that contingent odor-nursing exposure powerfully affects the development of odor hedonics. First, the mother's breast odor releases increasingly reliable positive head-turning as a function of suckling experience (Macfarlane,

1975). Second, even arbitrary artificial odorants can elicit typical approach/appetitive responses in neonates after their association with nursing (Schleidt & Genzel, 1990; Delaunay-El Allam, 2010).

Non-associative and associative processes result in a shift in the hedonic appraisal of the stimulus, as attested by non-random orientation to, exploration of, and apparent willingness to (orally) seize it. On a psychological level, one may suggest that they engage processes involving reduction of stimulus novelty, increase in familiarity and complexification in "knowledge", including predictive value of pending reward. For newborns, current evidence suggests that novelty of odor stimuli generally releases dislike. Emotion theorists (e.g., Delplanque et al., 2009) emphasize indeed that the assessment of novelty is the leading event, even before the assessment of hedonic value, in the appraisal of the pertinence of an odor stimulus. Motivation theorists (e.g., Berlyne, 1960) also suggest that organisms generally seek an optimal level of stimulation based on the novelty of stimuli and related surprise effects.

5.3 Exposure effects as constrained by an inherently biased neonatal brain?

At birth, when perceptual processes are not yet fully mature, perception of incoming odorants may be in part dictated by the state of neural architecture of the olfactory system or by the neurochemical stage of the brain. The chemosensory system is conceived as an assembly of "modules" constituted by the glomeruli in the olfactory bulb, where the OSNs synapse with the second-order mitral cells. Evidence from rodents indicates that such functional modules within olfactory bulbs process distinct categories of chemical compounds, as glomeruli drain axons from OSNs expressing a same olfactory receptor protein (e.g., Johnson & Leon, 2002; Kobayakawa et al., 2007). Although as yet we know little about the developmental dynamics of these modules, we appreciate that they are specified progressively and organized heterochronously. In the rat pup, for example, odor activation recruits only a limited set of zones in the olfactory bulbs during the first 3 postnatal days, but by day 15 odor-based bulbar activation increases to nearly adult levels (Astic & Saucier, 1982). Thus, the earliest appearing glomeruli are presumably linked with OSNs that become functionally mature in advance of others. Such earlier-developing OSNs may be caught up in detecting stimuli that mediate vital responses in newborn organisms. Thus, the spatiotemporal structure of the native olfactory system might determine which stimulus has access to the brain and to the initial guidance of behavior.

The above point extends to a single sensory system the principle of ordered onset of the whole sensorium noted by Gottlieb (1971). Later, Turkewitz, and

Kenny (1985) theorized this principle as an evolved strategy for progressivity in brain exposure to sensory activity and for reduction of competition between emerging neurosensory systems. Here, we propose the similar logic within a same sensory modality: a limited set of early functional processing modules may prevent informational overpower in the sense modality which supports actions that are critical to neonatal adaptation. In sum, apparent sensory specializations in olfaction might emerge from time-ordered neural development, leading certain chemostimuli to be more ready than others to be engaged in the control of behavior. This would imply that it is not so much the stimuli which are "innate", but that the receptive competence of the olfactory system imposes a constraint upon the nature of the stimuli that will first accede to the control of neonatal behavior.

Another early perceptual bias pertains to sensitive periods during which the brain is "expecting" sensory experience to engage the organism in species-typical development. For example, within the first postnatal hour, human infants acquire an arbitrary odor as attractive when tested 2–3 days later against a novel odorant. A similar odor exposure applied at the age of 12 hours post-birth does not come out with differential response in the same testing conditions (Romantshik et al., 2007). It appears that right after birth infants are in a neurochemical state of increased arousal due to heightened norepinephrine levels induced by labor-related activation of the locus coeruleus (Lagercrantz & Slotkin, 1986), both processes known to up-regulate odor learning in rat pups (Wilson & Sullivan, 1994). Thus, any environmental odor coinciding with this brain state may be acquired as positive. As the natural situation dictates that this odor normally originates from the mother, pups indeed acquire maternal odor cues to guide their initial vital actions. Another case of sensitive period evidenced in the rat (e.g., Moriceau & Sullivan, 2004) is based on a shift from an inability to form negative conditioning of an odor contingent with a painful treatment before day 10 into effective negative odor conditioning from day 10. This abrupt shift in the negative conditionability of odors is explained by the engagement of the amygdala (Barr et al., 2009). Thus, again, a certain state of maturation in the neural circuitry of learning can bias the way odorants are engaged into preferences.

5.4 Are some odorants unconditionally attractive/repulsive to newborns?

So far, very few odorants have been fully investigated in mammals to evaluate whether their hedonic value is unconditional, i.e., does not depend from previous direct exposure effects. Such unconditional perceptual activity can be either in positive or negative hedonic directions. On the negative side, there is evidence for avoidance of predator odors in rats and mice, despite these have not been exposed to such stimuli for generations. However, studies assessing the aversive status of predator odorants have rarely been undertaken in newborns. One study exposed mice pups to a component of fox feces during the preweaning period, and showed subsequent reduced fear-related responses (Hacquemand et al., 2010). This would suggest that trigeminal and novelty properties of predator odorants might be involved in what is considered to be "innate predator fear".

On the positive side, infant mammals approach odors associated with conspecific milk or other mammary skin secretions, even before they have been postnatally exposed to them or to any lactating female (Keil et al., 1990; Marlier & Schaal, 2005; Doucet et al., 2009; reviewed in Schaal, 2010). The chemical basis of this early attraction of milk has been studied in depth in the rabbit, where a monomolecular compound, 2-methyl-but-2-enal (MB), was found to elicit responses similar to that triggered by milk odor (Schaal et al., 2003). The selective activity of MB was ascertained not to be due to novelty or non-specific arousal effects (Coureaud et al., 2003). The species generality of the releasing potency of the MB was established in testing either pups from different breeds, pups from different colonies exposed to varied diets, or wild-type pups (Schaal et al., 2003; Coureaud et al., 2008). Finally, independence of the MB activity from prior exposure effects was verified. First, prenatal experience was ruled out in showing that the odors of amniotic fluid and of pregnant female's blood were behaviorally ineffective, and that MB is not chemically traceable in them (Schaal et al., 2003). Further, natal or postnatal learning was ruled out in testing pups deprived of previous MP exposure. Remarkably, the MB response of rabbit neonates remained fully functional despite deprivation in the days following birth (Coureaud et al., 2000).

Thus, at least in the case of rabbit newborns, there is evidence for odorants that can unconditionally release approach and oral capture in the minutes following birth. Functionally similar attractive volatile compound intervening in mother-infant interactions have been proposed in the rat. Dimethyl disulfide (DMDS) was shown to elicit nipple grasping in pups, but with moderate releasing potency (about 50%) relative to that of nipples themselves (Pedersen & Blass, 1981). But, in these studies, the rat pups were aged 3–5 days at testing, implying that the meaning of DMDS might derive from extensive exposure during nursing prior to testing. Thus, until DMDS is further assessed for responsiveness in newly born pups, its signal value might be attributable to postnatal experience.

5.5 Do some odorants act as primary reinforcers in newborns?

A stimulus that is rewarding without learning is termed a primary reinforcing stimulus (Rolls, 2011). It not only releases acceptance (or rejection) in itself, but additionally can contribute to convert an initially neutral stimulus into an attractive (or aversive) stimulus. Taste is considered mediating such primary reinforcing effects. Sweet taste releases indeed acceptance and alters the hedonic value of any contingent stimulus (e.g., Blass et al., 1984). It is unclear so far whether odorant compounds exist that would similarly mediate primary reinforcing effects (apart from pungent/irritant compounds that could support aversion learning). Some odor compounds are often cited as candidates – such as pheromones, or odorants derived from rotting food or fire (e.g., Rolls, 2011), but their "primary" reinforcing value has never been adequately examined.

One odorant compound may be considered as a candidate primary reinforcing agent: the MB of rabbit milk. As mentioned above, MB releases a stereotyped response right at birth, and this response is impervious to postnatal experience during the first postnatal days (Montigny et al., 2008). During this period, deprivation of MB does not alter the pups' response to it. Further, enforcing MB-deprived pups to learn conventional odorants up to day 6 has no effect on subsequent reactivity to MB. But, conversely, MB strongly affects the learning of conventional odorants. Any odorant made contingent with MB incorporates its strong releasing properties after a single, 15-second pairing episode (Coureaud et al., 2006). For a conventional odorant to become attractive in pups without MB appears comparatively slow. In sum, although both unconditional and plastic olfactory mechanisms are dissociable, they are functionally, but asymmetrically, linked. The primary odorant reinforcer appears to tutor plastic olfactory mechanisms in rabbit pups.

However, the tutoring potency of MB over olfactory learning is time-bound in several ways. First, MB-induced odor learning functions only during the first 4–5 postnatal days (Montigny et al., 2006), demarcating a sensitive period. Second, MB-induced response shifts from initial automatism in neonates to progressively become controlled by the metabolic needs of growing pups (Montigny, 2007). In sum, rabbit pups begin life with a powerful olfactory mechanism for bootstrapping their initial attempts in finding their way to the source of milk. MB constitutes an ontogenetic adaptation that potently increases the repertoire of meaningful olfactory cues.

It may be noted that the MB-response loop described in neonatal rabbits satisfies all eight innateness criteria proposed by Mameli and Bateson (see above). But decreeing that this stimulus-response system is "innate" or "inborn" does not really advance our understanding of the dynamic processes through which it comes up to act as an automatic releaser and a primary reinforcer. How, when, and at which level of the olfactory system do genes and experience interrelate to produce such a phenotypic trait? Which are the interactive processes in the embryonic environment and gene networks that selectively operate to produce sensory neurons in the olfactory placode that will express receptor proteins that capture MB? Are there MB-specific OSNs and do they have broad/restricted distribution in the olfactory mucosae? Do axons of these hypothetically MB-specific OSNs correspondingly target a wide array of glomeruli in the olfactory bulb or only a circumscribed set of glomeruli? Are OSN-glomeruli chains dedicated to process the MB developmentally ahead from other OSN-glomeruli chains? Beyond the olfactory bulb, which neural network is recruited, and when, in the generation of arousal, motivation, hedonic reward, coordinated motor response, and memory? And finally, when the MB-response loop wanes around weaning, what is the fate of its neural underpinnings?

We now have multiple examples of stimulus-behavior systems that are supported by dedicated populations of OSNs, mostly in invertebrates (e.g., Kurtovic et al., 2007; Semmelhack & Wang, 2009; Wisotsky et al., 2011). When these receptive structures are lesioned or genetically ablated, corresponding perceptual skills and behavior are severely perturbed or abolished. In vertebrates also cases of well-defined behaviors depend on subsets of OSNs that could be genetically targeted (e.g., Kobayakawa et al., 2007; Munger et al., 2010). In mice, a specialized group of OSNs expressing a certain type of receptor reacts to carbon disulfide, an odor compound emitted from the mouth of rodents and which promotes attraction and learning of associated food odors (Munger et al., 2010). Thus, in certain cases where hard issues can be rigorously addressed with new experimental tools, the genetic determinants of sensory processes and their relation to behavior can be established.

6. Conclusion

Through their rather radical claims on neonatal odor preferences as determined either by exclusive learning or by innate processes, T. Engen and J. Steiner raised important questions. After delving into the current literature in the field, this preliminary essay attempted to sort out the validity of these claims and their implications for the understanding of the earliest manifestations of olfactory hedonism. Ultimately both esteemed mentors were partially accurate: on the one hand, newborn brains are competent to support odor-based preferences and to actualize them in behavior from the very first minutes of aerial life; on the other hand, early odor preferences are learned, or, more broadly, depend on the chemosensory experience previously encoded in the fetal environment. But most significant are the questions raised in challenging Engen's and Steiner's views. First, in concert with many other scientists, they highlight newborn organisms as fascinating model systems for analyzing the development of perception and behavior. But neonatal skills do not emerge from "nothing", as implied by the empiricist view, nor do they emerge as bundles of reflexes from a genetically pre-specified plan, as implied by the nativistic view. Neonatal organisms are not "virgin" of previous experience, but are the transitory product of interweaving biological, including genetic, and experiential processes during development. Second, when a behavioral ability is well characterized at an early point in development, investigations should not rest on hollow concepts such as those mobilized by the nature-nurture dichotomy. One ideal illustration is human response to sweet and bitter tastes. Based on sucrose or quinine-elicited stereotyped responses in newborns, the general sentence of taste experts to explain hedonic responses to sweet/bitter taste is that they are "innate". This opinion recurs since the earliest observers (e.g., Kroner, 1882; Preyer, 1884) up to modern investigators (Steiner et al., 2000; Rolls, 2011), assigning a quasi-dogmatic status to the notion of taste innateness. However, any dogma is there to be put under test. Recent evidence indicates indeed that sweet taste is strongly influenced by exposure to sweetness during fetal life (e.g., Zhang et al., 2011). Finally, the notions discussed in this assay underscore the power of language in scientific tradition. They show that denominations can be helpful in the toolbox of science, as they can sometimes be distracting for long periods to engage further inquiry.

References

Alaoui-Ismaili, O., Vernet-Maury, E., Dittmar, A., Delhomme, G., & Chanel, J. (1997). Odor hedonics: connection with emotional response estimated by autonomic parameters. *Chemical Senses, 22*, 237–248.

Alberts, J. R. (1984). Ontogeny of social recognition: An essay on mechanism and metaphor in behavioral development. In E. S. Gollin (Ed.), *The comparative development of adaptive skills* (pp. 95–136). Hilldale, NJ: Erlbaum.

Alberts, J. R. (1987). Early learning and ontogenetic adaptation. In N. A. Krasnegor, E. M. Blass, M. A. Hofer, & W. P. Smotherman (Eds.), *Perinatal Development, A Psychobiological perspective* (pp. 12–38). Orlando, FA: Academic Press.

Alberts, J. R., & Decsy, G. J. (1990). Terms of endearment. *Developmental Psychobiology, 23*, 569–584.

Astic, L., & Saucier, D. (1982). Ontogenesis of the functional activity of rat olfactory bulb: Autoradiographic study with the 2-deoxyglucose method. *Developmental Brain Research, 2*, 1243–1256.

Barr, G. A., Moriceau, S., Shionoia, K., Muzny, K., Goa, P., et al. (2009). Transitions in infant learning are modulated by dopamine in the amygdala. *Nature Neuroscience, 12*, 1367–1369.

Bartocci, M., Winberg, J., Papendieck, G., Mustica, T., Serra, G., & Lagercrantz, H. (2001). Cerebral hemodynamic response to unpleasant odors in the preterm newborn measured by near-infrared spectroscopy. *Pediatric Research, 50,* 324–330.

Bateson, P. P. G. (2001). Taking the stink out of instinct. In H. Rose, & S. Rose (Eds.), *Alas poor Darwin: Arguments against evolutionary psychology* (pp. 157–173). London, UK: Vintage.
Berlyne, D. E. (1960). *Conflict arousal and curiosity.* New York: McGraw-Hill.
Berntson, G. G., Tuber, D. S., Ronca, A. E., & Bachman, D. S. (1983). The decerebrate human: associative learning. *Experimental Neurology, 81*, 77–88.
Bingham, P. M., Abassi, S., & Sivieri, E. (2003). A pilot study of milk odor effects on nonnutritive sucking by premature infants. *Archives in Pediatric and Adolescent Medicine, 157,* 72–75.
Blass, E. M. (1990). Suckling: Determinants, changes, mechanisms, and lasting impressions. *Developmental Psychology, 26,* 520–533.
Blass, E. M., Ganchrow, J. R., & Steinerm J. E. (1984). Classical conditioning in newborn humans 2–48 hours of age. *Infant Behavior and Development, 7,* 223–235.
Bolhuis, J. J. (1996). Development of perceptual mechanisms in birds: Predispositions and imprinting. In C. F. Moss, & S. J. Shettleworth (Eds.), *Neuroethological Studies of Cognitive and Perceptual Processes* (pp. 158–184). Boulder, CO: Westview Press.
Brunjes, P. C., & Frazier, L. L. (1986). Maturation and plasticity in the olfactory system. *Brain Research Reviews, 11*, 1–45.
Christie, I. C., & Friedman, B. H. (2004). Autonomic specificity of discrete emotion and dimensions of affective space: A multivariate approach. *International Journal of Psychophysiology, 51,* 143–153.
Chuah, M. I., & Zheng, D. R. (1987). Olfactory marker protein is present in olfactory receptor cells of human fetuses. *Neuroscience, 23,* 363–370.
Coureaud, G., Schaal, B., & Langlois, D. (2001). The interaction between pheromone-elicited and odour-elicited behaviour in the newborn rabbit. *Chemical Senses, 26*, 781.
Coureaud, G., Langlois, D., Perrier, G., & Schaal, B. (2003) A single key-odorant accounts for the pheromonal effect of rabbit milk: Further test of the mammary pheromone's activity against a wide sample of volatiles from milk. *ChemoEcology, 13,* 187–192.
Coureaud, G., Montigny, D., Dewas, M., Moncomble, A. S., Patris, B., & Schaal, B. (2005). *A sensitive period for pheromone-induced odour learning in newborn rabbits.* 24th International Ethological Conference, Budapest, Hungary, 20–27 August.
Coureaud, G., Moncomble, A. S., Montigny, D., Dewas, M., Perrier, G., & Schaal, B. (2006). A pheromone that rapidly promotes learning in the newborn. *Current Biology, 16,* 1956–1961.
Coureaud, G., Rödel, H., Kurz, C. A., & Schaal, B. (2008). Age dependent responsiveness to the mammary pheromone in domestic and wild rabbits. *ChemoEcology, 18*, 52–59.
Davis, L. B., & Porter, R. H. (1991). Persistent effects of early odour exposure on human neonates. *Chemical Senses, 16,* 169–174.
Delaunay-El Allam, M., Soussignan, R., Patris, B., & Schaal, B. (2010). Long lasting memory for an odor acquired at the mother's breast. *Developmental Science, 13*, 849–863.
Delplanque, S., Grandjean, D., Chrea, C., Coppin, G., Aymard, L., et al. (2009). Sequential unfolding of novelty and pleasantness appraisals of odors: Evidence from facial electromyography and autonomic reactions. *Emotion, 9*, 316–328.
Doty, R. L. (1975). An examination of relationships between pleasantness, intensity and concentration of 10 odorous stimuli. *Perception & Psychophysics, 17*, 492–496.
Doty, R. L., Brugger, W. E., Jurs, P. C., Orndorff, M. A., Snyder, P. I., & Lowry, L. D. (1978). Intranasal trigeminal stimimulation from odorous volatiles: Psychometric responses from anosmic and normal humans. *Physiology & Behavior, 20,* 175–185.

Doucet, S., Soussignan, R., Sagot, P., & Schaal, B. (2009). The secretion of areolar (Montgomery's) glands from lactating women elicits selective, unconditional responses in neonates. *PLoS One, 4*, e7579.

Doucet, S., Soussignan, R., Sagot, P., & Schaal, B. (2011). An overlooked aspect of the human breast: Areolar glands in relation with breastfeeding pattern, neotatal weight gain and dynamics of lactation. *Early Human Development* (on line, 16 August).

Döving, K. B., & Pinching, A. J. (1973). Selective degeneration of neurons in the olfactory bulb following prolonged odor exposure. *Brain Research, 52*, 115–129.

Engen, T. (1965). Psychophysical analysis of the odor intensity of homologous alcohols. *Journal of Experimental Psychology, 70*, 611–616.

Engen, T. (1974). Method and theory in the study of odor preferences. In I. W. Johnson, D. G. Moulton, & A. Türk (Eds.), *Human responses to environmental odors* (pp. 121–141). London: Academic Press.

Engen, T. (1982). *The perception of odors*. New York: Academic Press.

Engen, T. (1986). Children's sense of smell. In H. Meiselman, & R. S. Rivlin (Eds.), *Clinical measurement of taste and smell* (pp. 316–325). New York: Macmillan.

Engen, T. (1988). The acquisition of odor hedonics. In S. van Toller, & G. H. Dodd (Eds.), *Perfumery: The psychology and biology of fragrance* (pp. 79–104). London: Chapman & Hall.

Engen, T. (1991). *Odor sensation and memory*. New York: Praeger.

Engen, T., Lipsitt, L. P., & Kaie, H. (1963). Olfactory response and adaptation in the human neonate. *Journal of Comparative and Physiological Psychology, 56*, 73–77.

Engen, T., & Lipsitt, L. P. (1965). Decrement and recovery of responses to olfactory stimuli in the human neonate. *Journal of Comparative and Physiological Psychology, 59*, 312–316.

Engen, T., Cain, W. S., & Rovee, C. K. (1968). Direct scaling of olfaction in the newborn infant and the adult human observer. In N. Tanyolac (Ed.), *Theories of odors and odor measurement* (pp. 271–294). New York: Spartan Books.

Faas, A. E., Sponton, E. D., Moya, P., & Molina J. C. (2000). Differential responsiveness to alcohol odor in human neonates. Effects of maternal consumption during gestation. *Alcohol, 22*, 7–17.

Field, T. (2007). *The amazing infant*. Malden, MA: Blackwell.

Gaugler, C., Messer, J., Marlier, R., Soussignan, R., & Schaal, B. (2003). Olfaction in premature newborns: detection, discrimination and hedonic categorization. *Biology of the Neonate, 84*, 268.

Gottlieb, G. (1971). Ontogenesis of sensory function in birds and mammals. In E. Tobach, L. Aronson, & E. Shaw (Eds.), *The biopsychology of development* (pp. 67–128). New York: Academic Press.

Gottlieb, G. (1976). Conceptions of prenatal development: behavioral embryology. *Psychological Review*, 83, 215–234.

Gottlieb, G. (2007). Probabilistic epigenesis. *Developmental Science, 10*, 1–11.

Goubet, N., Rattaz, C., Pierrat, V., Alléman, E., Bullinger, A., & Lequien, P. (2002). Olfactory familiarization and discrimination in preterm and full-term newborns. *Infancy, 3*, 53–75.

Goubet, N., Rattaz, C., Pierrat, V., Bullinger, A., & Lequien, P. (2003). Olfactory experience mediates response to pain in preterm newborns. *Development Psychobiology, 42*, 171–180.

Goubet, N., Strasbaugh, K., & Chesney, J. (2007). Familiarity breeds content? Soothing effect of a familiar odor on full-term newborns. *Journal of Developmental and Behavioral Pediatrics, 28*, 189–194.

Gross-Isseroff, R., & Lancet, D. (1988). Concentration-dependent changes of perceived odor quality. *Chemical Senses, 13*, 191–204.
Hacquemand, R., Pourié, G., Jacquot, L., & Brand, G. (2010). Postnatal exposure to synthetic predator odor (TMT) induces quantitative modification in fear-related behaviors during adulthood without change in corticosterone level. *Behavioural Brain Research, 215*, 58–62.
Haddad, R., Weiss, T., Khan, R., Nadler, B., Mandairon, N., et al. (2010). Global features of neural activity in the olfactory system form a parallel code that predicts olfactory behavior and perception. *Journal of Neuroscience, 30*, 9017–9026.
Haller, R., Rummel, C., Henneberg, S., Pollmer, U., & Köster, E. P. (1999). The effect of early experience with vanillin on food preference later in life. *Chemical Senses, 24*, 465–467.
Hepper, P. G. (1995). Human fetal "olfactory" learning. *International Journal of Prenatal, Perinatal and Psychological Medicine, 7*, 147–151.
Horn, G. (2004). Pathways of the past: The imprint of memory. *Nature Reviews Neuroscience 5*, 108–120.
Irzhanskaia, K. N., & Felberbaum, R. A. (1954). Conditioned reflex activity in premature infants. *Fisiologikeski Zhurnal SSSR, 40*, 668–672.
Johnson, B. A., & Leon, M. (2002). Modular representations of odorants in the glomerular layer of the rat olfactory bulb and the effects of stimulus concentration. *Journal of Comparative Neurology, 422*, 496–509.
Johnson, M. H., & Morton, J. (1991). *Biology and cognitive development. The case of face recognition.* Oxford: Blackwell.
Jones, S. V., Choi, D. C., Davis, M., & Ressler, K. J. (2008). Learning-dependent structural plasticity in the adult olfactory pathways. *Journal of Neuroscience, 28*, 13106–13111.
Joseph, R. (2000). Fetal brain, behavior and cognitive development. *Developmental Review, 20*, 81–98.
Karmiloff-Smith, A. (2001). Why baby brains are not Swiss Army knives. In H. Rose, & S. Rose (Eds.), *Alas poor Darwin: Arguments against evolutionary psychology* (pp. 144–156). London, UK: Vintage.
Keil, W., von Stralendorff, F., & Hudson, R. (1990) A behavioral bioassay for analysis of rabbit nipple-search pheromone. *Physiology and Behavior, 47*, 525–529.
Khan, R., Luk, C. H., Flinker, A., Agarwall, A., Lapid, H., Haddad, R., & Sobel, N. (2007). Predicting odor pleasantness from odorant structure: pleasantness as a reflection of the physical world. *Journal of Neuroscience, 27*, 10015–10023.
Kobayakawa, K., Kobayakawa, R., Matsumoto, H., Oka, Y., Imai, T., et al. (2007). Innate versus learned odour processing in the mouse olfactory bulb. *Nature, 450*, 503–510.
Kroner, T. (1882). Über die Sinnesempfindungen des Neugeborenen. *Bresslauer Ärztliche Zeitschrift, 4*, 20–55.
Kurtovic, A., Widmer, A., & Dickson, B. J. (2007). A single class of olfactory neurons mediates behavioural responses to a Drosophila sex pheromone. *Nature, 446*, 542–546.
Lagercrantz, H., & Slotkin, T. A. (1986). The "stress" of being born. *Scientific American, 254*, 100–107.
Laing, D. G., & Panhuber, H. (1978). Neural and behavioral changes in rats following continuous exposure to an odor. *Journal of Comparative Physiology, 124*, 259–265.
Lehrman, D. S. (1953). A critique of Konrad Lorenz's theory of instinctive behavior. *Quarterly Review of Biology, 28*, 337–363.

Lewkowicz, D. J. (2011). The biological implausibility of the nature-nurture dichotomy and what it means for the study of infancy. *Infancy,* doi: 10.1111/j.1532-7078.2011.00079.x.

Lickliter, R., & Honeycutt, H. (2003). Developmental dynamics: Toward a biologically plausible evolutionary psychology. *Psychological Bulletin, 129,* 819–835.

Macfarlane, A. (1975). Olfaction in the development of social preferences in the human neonate. *Ciba Foundation Symposia, 33,* 103–113.

Mameli, M., & Bateson, P. (2006). Innateness and the sciences. *Biology and Philosophy, 21,* 155–188.

Mandairon, N., Poncelet, J., Bensafi, M., & Didier, A. (2009). Humans and mice express similar olfactory preferences. *PLoS ONE, 4,* e4209.

Marlier, L., Schaal, B., & Soussignan, R. (1998). Neonatal responsiveness to the odor of amniotic and lacteal fluids: A test of perinatal chemosensory continuity. *Child Development, 69,* 611–623.

Marlier, L., Schaal, B., Gaugler, C., & Messer, J. (2001). Olfaction in premature newborns: Detection and discrimination abilities two months before gestational term. *Chemical Signals in Vertebrates, 9,* 205–209.

Marlier, L., & Schaal, B. (2005). Human newborns prefer human milk: Conspecific milk odor is attractive without postnatal exposure. *Child Development, 76,* 155–168.

Mennella, J. A., Jagnow, C. P., & Beauchamp, G. K. (2001). Pre- and post-natal flavor learning by human infants. *Pediatrics 107,* 1–6.

Michel, G. F., & Moore, C. L. (1995). *Developmental Psychobiology, An interdisciplinary science.* Cambridge, MA: Bradford-MIT Press.

Mizuno, K., & Ueda, A. (2004). Antenatal olfactory learning influences infant feeding. *Early Human Development, 76,* 83–90.

Montigny, D. (2008). *Fonctions adaptatives immédiates et différées de la phéromone mammaire chez le lapereau.* Villetaneuse, France: Doctoral thesis, University of Paris 13.

Montigny, D., Coureaud, G., & Schaal, B. (2006). Shift from automatism to prandial control in the response of newborn rabbits to the mammary pheromone, *Physiology and Behavior, 89,* 742–749.

Moore, D. (2001). *The dependent gene: the fallacy of nature vs. nurture.* New York: W.H. Freeman.

Moriceau, S., & Sullivan, R. (2004). Unique neural circuitry for neonatal olfactory learning. *Journal of Neuroscience, 24,* 1182–1189.

Munger, S. D., Leinders-Zufall, T., McDougall, L. M., Cockerham, R. E., Schmid, A., et al. (2010). An olfactory subsystem that detects carbon disulfide and mediates food-related social learning. *Current Biology, 20,* 1438–1444.

Pedersen, P. A., & Blass, E. M. (1981). Olfactory control over suckling in albino rats. In R. N. Aslin, J. R. Alberts, & M. R. Petersen (Eds.), *Development of perception, Psychobiological perspectives, Audition somatic perception and the chemical senses* (pp. 359–381). New York: Academic Press.

Perl, E., Shay, U., Hamburger, R., & Steinerm, J. E. (1992). Taste- and odor-reactivity in elderly demented patients. *Chemical Senses, 17,* 779–794.

Pihet, S., Schaal, B., Bullinger, A., & Mellier, D. (1996). An investigation of olfactory responsiveness in premature newborns. *Infant Behavior and Development, ICIS Issue: 676.*

Pinker, S. (2002). *The blank slate: The modern denial of human nature.* New York: Viking Press.

Poncelet, J., Rinckm, F., Bourgeat, F., Schaal, B., Rouby, C., Bensafi, M., & Hummel, T. (2010). The effect of early experience on odor perception in humans: Psychological and physiological correlates. *Behavioural Brain Research, 208,* 458–465.
Porter, R., Makin, J. W., Davis, L. B., & Christensen, K. M. (1991). An assessment of the salient olfactory environment of formula-fed infants. *Physiology and Behavior, 50,* 907–911.
Preyer, W. (1884). *Die Seele des Kindes.* Leipzig: Grieben.
Rieser, J., Yonas, A., & Wilkner, K. (1976). Radial localization of odors by human newborns. *Child Development, 47,* 856–859.
Ripin, R., Hetzer, H., & Bühler, C. (1930). Frühestes Lernen des Säuglings in der Ernährungssituation. *Zeitschrift für Psychologie, 118,* 82–127.
Rolls, E. T. (2011). Chemosensory learning in the cortex. *Frontiers in Systems Neuroscience, 5,* 1–13.
Romantshik, O., Porter, R. H., Tillmann, V., & Varendi, H. (2007). Preliminary evidence of a sensitive period for olfactory learning by human newborn. *Acta Paediatrica, 96,* 372–376.
Ronca, A. E., Berntson, G. G., & Tuberm, D. S. (1985). Cardiac orienting and habituation to auditory and vibrotactile stimuli in the infant decerebrate rat. *Developmental Psychobiology, 18,* 545–558.
Rose, H., & Rose, S. (2001). *Alas poor Darwin: Arguments against evolutionary psychology.* London, UK: Vintage.
Rosenblatt, J. S. (1983). Olfaction mediates developmental transitions in the altricial newborn of selected species of mammals. *Developmental Psychobiology, 16,* 347–375.
Rovee, C. K. (1969). Psychophysical scaling of olfactory response to aliphatic alcohols in human neonates. *Journal of Experimental Child Psychology, 7,* 245–254.
Rovee, C. K. (1972). Olfactory cross-adaptation and facilitation in human neonates. *Journal of Experimental Child Psychology, 13,* 368–381.
Russell, M. J. (1976). Human olfactory communication. *Nature, 260,* 520–522.
Sarnat, H. B. (1978). Olfactory reflexes in the newborn infant. *Journal of Pediatrics 92,* 624–626.
Schaal, B. (1988). Olfaction in infants and children: Developmental and functional perspectives. *Chemical Senses, 13,* 145–190.
Schaal, B. (2005). From amnion to colostrum to milk: Odor bridging in early developmental transitions. In B. Hopkins, & S. P. Johnson (Eds.), *Prenatal development of postnatal functions* (pp. 51–102). London: Praeger.
Schaal, B. (2010). Mammary odor cues and pheromones: Mammalian infant-directed communication about maternal state, mammae, and milk. *Vitamins & Hormones, 83,* 83–136.
Schaal, B., & Orgeur, P. (1992). Olfaction in utero: can the rodent model be generalized? *Quarterly Journal of Experimental Psychology. Comparative and Physiological Psychology, 44B,* 245–278.
Schaal, B., Orgeur, P., & Rognon, C. (1995). Odor sensing in the human fetus: Anatomical, functional and chemo-ecological bases. In J. P. Lecanuet, N. A. Krasnegor, W. A. Fifer, & W. P. Smotherman (Eds.), *Prenatal Development, Psychobiological Perspective* (pp. 205–237). Hillsdale, NJ: Erlbaum.
Schaal, B., Marlier, L., & Soussignan, R. (2000). Human foetuses learn odours from their pregnant mother's diet. *Chemical Senses, 25,* 729–737.
Schaal, B., Coureaud, G., Langlois, D., Giniès, C., Sémon, E., & Perrierm G. (2003). Chemical and behavioural characterization of the rabbit mammary pheromone. *Nature, 424,* 68–72.

Schaal, B., Hummel, T., & Soussignan, R. (2004). Olfaction in the fetal and premature infant: Functional status and clinical implications. *Clinics in Perinatology, 31*, 261–285.

Schaal, B., Doucet, S., Soussignan, R., Rietdorf, M., Weibchen, G., & Francke, W. (2008). The human breast as a scent organ: Exocrine structures, secretions, volatile components, and possible function in breastfeeding interactions. In J. L. Hurst, R. J. Beynon, S. C. Roberts, & T. D. Wyatt (Eds.), *Chemical Signals in Vertebrates,* Vol. 11 (pp. 325–335). New York: Springer.

Schaal, B., Coureaud, G., Doucet, S., Delaunay-El Allam, M., Moncomble, A. S., et al. (2009). Olfactory mammary signalisation in females and neonatal odour processing: Ways evolved in rabbit and human. *Behavioral Brain Research, 200,* 346–358.

Schaal, B., & Durand, K. (2012). Olfaction in human multisensory development. In A. Bremner, D. Lewkowicz, & C. Spence (Eds.), *Multisensory development.* Oxford: Oxford University Press.

Schleidt, M., & Genzel, C. (1990). The significance of mother's perfume for infants in the first weeks of their life. *Ethology and Sociobiology, 11,* 145–154.

Schneirla, T. C. (1965). Aspects of stimulation and organization in approach/withdrawal processes underlying vertebrate behavioural development. In D. S. Lehrman, R. A. Hinde, & E. Shaw (Eds.), *Advances in the Study of Behaviour* (pp. 1–74). New York: Academic Press.

Semmelhack, J. L., & Wang, J. W. (2009). Select Drosophila glomeruli mediate innate olfactory attraction and aversion. *Nature, 459,* 218–223.

Smotherman, W. P., & Robinson, S. R. (1987). Psychobiology of fetal experience in the rat. In N. E. Krasnegor, E. M. Blass, M. A. Hofer, & W. P. Smotherman (Eds.), *Perinatal development: A psychobiological perspective* (pp. 39–60). Orlando, FA: Academic Press.

Soussignan, R., & Schaal, B. (2004). Emotional processes in human newborns: A functionalist perspective. In J. Nadel, & D. Muir (Eds.), *Emotional Development* (pp. 127–159). Oxford: Oxford University Press.

Soussignan, R., Schaal, B., Marlier, L., & Jiang, T. (1997). Facial and autonomic responses to biological and artificial olfactory stimuli in human neonates: Re-examining early hedonic discrimination of odors. *Physiology & Behavior, 62,* 745–758.

Spelke, E. S., & Newport, E. L. (1998). Nativism, empiricism, and the development of knowledge. In W. Damon, & R. M. Lerner (Eds.), *Handbook of child psychology,* Volume 1 Theoretical models of human development (pp. 275–340). New York: Wiley.

Spencer, J. P., Blumberg, M. S., McMurray, B., Robinson, S. R., et al. (2009). Short arms and talking eggs: Why we should no longer abide the nativist-empiricist debate. *Child Development Perspectives, 3,* 79–87.

Steiner, J. E. (1974). Innate, discriminative human facial expressions to taste and smell stimulation. *Annals of the New York Academy of Sciences, 237,* 229–233.

Steiner, J. E. (1977). Facial expressions of the neonate infant indicating the hedonics of food-related chemical stimuli. In J. M. Weiffenbach (Ed.), *Taste and Development. The genesis of sweet preference* (pp. 173–204). Bethesda, MD: US DHEW.

Steiner, J. (1979). Human facial expressions in response to taste and smell stimulations. In L. P. Lipsitt, & H. Reese (Eds.), *Advances in child development,* Vol. 13 (pp. 257–295). New York: Academic Press.

Steiner, J. E., Glaser, D., Hawilo, M. E., & Berridge, K. (2001). Comparative evidence of hedonic impact: Affective reactions to taste by human infants and other primates. *Neuroscience and Biobehavioral Reviews, 25,* 53–74.

Stirnimann, F. (1936). Versuche über Geschmack und Geruch am ersten Lebenstag. *Jahrbuch für Kinderheilkunde, 146,* 211–227.

Sullivan, R. M., Taborsky-Barba, S., Mendoza, R., Itano, A., Leon, M., Cotman, C. W., et al. (1991). Olfactory classical conditioning in neonates. *Pediatrics, 87,* 511–517.

Todrank, J., Heth, G., & Restrepo, D. (2011). Effects of in utero odorant exposure on neuroanatomical development of the olfactory bulb and odor preferences. *Proceeding of the Royal Society B, 279,* 1949–1955.

Turkewitz, G., & Kenny, P. A. (1985). The role of developmental limitations of sensory input on sensory/perceptual organization. *Journal of Developmental and Behavioral Pediatrics, 15,* 357–368.

Varendi, H., Porter, R. H., & Winberg, J. (1994). Does the newborn baby find the nipple by smell? *Lancet, 344,* 989–990.

Varendi, H., & Porter, R. H. (2001). Breast odor as the only maternal stimulus elicits crawling towards the odour source. *Acta Paediatrica, 90, 372*–375.

Wang, H., Wysocki, C. J., & Gold, G. (1993). Induction of olfactory receptor sensitivity in mice. *Science, 260,* 998–1000.

Wilson, D. A., & Sullivan, R. (1994). Neurobiology of associative learning in the neonate: Early olfactory learning. *Behavioral and Neural Biology, 61,* 1–18.

Wisotsky, Z., Medina, A., Freeman, E., & Dahanukar, A. (2011). Evolutionary differences in food preferences rely on Gr64e, a receptor for glycerol. *Nature Neuroscience, 14,* 1534–1541.

Yee, K. K., & Wisocki, C. J. (2001). Odorant exposure increases olfactory sensitivity: Olfactory epithelium is implicated. *Physiology & Behavior, 72,* 705–711.

Youngentob, S. L., & Kent, P. F. (1995). Enhancement of odorant-induced mucosal activity patterns in rats trained on an odorant identification task. *Brain Research, 670,* 82–88.

Youngentob, S. L., Kent, P. F., Sheehe, P. R., Molina, J. C., Spear, N., & Youngentob, L. M (2007). Experience-induced fetal plasticity: The effect of gestational ethanol exposure on the behavioral and neurophysiologic olfactory response to ethanol odor in early postnatal and adult rays. *Behavioral Neuroscience, 121,* 1293–1305.

Zhang, G. H., Chen, M. L., Liu, S. S., Zhan, Y. H., Quan, Y., et al. (2011). Effects of mother's dietary exposure to acesulfam-K in pregnancy and lactation on the adult offspring sweet preference. *Chemical Senses, 36,* 763–770.

CHAPTER 14

The acquisition of odour preferences via evaluative olfactory conditioning

Historical background and state of the art

Gesualdo M. Zucco
University of Padova, Italy

Odours are powerful in bringing back distant memories linked to emotional content. This inherent hedonic property of olfactory stimuli, together with poor accessibility to semantic processing and language, makes odour particularly suitable for studies on evaluative conditioning, a variant of Pavlovian classical conditioning, through which a neutral stimulus acquires the affective valence of another stimulus with which it is paired (Hermans & Baeyens, 2002; Levey & Martin, 1975, 1987). Investigations on olfactory evaluative learning have undergone a surge in recent decades, ranging from literature on taste aversions, emergence of phobias, unconscious and subliminal conditioning, and cross-modal conditioning (De Howver et al., 2001; Li et al., 2007; Rozin et al., 1998; Zucco et al., 2009), to the effect of odour conditioning on consuming attitude, cognitive performance, and somatic syndromes in humans (Schneider et al., 1999; Van den Bergh et al., 1999; Vermetten & Bremner, 2003; Zucco, Militello, & Doty, 2008). The aim of this chapter is, therefore, to examine the most relevant literature on preference acquisition via evaluative olfactory conditioning, from the earliest demonstrations to the most recent investigations.

Introduction

Several studies have supported Engen's assertion (1982, 1988) that the hedonic characteristic of odour dominates odour perception. Notably, available evidence shows that odour cues reliably differ from verbal and visual cues as they evoke more emotional, affective, old, and evocative personal memories (Chu & Downes, 2000; Herz, 2004; Larsson & Willander, 2009; Zucco, Aiello, Turuani, & Koester, 2012; cf., also Herz, this book). Consistent with these outcomes, physiological records during the recall of odour-evoked autobiographical episodes have shown significantly greater activation in the amygdala and hippocampal regions (respectively involved in emotion and memory) only by emotionally valenced olfactory stimuli, compared to other cues, therefore, providing support to behavioural data (Royet, Zald, Versace, et al., 2000; Herz, Eliassen, Beland, & Souza, 2004).

These properties make the sense of olfaction particularly suitable for studying associative learning (Wassermann & Miller, 1997), namely a phenomenon through which separated events come to be linked by experience. Indeed, the hedonic (pleasant-unpleasant) dimension of an odorant is usually due to an associative process between the odorant itself and the situation during which it was experienced first. On these grounds, a neutral odorant can acquire a disliking valence if it is paired with an unpleasant event (for example, smelling the odour of a food associated with nausea or emesis).

Within the field of cognitive psychology, evaluative conditioning is a process that better fits this form of associative learning through which an olfactory conditioning stimulus acquires hedonic valence. Such a process was mainly investigated in the visual and verbal sensory modalities. However, in recent years, a significant body of research has accumulated in the olfactory domain (e.g., Engen, 1988; 1991; Herz, 2001; Hermans & Baeyens, 2002; Levey & Martin, 1975, 1987; Rozin, Wrzesniewski, & Byrnes, 1998; Van Reekum, Van den Bergh, & Frijda, 1999).

This chapter focuses on the historical background of evaluative olfactory conditioning, on its domains of application within cognitive psychology, and on its relevance in better understanding the processes involved in the acquisition of odour preferences and formation of human emotions.

1. Classical and evaluative conditioning: General concepts

In the beginning of the 20th century, a Russian physiologist, Ivan P. Pavlov, developed an associative learning paradigm based on conditioned reflexes. This paradigm was later successful in explaining not only elementary acquired reflexives and autonomic responses in humans and other animals, such as salivary responses, but also cognitively-guided behaviours, such as beliefs and attitudes. Pavlovian conditioning was studied in depth for the first half of the 20th century by behavioural psychologists; later, in the 1960s, it was enriched both methodologically and in the number of phenomena that could be investigated by cognitive psychologists as well.

According to the model (Pavlov, 1927), the repeated association between a neutral stimulus (NS), such as a sound, and an unconditioned stimulus (US), such as food, creates an associative link between the neutral stimulus and the US. Thus, a second step, the presentation of the neutral stimulus alone (now termed conditioned stimulus, or CS) induces the salivary conditioned response (CR) previously elicited by the US as unconditioned response (UR).

Our daily-life is influenced by numerous situations in which this model is supposed to find its concrete application, for example, in the powerful impact of advertisement and consumer science in influencing human conditioned preferences, in the acquisition of odour/taste aversions, and, specifically, in the ways in which most phobias are acquired.

The earliest studies on classical conditioning focused on reflexive responses such as eye blink, salivary, and galvanic skin response. They allowed researchers to establish some principles and processes characterising classical conditioning, which can be summarised as follows: (1) Conditioning is most effective when the CS precedes the US or at least when both are presented simultaneously; (2) human beings and other animals respond passively when exposed to CSs (i.e., it is difficult to voluntarily avoid a CR, for example, salivation, after exposure to a CS light); (3) CRs are mediated mostly by the autonomic nervous system. Furthermore, once acquired, the CR is subject to other processes, one of which is the generalisation to stimuli that are similar to the original CS (Pavlov's dogs, for instance, also responded to sounds varying in frequency from those of the original CS). The intensity of the generalised response depends, however, on the distance in similarity between the original and generalised CS (i.e., the greater the distance, the weaker the response), a phenomenon known as "generalisation gradient."

Another process of classical conditioning is extinction, that is, the suppression of the acquired CR. It occurs as a consequence of several presentations of the CS no longer followed by the US; through this procedure, the CR gradually diminishes until it disappears. The earliest behaviourists largely used a technique known as flooding (i.e., an applied variation of extinction) in the treatment of phobias, before Wolpe (1969) developed the less traumatic "progressive desensitisation" method. According to flooding, a person who acquired a phobia for dogs is exposed to a harmless CS dog until his autonomic and cognitive conditioned reactions disappear. The cognitive-behavioural techniques based on extinction (i.e., flooding and progressive desensitisation) are of considerable value in the treatment of olfactory disorders, for instance, multiple chemical sensitivity (Spyker, 1995), a debilitating disease occurring in response to many chemically unrelated substances, often through a conditioning mechanism.

Discrimination is a further process. It is the reverse of generalisation. Here, to have a CR (e.g., salivation) elicited only by the original CS (e.g., a 1000-Hertz sound), this has to be re-presented followed by the original US (e.g., food), as done in the acquisition phase. Rethinking the previous example about phobia for dogs, an unethical discrimination procedure would consist of exposing the phobic person to the dog (CS) that bit him, allowing this dog to bite (US) him again, and to harmless dogs that will not bite him.

A last CC process is called second order conditioning. Here, in the first step, a neutral stimulus followed by the CS is presented; in the second step, the presentation of the NS alone (or second order CS) will induce, according to the paradigm, the CR. In other words, if a sound (CS) induces a salivary response (CR), then the repeated presentation of a light (NS) before the sound will create a new association, so that the presentation of the light alone will induce the salivary response. Young people often unconsciously apply second order conditioning when training their dogs; for example, when the leash (first CS) is preceded by "let's go for a walk" (second CS) and by snapping the fingers (third CS).

These processes have been addressed in detail as they are largely shared by a variant of Pavlovian CC, namely evaluative conditioning (EC). This paradigm applies to situations where a NS acquires the affective valence of another stimulus (US) with which it is paired with and induces the affective UR previously elicited by the US. A neutral odorant, for example, could acquire a positive or negative valence if paired with a pleasant or unpleasant event. Taste aversions are usually learned through this mechanism, where nausea is the CR to a taste or flavour that was incidentally associated with an US inducing nausea. The definition of EC was introduced by Martin and Levey (1978) following an earlier study (Levey & Martin, 1975) where they designed the prototypical "picture-picture" paradigm. In this study, they asked participants to evaluate a series of paintings for pleasantness. The two stimuli that each participant liked the most and the two stimuli that they liked the least were used as USs; those paintings rated as neutral served as CSs. In the acquisition phase, CS-US pairs were presented several times. The participants were then required to again judge the previously rated neutral paintings for pleasantness on a scale ranging from –100 (completely dislike) to 100 (completely like). The main result was that the neutral paintings associated with the liked paintings were rated more positively, whereas the reverse occurred with the neutral paintings associated with the negative ones.

Although evaluative conditioning can be seen as a variant of Pavlovian classical conditioning, most authors suggest that it is a unique form of affective associative learning, useful for explaining the acquisition of perceptual preferences and related emotions (e.g., Dawson et al., 2007; De Houwer, Thomas, & Baeyens, 2001; Field, 2000; Rozin, Wrzesniewsky, & Byrnes, 1998). There are two specific aspects of EC that seem to support a distinction between it and the classic form of Pavlovian conditioning. Firstly, evaluative conditioning is highly resistant to extinction (cf. for related experiments, Baeyens et al., 1988, 1995; De Houwver et al., 2000). Secondly, awareness of the relationship between CS and US (i.e., contingency awareness) and awareness of the predicted outcomes or purposes of a study (i.e., demand awareness) are not necessary conditions for the occurrence of EC (e.g., Hermans & Baeyens, 2002; Zucco, Paolini, & Schaal, 2009).

Resistance of EC to extinction can be easily observed in everyday life experiences. For example, people who exhibit an aversion to a food (because they have associated it with an illness) know very well how difficult it is to overcome the feeling of nausea induced by the food and to start to taste it again. The first unpleasant experience between foods (CSs) and negative events (USs) is thus difficult to forget.

Whether the importance of awareness in evaluative conditioning learning is, instead, a real problem or an artefact is difficult to ascertain. Although numerous papers have examined this subject, the conditions under which evaluative conditioning occur still remain debated (e.g., Dawson et al., 2007; De Houwver et al., 2001; Dickinson & Brown, 2007; Field, 2000; Hermans & Baeyens, 2002; Pleyer et al., 2007; Rozin et al., 1998; Wardle et al., 2007). So far, the most suggestive evidence for unaware learning based on evaluative conditioning comes from studies using flavours and subliminal or low intensity odorants as CSs (detailed below). However, the belief of some authors (e.g., Field, 2000) that evaluative conditioning does not occur without awareness may be also due to the artificiality of the laboratory environment, which may not favour genuine ignorance during the application of a CR procedure. Looking back to the psychological literature on emotions more than one century ago, James and Lange (James, 1884, 1890), demonstrated that physiological reactions come suddenly and automatically when one is exposed to an emotional event, and that it is only later that cognitive awareness can take place (see, also, Schacter & Singer, 1962). If an odour previously associated with a painful event can trigger similar physiological reactions before awareness (as it is), then the debate on contingency and demand awareness in EC appears rather unproductive.

2. Verbal and attitude conditioning: The earliest demonstrations of evaluative conditioning

Between 1930 and 1950, Gregory Razran carried out an extensive series of pioneering experiments on evaluative conditioning. Many of them used odorants or flavourings as USs and verbal and pictorial materials as CSs. Razran is known for coining the term "semantic conditioning" to refer to all conditioning situations in which the CSs are words, and for devising the so-called "free lunch technique" to study human conditioned preferences. Within the framework of semantic conditioning, one of Razran's (1949) studies was about the generalisation of the conditioned response and the role of semantic and phonological factors in conditioning.

In the first session, university college students, misinformed about the true aim of the experiment, were presented the words "style, urn, freeze, and surf" 15 times while they were eating lunch. In a second session, they were shown words that were related to the experimental CSs words either phonetically (i.e., "stile, earn, frieze, and serf") or semantically (i.e., "fashion, vase, chill, and wave"). Salivation was used as the measure of learning. Interestingly, it was observed that besides the appearance of a CR (i.e., salivation), when the participants were exposed to the initial experimental words, the gradient of generalisation was higher for the semantically-related words than for the phonetically-related words. Razran suggested that generalisation in humans is probably driven more by high level cognitive processes (such as language acquisition and semantic knowledge) than by low level cognitive processes (such as phonological similarity).

In another study (Razran, 1940), the "luncheon technique" was used to investigate whether political slogans as CSs (e.g., "America for Americans," "workers of the world unite," and "no other ism but Americanism") could change their affective valence if they are associated either with a free lunch or with unpleasant (putrid) odorants. On the acquisition session, the author asked 16 students to rate the slogans on a 7 point scale for (1) personal approval, (2) social effectiveness, and (3) literary value. The slogans were then divided into two sets. One set was presented while the participants were enjoying a free lunch and the other while they were required to inhale a number of unpleasant odorants. After about eight conditioning sessions, the original procedure was repeated with the subjects re-rating each slogan for approval, effectiveness, and literary value. It was observed that the slogans paired with the free lunch showed increases in ratings, while those combined with the unpleasant odorants showed decreases in ratings, demonstrating that the association of a political slogan (CS) with either a pleasant or unpleasant US could change the valence of the CS. Moreover, the author assumed that the conditioning process was unconscious since, according to post-experimental interviews, the participants had no knowledge about which slogans were combined with pleasant or unpleasant stimuli.

In a following experiment, again using the luncheon technique, Razran (1954) examined whether the conditioned stimuli could by themselves also induce special food or eating attitudes, namely, a kind of cognitively-driven conditioning. Razran exposed 20 male students (10 experimental and 10 controls) to 22 to-be-conditioned items, consisting of musical excerpts, unfamiliar paintings, slogans and quotations, and photographs of college girls. The materials were presented during six consecutive free luncheons to the participants, who were unaware of the real purposes of the experiment. They were just told that the experimenter was interested in examining how different stimuli are associated in the human mind and how eating affects such connections. One day after the last session, the participants were engaged in three tests (food-related free verbalisation, food-related rhyme finding, and speed of unscrambling food-related scrambled words) assessing variables of conditioned food attitudes. The tests were performed by the participants while they were again exposed to the conditioned material. Razran found that in all three parameters, the difference between the experimental and control groups (the latter was given no food) were highly reliable for the four kinds of conditioned stimuli that induced the evocation of food-related verbal reactions. He interpreted the outcome of this study as resulting from unconscious cognitive conditioning. Thus, Razran is a precursor in studying the effectiveness of olfactory stimuli to induce hedonic reactions even below the level of awareness.

Almost in the same period, and in the same current as Razran's study, Arthur Staats carried out important studies on human conditioning. Although these were related to the verbal system and not to olfaction, they are worth mentioning here since they also provided an early demonstration of evaluative conditioning and were forerunners of modern studies on attitude formation.

The basic paradigm (Staats & Staats, 1957, 1958) foresaw the presentation of to-be-conditioned verbal stimuli (e.g., nonsense syllables) paired with positively or negatively valenced words as USs, with the hypothesis that the response firstly elicited by these words would be, after conditioning, induced by the verbal stimuli. The paradigm worked well with a large class of materials, namely (besides the nonsense syllables), colours, familiar masculine name, and names of nations. One of the most impressive works carried out by Staats (cf., Staats & Staats, 1958) was the study that follows on the conditioning of affective response to national names.

Twenty-four students, unaware of the real purpose of the experiment, were presented with national names (CSs) and emotionally-valenced words (USs), with the instruction to pay attention to the visually presented names and to repeat aloud the words immediately after the experimenter pronounced them. The participants were also told that the main purpose of the experiment was to study how the two types of learning took place simultaneously. The national names (German, Swedish, Greek, Italian, Dutch and French) were each presented 18 times and paired with a different word. The names Dutch and Swedish were systematically and respectively paired with words having positive (e.g., honest, pretty) and negative (e.g., ugly, thief) evaluative meanings, while the other names were paired with non-emotionally valenced words (e.g., large, twelve). At the end of the first session, the participants engaged in a deceit memory test, which was followed by the rating of the six national names on a 7-points semantic differential scale. The latter task was deceptively proposed with the aim to ascertain participants' feeling about the words, since feelings can affect learning. The participants were also asked questions to determine their awareness about the purposes of the experiment. The results were unequivocal: The national names paired with positively-valenced words were evaluated as positive, while the reverse occurred for the national names paired with negatively-valenced words.[1] Similar results were also obtained using the galvanic skin response (Staats, Staats, & Crawford, 1962). The only criticism that can be addressed in these studies is that the author neglected to check the CSs and USs stimuli he used for their affective valence before the experimental sessions took place.

It would be interesting to examine whether odorants as USs would also be effective in altering the evaluative meaning of national names; a kind of study that has never been performed (although a similar study was performed by Allen and Janiszewsky (1989) using Norwegian words as CSs and positive reinforcement as USs).

As a last remark, I would like to point out that the EC mechanisms that explain the attitude changes towards national names plausibly mirror what happens in everyday life. Negative judgments on ethnicities, indeed, often occur in the absence of direct experiences with those ethnicities rather as a result of continuous exposure to US words or sentences associated with those ethnicities that have negative evaluative meanings. Examples of this process are confined to the olfactory domain and can be found in the anthropological literature (e.g., Classen, Howes, & Synnot, 1994). These authors point out how feelings of disliking towards a given social category (e.g., working-class), ethnicities, and even women may be based on their assumed foul smell. Therefore, evaluative conditioning appears to be a substantial component of prejudice.

3. Domains of evaluative olfactory conditioning

The pioneering studies performed by Razran, Staats, and their colleagues represent the base from which several researchers began to investigate a wide range of psychological phenomena framed within the evaluative conditioning paradigm. Although most studies focused on visual conditioning (cf., Levey & Martin, 1975; Martin & Levey, 1978), a significant body of research literature has also

1. The positive and negative words used by Staats as USs are not natural USs stimuli. For example, the words "bad" or "good" are unmeaning neutral stimuli to a child that is beginning to learn language; however, through a conditioning process, they will acquire a negative or positive valence since they will be paired with real USs (e.g., the mother's non verbal indicators). The process that allows these initially non-sense words to became meaningful is second order conditioning.

accumulated in the olfactory domain. The following sections will address these issues with an emphasis on the acquisition of human preferences and related emotions.

3.1 The flavour-flavour paradigm

Taste and smell are combined in the flavour evaluative conditioning paradigm. Accordingly, participants are requested to experience two flavours as CSs, one mixed with sugar and the other in plain water. After conditioning, the flavours are presented in unsweetened solutions. A clear preference ensues for those flavours previously paired with sugar. This paradigm was first proposed by Zellner, Rozin, Aron, and Kulish (1983), who asked their participants to drink 48 small samples of two kinds of flavoured tea, 24 with sugar added (CS+) and 24 unsweetened (CS). On the day of conditioning and one week later, the participants were tested for their preference for one of the two flavours. A preference enhancement was observed for the sweetened flavours presented during the conditioning session. This finding was observed by Zellner et al. in the three separate experiments of this study, which differed in a number of stimulus properties (e.g., sugar concentration, quality of the stimulus – sweetness or bitterness, and context). The efficacy of this paradigm was demonstrated by this same group in subsequent studies (for reviews, cf., Rozin & Zellner, 1985; Rozin & Fallon, 1987).

A variation of the Zellner et al.'s flavour-flavour paradigm was employed by Baeyens, Eelen, Van Den Bergh, and Crombez (1990).[2] The participants (72 high-school children), who were not informed about the real purposes of the experiment, were presented with artificial flavours mixed in either likeable or dislikeable solutions (as CSs+) or just plain water (as CSs–). The CSs+ flavours (i.e., orange, apricot, pear, and raspberry) were mixed with either sugar (as positive-liking US) or with Tween 20 (i.e, polysorbate 20), a substance with a soap-like taste (as negative-disliking US). Four flavourless coloured solutions were also presented to the participants to study whether the perceptual characteristic of a food could also acquire a liking or disliking value if paired with pleasant or unpleasant flavours. The authors hypothesised that, after conditioning, the CS+ sugar-added flavour would be rated as more liked than the matched CS– control flavour, and that the CS+ Tween-added flavour would be rated as less liked than the CS– control flavour. The second experimental session was carried out on the day of conditioning and one week later. Results showed a significant effect (both immediately and one

2. The group of Belgian psychologists led by Frank Baeyens is one of the most active and outstanding group of researchers investigating EC.

week after conditioning) only in the flavour-flavour negative condition. According to the authors, this was probably due to what they interpret as an "affective positive-negative asymmetry," that is, "a negative event often has more affective impact than equally intense positive events" (Baeyens et al., 1990, p. 451). This refers to the positive sugar USs compared to the negative Tween USs. The authors also found that the contingent presentation of the coloured unflavoured solutions did not induce any conditioning effect. Finally, none of the subjects in the flavour-flavour groups evidenced any explicit knowledge of the CS-US contingency relation, as evidenced by the post-experimental interviews. This latter finding is reminiscent of the debate mentioned above on whether or not contingency awareness (i.e., awareness relative to the relationship between CS and US) and demand awareness (i.e., relative to the predicted outcomes of a study) may play a role in the occurrence of evaluative conditioning. Undoubtedly, the studies using the favour-flavour paradigm represent the best suggestive evidence that evaluative learning does not require awareness. Field (2000) suggests that the reasons underlying the unaware learning of tastes can be due to their ecological relevance, as gustation mediates stimuli that are vital to avoid harmful substances and to ingest beneficial ones.[3] Accordingly 'it makes sense that it is important to learn about tastes in a spontaneous way and this may best be achieved by learning without conscious control' (Field, 2000, p. 24).

3.2 Cross-modal evaluative conditioning

The flavour-flavour paradigm examined above, as well the picture-picture paradigm devised by Martin and Levey (1978), are representative of evaluative conditioning studies where the CSs and the USs belong to a same modality. Numerous other studies have demonstrated that evaluative conditioning can also occur when CSs and USs belong to different sensory modalities.

For instance, Todrank, Byrnes, Wrzesniewski, and Rozin (1995) showed that pleasant and unpleasant odorants (as USs) associated with photographs of neutral human faces (as CSs) influence subsequent preference judgments about the people represented in the photographs. Indeed, in a test session, participants provided pleasantness ratings influenced by the valence of the odorant associated with the photograph to be judged in the first session. Similar results were obtained

3. The reader interested in knowing more on the debate about awareness in evaluative conditioning studies on flavours can refer to two investigations that recently appeared in the *Journal of Learning and Behaviour*. The first (Dickinson & Brown, 2007) demonstrated that flavour evaluative conditioning is substantially unaffected by awareness, while the second (Wardle, Mitchelll, & Lovibond, 2007) showed the converse (see reference list).

by Van Reekum, Van den Bergh, and Frijda (1999), who simultaneously paired either pleasant or unpleasant unconditioned odorants or sounds with hedonically neutral abstract paintings.

Hermans, Baeyens, and Natens (2000), instead, used four pictures of liquid-soap bottles as CSs and two odorants (lemon and civet) as the positive and negative US, respectively (the remaining two pictures served as controls with no odorants associated). The pleasant and unpleasant odorants were applied to a cotton swab and paired with one of the two experimental pictures in a transparent bag. The other two pictures served as controls (i.e., no odour was applied to the cotton swab). Eight acquisition trials were administered with the participants being required to look at the picture, open the bag, smell the cotton swab, close the bag, and look at the picture again. In the test session, the participants were asked to rate the four CSs for pleasantness. The CS picture of the bottle of liquid soap associated with the positive US (lemon) was rated as more pleasant than that associated with the negative US (civet). In addition, the participants attributed better quality to the first positive CS than to the negative and exhibited a higher propensity to buy it in a supermarket. This study reminds us of a study carried out by Razran (1954), where it was demonstrated that conditioned stimuli can also induce food attitudes, that is, a cognitively-driven conditioning.

A further cross-modal study was conducted by Rozin, Wrzesniewski, and Byrnes (1998), who presented to 20 participants USs *a priori* judged as negative (e.g., a real dead cockroach, a photo of Hitler) or as positive (e.g., a picture of a rainbow, a picture of Sean Connery and Michelle Pfeiffer), and paired with odorants as CSs. The participants were unaware of the real aims of the experiment, which was presented as a memory test for odour-object pairing. Following eight rounds of conditioning, each consisting of eight trials, the participants again rated the pleasantness of the odorants' CSs on a 200 mm visual analogue scale. The results indicated that the dead cockroach was the only US that influenced the emotional valence of the paired CS odorant. Nevertheless, it sounds a bit puzzling that the authors considered Hitler's and Connery-Pfeiffer's' photos as *a priori* strong negative and positive USs, respectively. It is not fully convincing that these stimuli induce unanimously negative vs. positive URs in younger university students, who did not live the World War II period. It may be presumed that the same experiment of Hitler's photo paired with an odorant would be more conclusive in elderly participants who experienced the concentration camps.

An interesting application of cross-modal evaluative conditioning learning can be found within the framework of the so-called motion-sickness paradigm, which illustrates how conditioned taste aversions can be experimentally induced (cf. Stockhorst, Enck, & Klosterhalfen, 2007, for review). Accordingly, nausea and vomiting (i.e., gastrointestinal URs to nauseogenic experiences) can also be induced using body rotation (i.e., a series of rotations administered to the participants while seated on a rotation chair). Kosterhalfen et al. (2000), for instance, noted that the presentation of a taste as CS (elderberry juice) immediately before the rotation onset as US provokes a conditioned aversion to the CS when it is presented alone (as demonstrated by the self-evaluated intensity of the experienced symptoms and quantity of juice consumed). The motion-sickness paradigm constitutes an interesting and potentially fruitful way to investigate how taste aversions can be experimentally acquired in humans, mirroring circumstances that are common in real-life situations.

The effectiveness of the cross-modal olfactory evaluative conditioning paradigm was also noted in developmental studies and in the emergence of somatic syndromes. For example, Hvastja and Zanuttini (1989) examined recognition memory and changes in odour preferences in three groups of children (aged 6 years 6 months, 8 years 9 months, and 10 years 5 months on average). The participants were presented with six unfamiliar odorants as CSs associated with slides depicting positive (e.g., a kitten) and negative (e.g., the open mouth of a bat) images as USs. One month after the learning session, the participants were asked to rate each odorant on a 9-point scale. The results showed that the odorants paired with the positive USs images were rated more pleasant than those paired with negative USs images, supporting the notion that odour hedonics is acquired through an associative process in childhood. Studies conducted at even earlier ages, in neonates and infants, also clearly show that odour preferences can derive in part from their association with pleasurable events perceived in other modalities (cf. Schaal & Durand, 2012; Soussignan, Delaunay, & Schaal, 2012, for reviews).

Furthermore, Zucco, Caldironi, and Shock (1989) demonstrated that olfactory evaluative conditioning can also emerge in severely multi-handicapped children. The participants were four congenitally-blind children (aged from 3 to 6 years) living in a specialised centre for multi-sensory diseases. In the pre-experimental session, four operators were required to place their faces and hands close to each child's nostrils for about 30–40 seconds. In the same session, the children were also presented with four odorants (essential oils) usually judged pleasant by adults. The entire session was videotaped. On both occasions, the children exhibited neither visible neurovegetative nor non-verbal reactions. In the following experimental session, each operator (the educator caring for each child) was required to sprinkle the parts of her body most exposed to daily contact with the child they had to care for (i.e., neck, hands, wrists) with one of the four odorants (as CS) and to repeat the procedure for 10 consecutive days. Therefore, each child was exposed to only one operator and to one odorant. At the end of the experimental session, the four children were tested again to assess whether the odorant

CSs worn by the operators acquired the same positive affective valence as the daily body contact (US) between each child and his/her operator. The children's reactions were videotaped for later analysis by six independent judges. The results indicated that 3 out the 4 children recognised the odorant worn by their own operator after smelling it, exhibiting clear non-verbal hedonic reactions (e.g., smiling). Conversely, such behaviours did not occur when they were exposed to the odorant worn by a non-familiar teacher.

A further impressive developmental study was performed by Bernstein (1978) on taste-learned aversions in children receiving chemotherapy. The participants were 41 children, aged 2 to 16 years, in treatment for neoplastic diseases. Some of them received chemotherapeutic drugs, toxic to the gastrointestinal tract and associated with nausea or emesis. The author tested whether such children would acquire an aversion to ice cream (as the CS) eaten prior to drug treatment (as the US). The experimental group was compared to a group of children receiving the same chemotherapy, but not in association with the ice cream, and to another group receiving chemotherapy not associated with gastrointestinal malaise but with receiving the ice cream. On the second session, about 4 months later, the children received the ice cream experienced in the first session and a novel ice cream. As compared to both control groups, the children from the experimental group exhibited significant preference for the novel ice cream over the one associated with malaise. This study is highly valuable since it shows that humans may acquire aversions to food through a single trial of conditioning, when such food is consumed before a treatment inducing discomfort or in association with nausea or vomiting. Although it may be considered ethically exceptionable, this study greatly contributed to the understanding of the causes of appetite loss experienced by cancer patients.

The studies reported above have demonstrated that evaluative conditioning is effective in inducing low-level or basic emotional conditioned responses in children (i.e., liking-disliking). Other authors have investigated whether odorants can affect cognitive performances. In particular, Epple and Herz (1999) showed that ambient odorants (as CSs) implicitly paired with frustrating tasks (as USs) negatively influence subsequent performance in cognitive tasks. The participants – forty-eight 5-year old children – underwent a failure/frustration task (US) in a room scented with an unfamiliar pleasant odorant (CS). They were then required to move a doll along a maze path to the centre under time constraints (5 minutes). They were also told that if they can reach the centre of the path in the allotted time, they would receive a prize. Unfortunately, this apparently simple task was rendered impossible, leading to feelings of frustration in the participants. After a break of about 20 minutes, following the unmanageable maze task, the children attended a second task in another room, the atmosphere of which was scented either with the same odorant as that of the maze room, with a different odorant, or with no odorant at all. In this second task, the children were asked to find and circle how many puppies pictured on a sheet were missing their tail (within 90 seconds). The entire session was videotaped. The children exposed to the same odorant in both sessions showed worse performance, circling significantly less puppies than the comparable control groups (i.e., no odour group and different odour group). According to the authors (see also, Herz, Schankler, & Beland, 2004), the experimental group probably experienced a decrease in motivation along the frustration task that yielded to a subsequent reduction in performance during the cognitive task. The odorant CS associated with failure was the core of this process. In the same line, Chu (2008) questioned whether an odorant could also increase performance in cognitive tasks. Accordingly, he used an adaptation of Epple and Herz's procedure, which pairs odorants CSs (fruit teas suffused in a test room) to success in cognitive tasks (as the US). In the first step, the participants (76, 11–13 year old, underachieving children) were requested to either look at a target picture and pick it out among others or to select the picture that was different than the set of other pictures. These tasks were performed in absence of ambient odorants. In a second experimental step, the children were asked to find two identical pictures among a set of several pictures while they were in the scented room. At variance from Epple and Herz' study, the latter task was very easy to perform, although presented by the experimenters as suitable for older children. Finally, in the third step, the children performed cognitive tasks similar to those realised in the first step, but in presence of the odorant used in the second step or in presence of a different odorant. Results clearly showed an increase in performance from the first to the third step, but only in children who were exposed twice to the same ambient odorant. According to the author, the odorant CS acted here as a cue that evoked the positive emotional state experienced with success in step 2, boosting motivation and improving subsequent cognitive performance.

Evaluative odour conditioning can also play a role in the emergence and maintenance of different kinds of somatic diseases. For example, odorants acting as CSs can trigger somatic symptoms such as those that prevail in multiple chemical sensitivity (Siegel, 1999; Van den Bergh et al., 1999; Zucco, Militello, & Doty, 2008), or can be shaped in cues reminiscent of traumatic a experience in patients affected by post-traumatic stress disorders (PTDS, Vermetten & Bremner, 2003).

Accordingly, Van den Bergh et al., exposed 64 students to two odorants CSs paired with either USs CO_2 enriched air (assumed to induce hyperventilation) or with normal air, and then measured their breathing patterns and somatic symptoms. After the acquisition phase, consisting of three conditioning trials, the participants exhibited increased respiratory behaviour and more symptoms to the odorant associated with CO_2 than to the odorant associated with air, when presented alone. Likewise, patients affected by multiple chemical sensitivity exhibit somatic symptoms in response to odorants previously experienced together with aversive events (e.g., a hyperventilation episode; see Labarge & McCaffrey, 2000, for review).

Furthermore, Vermetten and Bremner (2003) examined the role of odorants as cue reminders of past traumatic experiences in cases of PTSD. In all reviewed cases, the authors observed that odorants previously associated with the trauma became powerful cues of the memory related to the trauma itself and evoked symptoms of PTDS, as foreseen by the evaluative conditioning model.

Odorants can also induce anxiety symptoms. This was demonstrated by Schneider et al. (1999), who presented unpleasant odorant and pure air (as USs) and neutral faces (as CSs) to 12 patients suffering from social phobia and to 12 healthy controls. After conditioning, the faces paired with the unpleasant odorant were rated as more negative than those paired with pure air, demonstrating that conditioning occurred in both groups. This study also investigated the cortical and sub-cortical correlates of conditioned emotional learning by means of functional magnetic resonance imaging (fMRI). The presentation of a CS associated with a negative odorant led to the activation of the amygdala, with a decrease in activity in the group of normal subjects and an increase in the group of patients. Other authors (e.g., Zald & Pardo, 1997) also observed that the amygdala is involved in the elaboration of aversive odorants (namely, a condition in which humans exhibit highly negative emotional reactions). Indeed, only these stimuli (at variance from pleasant odorants) provoked significant amygdala activation (see also Buchel & Dolan, 2000; Herz et al., 2004; Gottfried, 2006).

Finally, Stockhorst et al. (1999) demonstrated that odorant CSs paired with insulin injection (as the US) at acquisition had the potential, when paired with a placebo injection during the session, to reduce glycaemia in healthy subjects. In all of the studies reported in this section, evaluative conditioning has shown to be a suitable and promising field of investigation of associative learning in which CSs and USs belong to different sensory modalities.

3.3 Ecological studies on evaluative conditioning

The evaluative conditioning paradigm has also demonstrated its effectiveness in contexts mirroring real-life experiences. For instance, Baeyens, Wrzesniewski, De Houwer, and Eelen (1996) examined odour conditioning in two ecological studies. In the first, unaware participants used toilet rooms (as USs) that were sprayed (CSs) either with lavender (group 1) or pine odour (group 2) for several days. For group 1, pine was the control odorant, while for group 2, it was lavender. In the test session, the participants were asked to evaluate both odorants on semantic differential scales. The results provided support for olfactory evaluative conditioning in real life contexts. Both odorants were, indeed, rated to be more positive than the control odorant, especially in those participants who used toilets as a relaxing break, while the reverse was observed for those participants who used the toilets out of mere necessity. Further, post-experimental interviews indicated that the participants were not aware of the manipulations, since only a negligible number of them realised that the odorant CS was the same than the one they experienced in the toilet room. In the second study, two distinct etheric oils (as CSs) were suffused in the air around participants while they were receiving a massage by a physiotherapist (USs), either in a relaxing, soft way or in a harsh, uncomfortable way. The participants exposed to the soft massage rated the CS more positively than the control odorant. However, the converse effect did not occur in the other condition. The authors explained this failure by the impossibility of administering a truly negative massage. Also in this study, most participants did not exhibit conscious recognition of the experimental odorant.

Further studies with ecological value have focused on marketing and consumer science, where emotional processes are prominent in decisions.

For example, Stuart, Shimp, and Engle (1987) demonstrated the occurrence of EC under natural advertising conditions. Through four experiments, they examined the amount of conditioning in the trials (1, 3, 10 and 20), the occurrence of latent inhibition (i.e., the participant's pre-exposition to the CS alone before pairing it to the US), and the effect of presentation procedure (i.e., forward – CS followed by US, and backward – US followed by CS). More than 200 students participated in each experiment. The CS consisted of a fictitious brand of toothpaste, while the USs consisted of four emotionally positive pictures selected out of 50, according to scales for the evaluation of affective reactions. Both CS and USs were presented on a coloured screen. The attitude of the participants towards the CS brand was measured by means of four types of evaluative scales. The outcomes indicated that the number of trials did not significantly affect the amount of conditioning, since it occurs effectively with a single CS-US pairing. However, latent inhibition appeared to affect the results. Indeed, the participants who were exposed to the CS alone, prior to pairing with the US, exhibited less positive attitudes towards the brand, as compared to the participants who were exposed to the standard CS-US procedure. This outcome is relevant from an applied perspective, suggesting that brands already familiar to the consumer before any advertising campaign (i.e., before being paired with a positively valenced US) lower the possibility that the consumer's attitude towards the brand becomes conditioned. Finally, consistent with the literature, backward has less, if any, conditioning effect compared to forward conditioning (cf., Domjan & Burkhard, 1985).

In a following study, Stuart, Shimp, and Engle (1991) performed a more in-depth evaluation of the influence of brand familiarity on the strength of conditioning. They exposed the participants to emotionally positive pictures (as USs) of actual brands of cola with different degrees of familiarity (as CSs). The novelty of the CS appears to be crucial in advertising, since conditioning effects were less significant for well-known brands (Coke and Pepsi) than for unknown or moderately known brands. Therefore, an excess of familiarity (i.e., a repetitive exposition to the CS) acts as an extinction factor, lowering or nullifying the CR. In any case, as the authors maintain, consumers' feelings and thoughts about objects are inevitably due to associative learning, whether they are mindlessly or consciously acquired.

In the same line of marketing and consumer science research, Rozin et al. (1988) asked two groups of participants (hair-washing liking vs. hair-washing neutral) to rate 7 shampoos' odour for pleasantness and to use the one evaluated as neutral for two weeks. In the test phase, the participants were required to again rate the 7 shampoos for pleasantness. Both groups showed only a slight improvement in liking for the shampoo they used, but unfortunately, no rating difference was found between the two groups for the target shampoo. The authors attribute this weak result to the fact that hair-washing is not a strongly liked activity.

A last related anecdotal example of the conditioned responses to a brand – although on the negative side – comes from a real experience reported by a student, who noticed that her favourite perfume had become suddenly disgusting. Later, it emerged that this was due to her habit of smelling a paper towel imbued with this scent to avoid the offensive odour of corpses during the practical session of anatomy courses. Consumer scientists work to obtain the opposite effect, pairing products CSs (e.g., perfumes) with well-known artists (actors, musicians, showgirls), to transfer the positive attributes of the latter to the product itself.

3.4 Low-intensity and subliminal odorants: The unconscious side of evaluative odour conditioning

In 1983, Kirk-Smith, Van Toller, and Dodd published a pioneering study demonstrating that a low intensity odorant paired with a stressful task influenced people's mood and attitudes without their awareness when they later re-experienced the odorant separately. This study was particularly important in highlighting unconscious odour learning processes as they may happen in everyday life conditions.[4]

4. For a discussion of consciousness in olfactory experiences see, e.g., Stevenson (2011) and Zucco and Job (2011).

In the first session, participants (12 males and 12 females) were given a stressful task (US) involving the completion of complex block patterns under strong time-constraint. This task consisted of assembling nine wooden blocks taken from the Wechsler adult intelligence scale (WAIS). The blocks were reduced in size from the original WAIS test and were presented in a diamond shape to increase the task's level of difficulty. Furthermore, to make the task more stressful, the participants were told at completion that their performance was slightly below average and that the last participant on the same task was a bit better. For half of the participants, the stressful event was experienced in the presence of a very low-intensity odorant (trimethyl undecylenic aldehyde, TUA). In the second session, 3–4 days later, participants were asked to complete mood evaluation questionnaires and to assess photographs of people on differential scales in the presence of TUA. No difference in mood ratings was found between male and female participants who had experienced the odorant on both sessions. However, a conditioning effect was found for the female participants (i.e., those who experienced TUA in session 1 reporting increased anxiety ratings and more negative judgment of people's photographs compared to controls). Since the participants reported no awareness of the odorant at both sessions, the authors hypothesised that unconscious emotional conditioning is based on the association between an odorant and emotional arousal. However, this study was criticised by Black and Smith (1994), who noticed several shortcomings, such as (i) the lack of appropriate controls; (ii) no verification that the block task actually elicited stress; and (iii) claiming a significant difference for the group of females without statistical corroboration. Further, Black and Smith ran statistical analyses indicating that mood ratings did not differ between odour-exposed females and non-odour-exposed females, thus disconfirming the conditioning of an unconsciously perceived odour.

Recently, Zucco, Paolini, and Schaal (2009) replicated Kirk-Smith et al.'s investigation, taking into account Black and Smith's observations. Additional checks were performed to improve the validity of the experiment, for instance: (1) The introduction of two further groups of participants to ascertain whether those who had to execute the task with or without the stress-inducing procedure reported differential stress and frustration feelings; (2) a procedure to select an odorant and photographs that were neutral; (3) an evaluation of the participant's olfactory function; and (4) the involvement of two examiners who were blind regarding experimental conditions, as well as of two clearly distinct environments for sessions 1 and 2.

The experimental setting devised by Kirk-Smith et al. was modelled (see Figure 1), but their procedure was simplified. The design of the main experiment (Experiment 1) followed a double factorial that crossed odour presence (a just

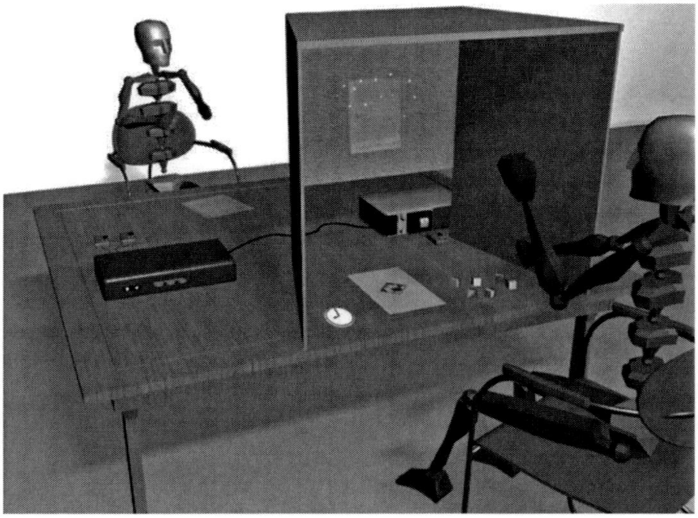

Figure 1. The experimental setting of the first session (from Zucco et al., 2009, reproduced by permission; adapted from Kirk-Smith et al., 1983).

noticeable concentration of essential oil of Neroli) with exposure to a stress-induction procedure. Thus, four conditions were compared, each comprising of 12 university student participants: (a) stress-related odour present in both the sessions 1 and 2 (present-present); (b) stress-related odour absent in both sessions (absent-absent); (c) odour present at session 1, but absent on session 2 (present-absent); and (d) odour absent on session 1, but present on session 2 (absent-present). Evidence for unconscious odour conditioning can only be found if condition "a" significantly differed from the other three conditions in the mood ratings on session 2 (the latter conditions resulting in similar ratings).

The results clearly supported Kirk-Smith et al.'s (1983) conclusions that odorants can induce unconscious hedonic reactions through a conditioning process. The participants exposed to the experimental odorant in both sessions (Experiment 1) assessed their mood as being significantly more negative and less positive in session 2 than the three control groups, who gave similar ratings (see Table 1). At variance by Kirk-Smith and colleagues, however, no gender differences were observed.

This pattern also occurred in a second experiment that was performed to rule out the argument that two exposures to the odour (i.e., on condition "a") would imply more chance to learn it (as opposed to conditions "b," "c," and "d," which involved only one or no exposure at all) favouring the differences found in ratings (see, p. 372). Furthermore, in both experiments, odour conditioning took place

Table 1. Self-rating (means and SDs) of positive and negative mood attributes of the Profile of Mood States (PMS) and of ratings of photographs using the Multiple Affect Checklist (MAC) as a function of the four conditions (Present–Present; Present–Absent; Absent–Present; Absent–Absent). From Zucco et al. (2009) (reproduced by permission).

Groups	Ratings					
	PMS positive attributes		PMS negative attributes		MAC assessment of photographs	
	Mean	SD	Mean	SD	Mean	SD
Present–Present	50.16*	8.91	25.08*	7.40	4.08*	0.44
Present–Absent	69.50	12.33	12.91	6.34	4.17	0.36
Absent–Present	66.58	8.39	17.58	6.52	4.27	0.56
Absent–Absent	65.50	11.24	15.08	7.24	4.09	0.40

* In millimetres.

without awareness of the experimental odorant, as demonstrated by the participants' answers to the post-experimental interview.

As a next step, it would be interesting to study whether a similar effect could be observed when relaxing and positive USs are paired with a low-intensity odorant. Moreover, it would be of interest to assess whether the temporal retention of the conditioned response is related to the intensity of the original discomfort, stress, or trauma received and whether stronger emotions linked with a neutral event imply longer sessions for the extinction of the CS.

An area related to that of unconscious odour conditioning is subliminal evaluative olfactory conditioning. The paradigm used is relevant in demonstrating that olfactory USs presented under threshold levels can drive implicit preferences and attitudes. A significant body of research exists on verbal and visual subliminal EC (e.g., De Houwer, Baeyens, & Eelen, 1994; De Houwer, Hendrickx, & Baeyens, 1997; Krosnick, Betz, Jussim, & Lynn, 1992; Niedenthal, 1990), while studies on olfactory subliminal conditioning are scarce. According to a general procedure during the learning phase a very brief (e.g., 10 msec), or subthreshold, presentation of a positive or negative US is followed by a CS whose valence is supposed to be changed by the undetectable US. Later, the participants are required to rate the CS target on several dimensions. Subliminal conditioning occurs whether the participants exposed to the positive US implicitly rate the target more positively than the participants exposed to the negative US.

One relevant study on subliminal odour EC was recently carried out by Li, Moallem, Paller, and Gottfried (2007). The authors investigated whether the affective valence of subliminally presented odorants can change preference judgments of human faces. In each trial, 31 students were delivered a bottle containing a pleasant (lemon), neutral (anisole), or unpleasant (sweat) odorant USs at a concentration below their specific detection thresholds (which were measured prior to the study). Each participant was asked to indicate whether or not the bottle contained an odorant. Immediately after a face with a neutral expression (as the CS) was displayed for about 1200 msec; each participant was required to rate the face's likeability on a visual analogue scale (ranging from "extremely unlikeable" to "extremely likeable"). Respiration and heart rate were also measured. The results clearly indicated that the faces were rated less or more likeable if they were preceded by the unpleasant or pleasant odorant, respectively, although the participants exhibited no awareness about the hedonic valence of the odorants. The heart rate was also substantially affected by the valence of the odorants with accelerative heart beats following the unpleasant odorant as compared to the pleasant odorant. This paradigm provided evidence that odorants presented subliminally can influence human likeability judgments and autonomic responses below the level of awareness.[5]

4. Summary and conclusion

The series of studies examined in the present chapter show the robustness of evaluative olfactory conditioning and provide information on how human preferences and the associated emotions may develop. They also confirm Trygg Engen's pioneering studies on the role of associative learning in the acquisition of olfactory responses.

The efficacy of evaluative conditioning in inducing affective conditioned reactions in humans has been exploited in several domains. Studies on the acquisition and maintenance of somatic diseases and phobias, on food aversions, and on influencing consumer choice decisions are particularly interesting. These represent examples of everyday conditions where odour-based evaluative conditioning processes massively occur to affect – mostly unconsciously – mood states, attitudes and effective behaviour. The evaluative olfactory conditioning paradigm has shown its efficacy also in developmental contexts as a mechanism capable of inducing whether basic hedonically valenced (liking-disliking) responses and influence cognitive performance and behaviours. These studies are relevant as they may contribute to help educators to favour learning in children.

---

5. This paradigm could be fruitful – if odorants are used as USs – to replicate Staats's (1958) study on the conditioning of affective responses to national names.

The present chapter is partially based on Zucco (2012), Conditionnement olfactif évaluatif. Lorsque les odeurs se transforment en émotions. In B. Schaal, C. Ferdenzi, & O. Wathelet (Eds.), *Olfaction et Émotion. Du culturel au neuronal.* Dijon: Edition Universitaires.

Furthermore, various techniques and procedures have successfully contributed to a better understanding of the association between odours and human preferences. Among others, low intensity stimulation, subliminal presentation, and motion-sickness procedures were revealed to be fruitful for the study of olfactory evaluative conditioning as they mimic, in the lab, real-life contexts. Evaluative Conditioning, therefore, still remains one of the best and promising paradigms to investigate the amazing world of olfaction.

References

Allen, C. T., & Janiszewsky, C. A. (1989). Assessing the role of contingency awareness in attitudinal conditioning with implications for advertising research. *Journal of Marketing Research, 26*, 30–43.

Baeyens, F., Crombez, G., Van den Bergh, O., & Eelen, P. (1988). Once in contact always in contact: Evaluative conditioning is resistant to extinction. *Advances in Behaviour Research and Therapy, 10*, 179–199.

Baeyens, F., Eelen, P., & Crombez, G. (1995). Pavlovian associations are forever: On classical conditioning and extinction. *Journal of Psychophysiology, 9*, 127–141.

Baeyens, F., Wrzesniewski, A., De Houwer, J., & Eelen, P. (1996). Toilet rooms, body massages and smell: Two fields study on human evaluative odor conditioning. *Current Psychology: Developmental Learning and Personality, 15*, 77–96.

Baeyens, F., Eelen, P., Van Den Bergh, O., & Crombez, G. (1990). Flavor-flavor and color-flavor conditioning in humans. *Learning and Motivation, 21*, 434–455.

Bernstein, I. L. (1978). Learned taste aversions in children receiving chemotherapy. *Science, 16*, 1302–1303.

Bierly, C., McSweeney, F. K., & Vannieuwkerk, R. (1985). Classical conditioning of preferences for stimuli. *Journal of Consumer Research, 12*, 316–323.

Black, S. L., & Smith, D. G. (1994). Has odor conditioning been demonstrated? A critique of unconscious odor conditioning in human subjects. *Biological Psychology, 37*, 265–267.

Buchel, C., & Dolan, R. (2000). Classical fear conditioning in functional neuroimaging. *Current Opinion in Neurobiology, 2*, 219–223.

Classen, C., Howes, D., & Synnot, A. (1994). *Aroma: The cultural history of smell.* New York: Routledge.

Chu, S. (2008). Olfactory conditioning of positive performance in humans. *Chemical Senses, 33*, 65–71.

Chu, S., & Downes, J. J. (2000). Odour-evoked autobiographical memories: Psychological investigation of proustian phenomena. *Chemical Senses, 25*, 111–116.

Dawson, M. E., Rissling, A. J., Schell, A. M., & Wilcox, R. (2007). Under what conditions can human affective conditioning occur without contingency awareness? Test of evaluative conditioning paradigm. *Emotion, 7*, 755–766.

De Houwver, J., Baeyens, F., & Eelen, P. (1994). Verbal evaluative conditioning with undetected US presentations. *Behaviour Research and Therapy, 32*, 629–633.

De Houwer, J., Hendrickx, H., & Baeyens, F. (1997). Evaluative learning with "subliminally" presented stimuli. *Consciousness and Cognition, 6*, 87–107.

De Houwver, J., Baeyens, F., Vansteenwegen, D., & Eelen, P. (2000). Evaluative conditioning in the picture-picture paradigm with random assignment of conditioned stimuli to unconditioned stimuli. *Journal of Experimental Psychology, ABP, 26*, 237–242.

De Houwer, J., Thomas, S., & Baeyens, F. (2001). Associative learning of likes and dislikes: A review of 25 years of research on human evaluative conditioning. *Psychological Bulletin, 127*, 853–869.

Dickinson, A., & Brown, K. J. (2007). Flavor-evaluative conditioning is unaffected by contingency knowledge during training with color-flavor compounds. *Learning and Behavior, 35*, 36–42.

Domjan, M., & Burkhard, B. (1985). *The principles of learning and behaviour*. Monterey, CA: Brooks/Cole.

Engen, T. (1982). *The perception of odors*. New York: Academic Press.

Engen, T. (1988). The acquisition of odor hedonics. In S. Van Toller, & G. H. Dodd (Eds.), *Perfumery: The psychology and biology of fragrance*. New York: Chapman & Hall.

Engen, T. (1991). *Odour sensations and memory*. New York: Praeger.

Epple, G., & Herz, R. S. (1999). Ambient odour associated to failure influence cognitive performance in children. *Developmental Psychobiology, 35*, 103–107.

Field, A. P. (2000). I like it, but I'm not sure why: Can evaluative conditioning occur without conscious awareness? *Consciousness and Cognition, 9*, 13–36.

Gottfried, J. A. (2006). Smell: Central nervous processing. *Advances in Oto-Rhino-Laryngology, 63*, 44–69.

Hermans, D., & Baeyens, F. (2002). Acquisition and activation of odor hedonics in everyday situations: Conditioning and priming studies. In C. Rouby, B. Schaal, D. Dubois, D. R. Gervais, & H. Holley (Eds.), *Olfaction taste and cognition* (pp. 119–139). New York: Cambridge University Press.

Hermans, D., Baeyens, F., & Natens, E. (2000). *Evaluative conditioning of brand CS with odor USs: a demonstration using the affective priming technique*. Unpublished internal research report, University of Leuven.

Herz, R. S. (2001). Ah, sweet skunk: Why we like or dislike what we smell. *Cerebrum, 3*, 31–47.

Herz, R. S., & Cupchik, G. C. (1995). The emotional distinctiveness of odour-evoked memories. *Chemical Senses, 20*, 517–528.

Herz, R. S., & Engen, T. (1996). Odor memory. Review and analysis. *Psychonomic. Bulletin and Review, 3*, 300–313.

Herz, R. S., Schankler, C., & Beland, S. (2004). Olfaction, emotion and associative learning: Effects on motivated behaviour. *Motivation and Emotion, 28*, 363–383.

Herz, R. S., Eliassen, J., Beland, S., & Souza, T. (2004). Neuroimaging evidence for the emotional potency of odor-evoked memory. *Neuropsychologia, 42*, 371–378.

Hvastja, L., & Zanuttini, L. (1989). Odour memory and odour hedonics in children. *Perception, 18*, 391–396.

James, W. (1884). What is an emotion? *Mind, 9*, 188–205.

James, W. (1890). *Principles of Psychology*. New York: Holt.

Kirk-Smith, M. D., Van Toller, C., & Dodd, G. H. (1983). Unconscious odour conditioning in human subjects. *Biological Psychology, 17*, 221–231.

Klosterhlfen, S., Kellermann, S., Stockhorst, U., et al. (2000). Pavlovian conditioning of taste aversion using a motion-sickness paradigm. *Psychosomatic, Medicine, 62*, 671–677.

Krosnick, J., Betz, A., Jussim, L., & Lynn, A. (1992). Subliminal conditioning of attitudes. *Personality and Social Psychology, Bulletin, 18*, 152–162.

Labarge, A. S., & McCaffrey, R. J. (2000). Multiple chemical sensitivity: A review of the theoretical and research Literature. *Neuropsychology Review, 10*, 183–211.

Larsson, M., & Willander, J. (2009). Autobiographical odour memory. *Annals of the New York Academy of Sciences, 23*, 318–323.

Levey, A. B., & Martin, I. (1975). Classical conditioning of human evaluative responses. *Behaviour Research and Therapy, 4*, 205–207.

Levey, A. B., & Martin, I. (1987). Evaluative conditioning A case for hedonic transfer. In H. Eysenck, & I. Martin (Eds.), *Theoretical foundations of behavioural therapy* (pp. 113–131). New York: Plenum Press.

Li, W., Moallem, I., Paller, K., & Gottfried, J. A. (2007). Subliminal smells can guide social preferences. *Psychological Science, 18*, 1044–1049.

Marinkovic, K., Schell, A. M., & Dawson, M. E. (1989). Awareness of the CS-UCS contingency and classical conditioning of skin conductance responses with olfactory CSs. *Biological Psychology, 29*, 39–60.

Martin, I., & Levey, A. B. (1978). Evaluative conditioning. *Advances in Behaviour Research and Therapy, 1*, 57–102.

Moore, A. B., & Murphy, C. (1999). A demonstration of classical conditioning of the human eyeblink stimulus. *Physiology and Behaviour, 66*, 689–693.

Pavlov, I. P. (1927). *Conditioned reflex* (Translated by G.V. Anrep). London: Oxford University Press.

Pleyers, G., Corneille, O., Luminet, O., & Yserbyt, V. (2007). Aware and (dis)liking: Item-based analyses reveal that valence acquisition via evaluative conditioning emerges only when there is contingency awareness. *Journal of Experimental Psychology: Learning, Memory, and Cognition, 33*, 130–144.

Razran, G. (1940). Conditioned response changes in rating and appraising socio-political slogans. *Psychological Bulletin, 37*, 481.

Razran, G. (1949). Stimulus generalization of conditioned response. *Psychological Bulletin, 46*, 337–365.

Razran, G. (1954). The conditioned evocation of attitudes (Cognitive conditioning?). *Journal of Experimental Psychology, 48*, 278–282.

Rozin P., & Fallon, A. (1987). A perspective on disgust. *Psychological Review, 94*, 23–41.

Rozin, P., & Zellner, D. A. (1985). The role of Pavlovian conditioning in the acquisition of, food like and dislikes. *Annals of the New York Academy of Sciences, 443*, 189–202.

Rozin, P., Wrzesniewski, A., & Byrnes, D. (1998). The elusiveness of evaluative conditioning. *Learning and Motivation, 29*, 397–415.

Royet, J. P., Zald, D., Versace, R., et al. (2000). Emotional responses to pleasant and unpleasant olfactory, visual, and auditory stimuli: A PET study. *Journal of Neuroscience, 20*, 7752–7759.

Schacter, S., & Singer, J. E. (1962). Cognitive, social and physiological determinants of emotional states. *Psychological Review, 69*, 379–399.

Schneider, F., Weiss, U., Kessler, C., Mueller-Gaertner, H. W., Posse, S., Salloum, J. B., et al. (1999). Subcortical correlates of differential classical conditioning of aversive emotional reactions in social phobia. *Biological Psychiatry, 45*(7), 863–871.

Shettlelworth, S. J. (1998). *Cognition evolution and behavior*. Oxford: Oxford University Press.

Siegel, S. (1999). Multiple Chemical Sensitivity as conditional response. *Toxicology, and Industrial Health, 15*, 323–330.

Schaal, B., & Durand, K. (2012). The role of olfaction in human multisensory development. In A. Bremner, D. Lewkowicz, & C. Spence (Eds.), *Multisensory Development*. Oxford, UK: Oxford University Press (in press).

Soussignan, R., Delaunay, M., & Schaal, B. (2012). L'épigenèse de l'hédonisme olfactif et gustatif chez l'enfant. In B. Schaal, C. Ferdenzi, & O. Wathelet (Eds.), *Olfaction et Emotion, du culturel au neuronal.* Dijon: Edition Universitaires de Dijon (in press).

Spyker, D. A. (1995). Multiple chemical sensitivities, syndrome and solution. *Clinical Toxicology, 33*, 95–99.

Staats, C. K., & Staats, A. W. (1957). Meaning established by classical conditioning. *Journal of Experimental Psychology, 1*, 74–80.

Staats, A. W., & Staats, C. K. (1958). Attitude established by classical conditioning. *Journal of Abnormal and Social Psychology, 57*, 37–40.

Staats, A. W., Staats, C. K., & Crawford, H. L. (1962). First-order conditioning of meaning and the parallel conditioning of a GSR. *The Journal of General Psychology, 67*, 159–167.

Stevenson, R. J. (2011). Olfactory illusions: Where are they? *Consciousness and Cognition, 20*, 1887–1898.

Stockhorst, U., Gritzmann, E., Klopp, K., Schotenfeld-Naor, Y., Huebinger, A., Berresheim, H. W., et al. (1999). Classical conditioning of insulin effects in healthy humans. *Psychosomatic Medicine, 61*, 424–435.

Stuart, E. W., Shimp, T. A., & Engle, R. W. (1987). Classical conditioning of consumer attitudes: Four experiments in an advertising context. *Journal of Consumer Research, 14*, 334–351.

Shimp,T. A., Stuart, E. W., & Engle, R. W. (1991). A program of classical conditioning experiments testing variation in the conditioned stimulus and context. *Journal of Consumer Research, 1*, 1–12.

Todrank, J., Byrnes, D., Wrzesniewski, A., & Rozin, P. (1995). Odour can change preferences for people in photographs: A cross modal evaluative conditioning study with olfactory USs and visual CSs. *Learning and Motivation, 26*, 116–140.

Van Den Bergh, O., Stegen, K., Van Diest, J., Raes, C., Stulens, P., Eelen, P., et al. (1999). Acquisition and extinction of somatic symptoms in response to odours: A Pavlovian paradigm relevant to Multiple Chemical Sensitivity. *Occupational and Environmental Medicine, 56*, 295–301.

Van Reekum, C. M., Van den Berg, H., & Frijda, N. H. (1999). Cross-modal preferences acquisition: Evaluative conditioning of pictures by affective olfactory and auditory cues. *Cognition and Emotion, 13*, 831–836.

Vermetten, E., & Bremner, J. D. (2003). Olfaction as a traumatic reminder in posttraumatic stress disorders: Case report and review. *Journal of Clinical Psychiatry, 64*, 202–207.

Wardle, S. G., Mitchell, C. J., & Lovibond, P. F. (2007). Flavor evaluative conditioning and contingency awareness. *Learning and Behavior, 35*, 233–241.

Wasserman, E. A., & Miller, R. (1997). What's elementary about associative learning? *Annual Review of Psychology, 48*, 573–607.

Wasserman, E. A., & Zentall, T. R. (2006). *Comparative cognition: Experimental explorations of animal intelligence*. New York: Oxford University Press.

Wolpe, J. (1969). *The practice of behaviour therapy*. New York: Pergamon Press.

Zald, D. H., & Pardo, J. V. (2000). Functional neuroimaging of the olfactory system in humans. *International Journal of Psychophysiology, 36*, 165–181.

Zellner, D. A., Rozin, P., Aron, M., & Kulish, C. (1983). Conditioned enhancement of human's liking for flavor by pairing with sweetness. *Learning and Motivation, 14*, 338–350.

Zucco, G. M. (2003). Anomalies in cognition: Olfactory memory. *European Psychologist, 8, 77–86.*

Zucco, G. M., & Job, R. (2011). Invariance of perception: The boundary between illusion and ambiguity in olfaction. *Consciousness and Cognition* (on line, 26 October).

Zucco, G. M., Caldironi, P., & Shock, V. (1989). Preferenze olfattive e comunicazione olfattiva in soggetti ciechi pluri handicappati. *Contributi di Psicologia, 2*, 131–144.

Zucco, G. M., Militello, C., & Doty, R. L. (2008). Discriminating between organic and psychological determinants of multiple chemical sensitivity. *Neurocase, 14*(6), 485–493.

Zucco, G. M., Paolini, M., & Schaal, B. (2009). Unconscious odour conditioning 25 years later: Revisiting and extending 'Kirk-Smith, Van Toller and Dodd'. *Learning and Motivation, 40*, 364–375.

Zucco, G. M., Aiello, L., Turuani, L., & Koester, E. P. (2012). Odor-evoked autobiographical memories: Age and gender differences along the life span. *Chemical Senses, 37*, 179–189.

DEDICATION

Writings in remembrance of Professor Trygg Engen

Birgitta Berglund

Trygg Engen: The scientist who made colleagues to grow

To be "trygg" means in Swedish to be secure, or feel safe and happy; "engen" translates to a lovely environment, the meadow, and is also the base of the verb "engender". Trygg engendered delight in students and researchers alike, and he contributed generously to the creation of the Stockholm School (Engen, 1995) – to my own and my students' scientific development.

In 1965, the chairman of the Psychological Laboratories of Stockholm University, Gösta Ekman, became my mentor. I wanted to produce a PhD thesis in psychophysics, delimited to sensory processing. After equipment building and experiments in warmth, audition and vibration, I turned to olfaction and olfactometers and to the adaptation to odors (Ekman et al., 1967). I then discovered Trygg Engen's research at Brown University. In 1961, he had been a Fullbright Fellow at our department and during this sabbatical, seven papers had been produced (e.g., Ekman et al., 1964). I voiced my surprise to Ekman, who had co-authored two of the papers. He responded that, during 1968, Trygg would spend a second sabbatical with us!

Trygg arrived with his wonderful family. Apart from our research together, he spent time composing the two in-depth educational and outstanding chapters on psychophysical methods for Kling & Riggs's *Woodworth and Schlosberg's Experimental Psychology*. He also drafted chapters on olfactory psychophysics and on human odor perception for two other handbooks. Trygg was also to become my lifetime mentor. Ekman became seriously ill in 1969 and died 1971. Trygg became a good friend and helped generously to improve my English, scientific writings, and career confidence. In 1972, he and his wife Betty took great care of me and my ex-husband Ulf Berglund during our Sloan Foundation post docs with the Teghsoonians at Smith College.

Three decades of research cooperation were to follow, with 1–2 month guest visits at Stockholm University. In 1982, Trygg also spent a third sabbatical with us as a Fullbright Fellow. The research in odor psychophysics continued (Berglund & Engen, 1993), but was extended to include clinical and environmental applications (Engen, 1986; Berglund et al., 1992); in all 10 publications. Among these was an experiment conducted in 1968 on emotions expressed in photographs and words. This paper was rejected by three journals and printed as an internal report in 1982. The photographs were Harold Schlosberg's set in which emotional expressions were enacted by Miss Lightfoot. Amazingly, at several occations in the passing years, the main reviewer, who rejected the paper, wrote Trygg and asked where that 'interesting paper' had been published!

In 1990, Trygg was honored with a Ph.D. (*Honoris Causa*) at Stockholm University. Trygg's research always had its own stance of new thought. I had the honor to co-work with him during his basic and innovative research period on problems of measurement and psychophysics and in environmental psyhology. Already in the late 70's, Trygg shifted his main theoretical interest from odor psychophysics to odor memory, whereas my interest went to the science of measurement. In 2002, Trygg honored me with his last research visit to my newly built Gösta Ekman Laboratory for Sensory Research at Stockholm University.

It is easy to decribe Trygg Engen as a person. He was always in a good mood and engaged readily in casual, social and professional interactions. I found him to be sensitive and considerate towards students and colleagues, and he came through as a warm person and a great mentor, who always reminded you about his students' accomplishments. He owned the rare quality of making you feel competent and encouraged you to pursue your ideas. There was also often some published material mentioned that you might like to look into. Trygg was also a rare family man who, after the children had left home, encouraged and supported his wife Betty to get a university degree and a profession. Privately, he was also a dedicated sailor with his own sailing boat. Once, during a summer visit, he and Betty rented a sailing boat to live in and to explore the Stockholm archipelago. This is how I will remember Trygg.

References

Berglund, B., Berglund, U., & Engen, T. (1992). Can sick buildings be assessed by testing human performance in field experiments? *Environment International, 18*, 221–229.

Berglund, B., & Engen, T. (1993). A comparison of self-adapataion and cross-adaptation to odorants presented singly and in mixtures, *Perception, 22*, 103–111.

Ekman, G., Berglund, B., Berglund, U., & Lindvall, T. (1967). Perceived intensity of odor as a function of time of adaptation. *Scandinavian Journal of Psychology, 8*, 177–186.

Ekman, G., Engen, T., Künnapas, T., & Lindman, R. (1964). A quantitative principle of qualitative similarity. *Journal of Experimental Psychology, 68*, 530–536.

Engen, T. (1986). Perception of odor and irritation. *Environment International, 12*, 177–187.

Engen, T. (1995). From classical to modern psychophysics at "the Stockholm School". In G. Neely (Ed.), *Perception and Psychophysics in Theory and Application*. Stockholm: Stockholm University (pp. 49–52).

William S. Cain

Trygg Engen: Rare mentor, rare human being

When I entered the psychology department at Brown University in September 1963, I figured that I would perform my research on human learning. Harold Schlosberg, the legendary chair of the department told us that we needed to have a mentor by Thanksgiving Day, an American holiday near the end of November. I could never get hold of the expert in human learning, so I pondered alternatives. Unlike any other person in my class, I had previously performed no research. Allergies prevented me from pursuit of animal research at Brown. My own insecurities guided me as well. The members of the vision and taste groups intimidated me. Trygg Engen did not.

At the first meeting of the academic year, Harold (Schlosberg) told all the newcomers that everyone in this small but powerful department should call everyone else by first name. He forbade competition among us and pointed to a wall with portrait photos of our predecessors. "They set the standard, compete with them," Harold said, knowing that the density of stars in the array left us awed. Although comforted by the noncompetitive attitude, I still gravitated toward that young, kind Norwegian pipe-smoker who entered our common room daily at 10:30 AM and chatted over a cup of coffee. He had only one student and one visiting fellow in his little group.

My only problem with potentially tying myself to Trygg was his area of research. Psychophysics had seemed dry as dirt to me in college and smell seemed remote and obscure. I reminded myself that my commitment to a mentor need only last through the academic year. If the work seemed impossibly dull, I could go shopping the following summer. I never did that. I found in psychophysics a discipline that mixed the quantitative with the behavioral in a proportion that seemed just right for me. As for the pursuit of olfaction, I liked the way it crossed boundaries into biology, food science, perfumery, and even air pollution. Furthermore, I had a dirty secret; I liked knowing a subject better than any of my fellow students, a tacit violation of Harold's dictum. Brown had produced only one former graduate in olfaction, Max Mozell, a rising star who had studied with Carl Pfaffmann, our taste guru. I could hardly know that four decades later I would receive an award with Max's name on it.

Trygg visited my office every day just after lunch. Whether I understood it fully or not, for I knew little about protocol, this meant that he cared deeply about my welfare. We talked about everything, including my coursework, research, and concerns. I always had his ear. It comes as no surprise that many years later even informal students would say the same about this mentor.

I had little sense at first of Trygg's creativity in the field of olfactory psychophysics. I felt obligated read everything he had written, an obligation I never wished to drop. Before I came to Brown, Trygg had collaborated with Carl (Pfaffmann) on information transmission in olfaction. While I was there, he pursued direct scaling of odor intensity and quality. After I left, he virtually created the field of odor memory, which led me to tell my own undergraduate student Harry Lawless at Yale to pursue graduate studies at Brown. Trygg had a hand in virtually every development in olfactory psychophysics, the study of infants and later of children, of odor preference, of adaptation, of odors and safety, and, including his work on memory, of odors and cognition. Then, as his career matured, he took the time to write his adventures in olfaction into a book, The Perception of Odors (Engen, 1982). Trygg proffered no pretense about what initiated olfaction. He had no theory of odor, all of which fell short anyway. His account began after the fact, after an "odor" got into the nervous system. Then he had strong opinions, some right and some wrong in my opinion, but more often right.

At the end of his career, he summed up his knowledge in another book, entitled Odor Sensation and Memory (Engen, 1991). When reviewing it, as I had the previous volume, I said,

> Olfaction is a topic that often makes instructors in general psychology and in sensation and perception uncomfortable. Textbooks often treat the modalitv superficially, even schematically. The accounts may dispense some facts, but few ideas. For instructors in search of ideas, Engen's book will oblige. This short, thematic narrative could, without making large demands on time, give considerable material to instructors who crave to give an idea-filled lecture on smell. The book could also give students three to four hours of readable narrative on a topic they may never again encounter outside the popular press and may therefore never otherwise come to know. (Cain, 1993)

Two decades later, the value of that little book has not changed, but alas, Trygg has departed. Only in recent years have I come to know how supportive he had become of young researchers at the end of his career. I have never heard more genuine expressions of affection for a person not formally a mentor, but effectively so. This man, whom I had come to love in his thirties, evoked the same feelings in students in his sixties and beyond. His legacy lives on in them and, if they truly absorbed his priorities, in their own students.

References

Cain, W. S. (1993). Redolence revealed. *Contemporary Psychology*, *38*, 529–530.
Engen, T. (1982). *Perception of odors*. San Diego, CA: Academic Press.
Engen, T. (1991). *Odor sensation and memory*. New York: Praeger.

Pamela Dalton

To begin to understand the impact of Trygg Engen's influence on my development as an olfactory scientist one only has to look at the broad scope of his impressive list of publications. As a naive graduate student, a postdoctoral fellow and a new faculty member, I often believed that I was asking truly original questions in the field of human odor perception. To my continued surprise then (and even to this day!) a literature search nearly always revealed that Trygg or one of his many students had already asked, and frequently answered, many of the most important questions. From basic research studies on sensory perception in neonates or the fundamentals of olfactory adaptation to questions in applied research regarding whether olfactory and taste cues can be used as deterrents among children, his work inspired me to think creatively and broadly. Indeed, much of my work in human odor perception – from understanding the effects of long-term adaptation to mapping the ways in which beliefs and expectations influence perception – continues avenues of research he earlier established. But his influence on my thinking and my career didn't stop with his writing; he was always a generous teacher when answering questions, filling in details where gaps existed in methods as described in the literature or simply providing the reassurance that my ideas were important and worthwhile. In 1995, the faculty of Monell recognized his enduring contributions to the advancement of chemosensory science by presenting him with the annual Manheimer Lectureship Award. While this was but one small tribute among many distinguished international awards he received in his lifetime, his graciousness and humble nature when accepting the award, coupled with the enthusiasm with which he delivered his scientific and informal lectures on that occasion remain indelibly etched in my memory. With gratitude for the influence he has had on my career, I have been delighted to be invited to contribute to this volume in his honor.

Richard L. Doty

Trygg Engen: Olfactory pioneer

When I took my first job in human olfaction, Trygg Engen was *the* human olfactory psychophysicist. As I noted elsewhere in this volume, his initial impact on my career was via his review of psychophysical methods in the third edition of *Woodworth and Schlossberg's Experimental Psychology* (Kling & Riggs, 1971). Although I had gained some familiarity with threshold testing after completing my master's thesis on vestibular psychophysics at Ames Research Center, Moffett Field, California, I was unaware of the breadth of psychophysical methodologies that were generally available. Trygg's lucid and easy-to-understand review was just what I needed to begin my studies on human olfaction. His early pioneering work on ratio scaling, as well as on odor memory, was very helpful to me and spawned whole new fields of academic activity.

Although I did not have the good fortune to study under Trygg at Brown University, I was fortunate to benefit from his presentations at the Association for Chemoreception Science (AChemS) meetings in Sarasota, Florida. I also had the opportunity to visit him at Brown on several occasions. At that time I was just beginning my career and Trygg was very gracious, going out of his way to help me understand the basics. We continued to maintain a professional relationship for a number of years thereafter. I have immensely benefitted indirectly from Trygg's influences via a number of his students, most notably Robert Mair, who encouraged me to submit data our group had collected on age and olfaction to *Science* magazine – a submission that proved extremely valuable to my entire career. Clearly Trygg's legacy lives on through his numerous students, as demonstrated by such fruitful interactions and this unique volume.

Trygg marched to the beat of a different drummer. Until reading various obituaries, I was unaware of his trials and tribulations during the Nazi occupation of Norway and his subsequent challenges within American academia. His research on applied problems with his collaborators at the University of Stockholm was trend setting at the time and provided a strong model for how psychophysics was relevant to practical affairs. His appearance on CBS television's "What's My Line" showed a willingness to go beyond the stereotypic and stodgy Ivy League college professor and to have fun with his profession and with others. His broad approach to life was refreshing to me as I am sure it was to those who knew him well.

Richard Gracely

I have been fortunate to know Trygg Engen as a psychology mentor and friend.

Trygg guided me through my first experience as a teaching assistant, ultimately agreed to be my PhD thesis advisor, and facilitated my first job that resulted in a 28-year career at the National Institutes of Health. After two years of animal research he lured me away from it to develop methods for medical psychology. He supported me in an unconventional project that began as the first dissertation outside of the physical confines of the psychology building in a nearby hospital and finished farther away at NIH. In a novel program he helped fashion, this project developed and applied psychophysical methods to the challenging problem of measuring sensory and motivational dimensions of human pain. This project and, more importantly, Trygg's guidance helped set me on a path toward a successful, fulfilling career in science. While I appreciate this, my fondest memories extend far beyond the academic walls.

I was privileged to be part of Trygg's wider family and to share in his optimistic view of the world. The once young Norwegian ski jumper brought a humorous and European viewpoint that was way ahead of its time. He walked or rode his bike to work, and often proclaimed that the only transportation a person ever needed was a bike, a pair of cross-country skis, and a sailboat. He was at home on the helm of his boat.

On the day before my dissertation defense, he asked me to help him bring his boat from its summer berth to Bristol, R.I. to be hauled out for the winter. Although I suspect his real purpose was to help me avoid pre-defense jitters, I agreed and found myself in very blustery weather with some sandwiches and beer. "What a glorious day!" he proclaimed, as we scanned the tall waves looking for other boats realizing that we were the only boaters crazy enough to brave the day. Later we approached the narrow entrance channel, marked by a series of buoys in the open shallow water. Trygg, with a permanent grin, reached down to start the motor. Instead, he paused, looked up, and shouted above the din, "our forefathers didn't have engines, LETS SAIL IT IN!" And so we did. Scanning backward to correct for drifting out of the narrow path with the waves and current pushing us sideways, we made it into the harbor, flying toward a crowd of boats moored and waiting to be hauled. We had one chance to grab our mooring before running into a sea wall. Trygg was ever confident and we did, indeed, grab it and avoid disaster. I will always remember him moments later as we sat below in the snug cabin. He was in his yellow bib overalls over a white Irish knit sweater, smiling with a twinkle in his eye as he poured glasses of rum for a toast to a very fine day. He was an inspiration for life, and I miss him.

Rachel S. Herz

I first met Trygg in the autumn of 1990, though I felt I already knew him from being a devotee of his writings and especially his ground-breaking book, *The Perception of Odors*. Earlier that year, I had begun research for my dissertation on the relationship between odor and emotional memory and as there was no one in Canada who could lead me in this topic I went on a quest to find and meet those who could.

It was a mid-October day when I met Trygg at his office at Brown University. We talked for about an hour and I was beyond thrilled to finally be able to share my excitement about olfaction with someone else, especially someone so important who was at the very heart of the field. By the end of our meeting I felt supported and inspired and with redoubled energy I carried out a stack of articles that Trygg had given me to read. From that moment on Trygg became my shining light and guide in all things olfactory.

Trygg and I were scheduled to meet again in 1992, as he was the designated external examiner for my PhD defense but when the date arrived, extenuating circumstances prevented him from coming and two faculty members from a different department were called to fill his spot at the last minute. Although we continued to correspond and work together on a paper, it was not until 1995 that I met Trygg again in person.

In December 1995 Trygg came to the Monell Chemical Senses Center in Philadelphia, where I was a junior faculty member, to give a talk and receive the prestigious Mannheimer award that he was being honored with. From that day forward we were in close and regular contact. In 1999, I moved to Rhode Island and through my connection with Trygg, was able to join the faculty of the Psychology department at Brown University as a Visiting Assistant Professor. By this time Trygg was long retired but he regularly came to the department and we had many very productive and helpful (to me) discussions about olfactory cognition and the mysteries that had yet to be resolved. In 2003, circumstances surrounding my appointment as a tenured faculty member (which did not come to pass) led Trygg to leave his department office. From then on our time together moved to his home or mine and we frequently enjoyed vigorous conversations and hearty dinners and lunches. Trygg and his wife, Elizabeth, became part of my family and I of theirs.

I am grateful that among my fondest memories of Trygg, was the last time that I saw him. It was an evening in late July 2009 and my husband and I had stopped over for a "quick" cocktail before going out to dinner in Providence. Recent problems that Trygg was suffering with his legs had considerably compromised his mobility, and for such an athletic, active and able man I can only imagine how frustrating it must have been – but he never let it show. In fact that night his dynamic presence danced around the room, his eyes sparkled, he laughed joyfully and he regaled us with many vivid and exciting stories from his past. The "quick" drink turned into a visit that lasted well over two hours and it would have lasted longer had our stomachs not been insistent. I now deeply wish that we would have stayed all evening. When we said our goodbyes that night I had no idea that I would never see him again.

As a graduate student, well before I ever met Trygg, he became my intellectual inspiration and mentor. Over the years as I got to know him better and better he became a father-like figure to me both professionally and personally and like an eager child I strove to make him proud in every way that I could. I owe the way that I think about olfaction to Trygg and I will continue to do everything I can to honor the scientist and the man that he was.

References

Engen, T. (1982). *The Perception of Odors.* New York: Academic Press.
Herz, R. S., & Engen, T. (1996). Odor memory: Review and analysis. *Psychonomic Bulletin and Review*, 3, 300–313.

Lew Lipsitt

Trygg Engen was among those fellow faculty members whom one could truly call collegial. When talking to him of a new idea, or something exciting that I had recently read, I could always count on Trygg's first reaction: "Yah. Yah." This was a deceptive first response as it could mean he endorsed the idea (or didn't), or might indicate he had read the same thing (or didn't). What it really meant was that Trygg was pondering what his genuine reply was going to be, and he was taking his time to articulate it. He was always receptive, always ready to spend his time with you on the subject and, if you just gave him some moments to personally reflect on it, he would give you a very thoughtful, often highly creative response. This was how it was in many aspects of our personal association and friendship. For present purposes, it is exactly how it was in our collaborative work on habituation in newborns. When he pondered the possible objection to some of our earlier work, on grounds that the seeming olfactory "habituation" effect might simply be a matter of "neural fatigue," he came up with the prospect of delivering multiple odorants simultaneously and testing for recovery with individual odorant components selectively. In describing the procedure we would ultimately use successfully, his description of the methodology was conveyed to me with excitement – with the passion of a scientist on the cusp of an important breakthrough. When he was sure I understood how this would solve "the problem," we went home to celebrate with a martini, which each of us knew just how to mix for the other. Trygg was an incomparable colleague – honest, highly principled, creative, and the best of friends.

Robert Mair

Trygg Engen: Mentor

I met Trygg in late September 1969 when I signed up for his Introduction to Psychology course. He was fresh from a sabbatical in Europe with an enthusiasm for psychology and an interest in knowing each of his students personally that made the course come alive. I got to know him better two years later in his role as faculty advisor for undergraduate honors students. Although I was working in another lab on a project related to language development he always seemed genuinely interested in what I was doing and welcomed discussions at office hours. Without Trygg's warmth, support, and enthusiasm I doubt that I would have given serious thought to a career in experimental psychology.

A year after I graduated, midway through my Conscientious Objector service, Trygg agreed to take me on as a part-time graduate student. As a scholar Trygg was driven by his interests and was undeterred when they didn't correspond to popular trends or funding priorities. In the end this led to thoughtful and original insights into once-neglected issues that fill the pages of this book. As a mentor Trygg gave his students unqualified support and encouragement to do the same. His constant admonishment was not to settle for the obvious. Looking back, I am amazed at his enthusiasm and encouragement for my fits and starts as an experimental psychologist: studying olfactory perception with neurological patients at the Boston VA, setting up a gas chromatograph in the lab to model the behavior of odorant molecules, building odd nosepieces to analyze birhinal olfactory interactions, raising rat pups in constant odor environments to examine neuroplasticity in olfactory pathways, building an electrophysiology lab to record odor-evoked activity of olfactory bulb neurons, and then working on a "side project" studying the neurochemistry and pharmacology of amnesia in Korsakoff syndrome patients (while working on a dissertation on adaptation of neuronal responses in the rat olfactory bulb).

Ultimately each of these projects was successful in one way or another. Some produced interesting results and published papers, others valuable learning experiences. Trygg's unwavering confidence in me helped me to develop confidence in myself. His emphasis on experimental rigor and fresh new thinking continues to inspire me in these directions as I work with undergraduate and graduate students in my lab at UNH. My work with him led to a postdoctoral fellowship with Bob Gesteland at Northwestern focusing on cellular aspects of olfactory receptor function, a return to the Providence VA (and Brown) for more work on amnesia (and collaborations with Trygg on the neuropsychology of olfaction), and a career at UNH studying memory and related cognitive functions, primarily in animal models.

Most importantly to me, Trygg was a model of balance: a successful career as a teacher and a scholar, a life fully lived, and a thoroughly decent man. He remains an inspiration.

~ ~ ~

Donald McBurney

Trygg was a young assistant professor when I was a graduate student at Brown. He was about 10 years older than I, and so I was able to relate to him more easily than to my major professor, Carl Pfaffmann. Carl had an imposing manner that intimidated his students, myself included. Carl went on leave twice during my time as a grad student, and left me in Trygg's care, which I recall as benign, but not quite benign neglect. Trygg's more easy going manner provided an anchor for me. Along with all graduate students in the department at the time, I took Trygg's course in psychophysics. I remember commenting to a fellow student that psychophysics was okay, but I couldn't imagine specializing in it. As it turned out, I was the only Pfaffmann student at the time (and perhaps ever) who never did electrophysiology, and only did psychophysical work, then and since.

I remember Trygg as an example of one who kept a sense of balance between work and the rest of life.

~ ~ ~

Mats J. Olsson and Steven Nordin

When we came to work with odor perception for a bachelor degree thesis with Trygg's close friend professor Birgitta Berglund at Stockholm University in the late 80's, we both belonged to those who had gone through life without paying attention to the sense of smell. Our work on the psychophysics of odor sensation and perception did certainly capture our interest. However, it was during Trygg's several visits to the Berglund lab during our graduate years that we were introduced to a wider perspective to the topic – the cognitive science approach to odor perception – which made our studies even more rewarding. In several ways Trygg's presence meant a lot to the graduate students at the lab. His solid knowledge, the international atmosphere, and his humbleness, altogether made him a deeply respected role model for us. A good indication that one has been deeply influenced by a person may be when one realizes, as we did, that one moves across the stage in the same characteristic manner as that person.

To the potential young scientist in the field that encounters this book we would like to pass on Trygg's legacy. Defy the short-sighted chase for new results and publications, and adopt a sense of urgency for reading. This will help you become a good scholar in your field, just as Trygg truly was. Rest in peace, Trygg. Thank you so much for sharing your time and wisdom with us.

~~~

Carolyn Rovee-Collier

A personal debt to Trygg Engen, in grateful admiration

When I entered graduate school in 1962 at Brown University, I examined the relation between newborn activity and nonnutritive sucking in Lew Lipsitt's Newborn Lab at the Providence Lying-In Hospital. When my master's thesis advisor left Brown after my first year, I wondered what my future might hold. Lew wasn't taking new students who would finish while he would be on sabbatical, in my fourth year of study. The only other faculty member of the Psychology Department who could direct my dissertation research with infants was Trygg Engen, but I had struggled in his graduate Psychophysics course – which was totally new to me – and doubted that he'd want to take me on. Fortunately, however, he agreed to direct my dissertation research – and forever changed my approach to research.

Trygg thought that the human newborn is an ideal psychophysical observer, unencumbered by social demands, language, or expectations, and he taught me to
~~~

think about infant data from the same perspective. This is one of the most important lessons I learned in graduate school and the one that has had the greatest impact on my research. As a result of the insight that infants respond to a *subjective stimulus or temporal interval*, my students and I have always selected test stimuli and temporal intervals that vary on a ratio scale and have usually plotted their responses on a log (sensory) scale. Karen Hildreth's dissertation, for example, documented the longest delay after forgetting that a brief reminder could still recover the forgotten memory. The difference in how the data were represented was dramatic. When she expressed the data in terms of the absolute delay (weeks) since forgetting, the function increased linearly with infant age. When she expressed the delay relative to how long it took infants of each age to forget, however, the *function was flat* (a constant ratio of 4:1) between 2 and 18 months of age! This "ratio rule" also characterizes the retention data of human adults and suggests, contrary to popular wisdom, that the fundaments of memory processing are developmentally invariant. Trygg would have been pleased.

Trygg was a Renaissance scholar in the field of olfaction. His interests were eclectic but deep-seated. He loved data and was genuinely excited by new findings – anyone's – and transmitted this same love and excitement to his students. I can't begin to count the number of times he'd stop me in the hall, waving some paper and saying, "Hey, so-and-so got these really neat data ..." and invariably ending with, "Now isn't that neat!". He pondered not only the "what?" but also the "how?" and "why?" of data ... and then designed impeccable experiments to answer these questions. Trygg was the mentor of many grateful and devoted students, and always took great pleasure in their success. He could well have authored the following passage from Walt Whitman's *Song of Myself* (in *Leaves of Grass*, 1855):

> *I am the teacher of athletes;*
> *he that by me spreads a wider breast than my own,*
> *proves the width of my own;*
> *he most honors my style*
> *who learns under it to destroy the teacher.*

Personally, Trygg was an old-world country gentleman – thoughtful, gracious, generous, kind, not a mean bone in his body. He was enormously proud of his family and students. I count my lucky stars for the wonderful memories of Trygg and his lovely wife Betty that are etched in my brain.

Benoist Schaal

In 1978, when entering the field of human ethology and behavioral biology as a fresh graduate student, my supervisor, H. Montagner, came up with several possible research topics, such as endocrine-behavior interrelations in children or the communicative functions of odors. This second theme became the focus of my dissertation on the adaptive roles of odors in human mother-infant relationships. At the time, the literature on infant olfaction was narrow and that on human mothers' olfaction was inexistent. As a starter (no Internet search at that times), my supervisor handed me over two chapters, one by Trygg Engen and one by Aidan Macfarlane. Trygg's chapter on *Method and theory in the study of odor preferences*[1] guided me into core issues in the experimental analysis of the development of odor hedonics, and revealed to me the pioneering studies on neonatal olfaction by the Providence school of experimental psychology. In this way, Trygg and his close colleague at Brown University, Lew Lipsitt, became intellectual mentors, opening my mind to theoretical approaches and creative methods then unknown to me (being trained as a biologist). Their papers certainly were those I read and re-read most at that time, and they remain inspiring nowadays. Several years later, in April 1986, at the occasion of a postdoctoral year in Montréal, I travelled down to Providence to meet Lew Lipsitt and Trygg Engen. There, delightfully welcomed by Lew, I could discuss with both of them about their ongoing work, get access to unpublished results, meet research assistants, visit their famous baby lab in the hospital. From the discussion with Trygg, I bear in mind a kind-hearted professor acutely listening to a novice and uttering an enthusiastic cascade of provocative questions on olfaction and the processes of its development in emotion and cognition. The origin(s) of preferences was one of his endeared issues. Holding that newborns had no preferences in the realm of odors, considering their perceptual capacities as a *tabula rasa* on which biases were to be impressed by familiarization, association or teaching, he was a radical empiricist. I was steadily intrigued in Trygg's position, there and in his writings. This led my colleagues, students, and me to analyze neonatal responses to a range of odorants chosen to be hedonically contrastive: newborns expressed clear-cut odor preferences. Later, some of these preferences could be traced back to fetal experience, which does not contradict Trygg's view but just pushed upward in development the formative impact of the environment on olfactory processes. But the issue remains open in humans on whether an odorant could work like sucrose from the very beginning of life, *i.e.*

1. Engen, T. (1974). Method and theory in the study of odor preferences. In A. Turk, J. W. Johnston, & D. G. Moulton (Eds.), *Human responses to environmental odors* (pp. 121–141). New York: Academic Press.

as an unconditional releasing stimulus and as a primary reinforcing agent. Recent evidence stemming from nonhuman mammals suggests that, in certain species, neonates are born with a partially carved *tabula*, in that they bear apparently unlearned olfactory "representations" (at least one) that are functionally connected with strong incentive potency and motor effectors. Whether such predisposed perceptual processes are generalizable to the newborns of our own species awaits further investigation. Trygg would certainly be dubitative, as firmly stated in his last book.[2] It is the role of mentors to articulate theories and sprinkle novice minds with unsolved problems. In that sense, Trygg Engen's work and thinking has been, and still is, influential on my own work.

Richard Stevenson

As a graduate student in the late 1980's my introduction to the study of olfaction was Professor Engen's book "*The Perception of Odors*". It is hard to stress how influential this book was for me and I would like here to briefly explain why. Having originally read this book over 20 years ago, I took the opportunity afforded by writing this chapter to re-read this work. I was struck by how prescient it was. Not only did he note that discovering the nature of the olfactory code would lead to a Nobel Prize (see p. 6) – beautifully anticipating Buck and Axel's triumph, for me it was the various hints as to where interesting discoveries might lie and, especially, the focus on learning and memory. This focus on learning and memory, which is also reflected in Professor Engen's extensive publication record, concerned the influence of experience on odor hedonics and the many apparently unusual aspects of olfactory memory – which led to a later an equally erudite book on just this topic. *The Perception of Odors* planted the germ of an idea in my mind, namely that learning and memory might be more deeply involved in olfactory perception than even Professor Engen had suspected. His book also stressed the totality of the perceptual process, noting that he had deliberately chosen to use the word *Perception* in the title because it dealt with "... whole objects" – the first mention that I had seen of an object-like concept in relation to smell. Similarly, he also seemed to recognize the limitations imposed by a structure-receptor-quality approach, noting the potential of pattern based coding. All of this insight, and the wealth of empirical data in his book – much coming from his lab and that of his many productive students – the focus on hedonics and experience, and odor

2. Engen, T. (1991). *Odor sensation and memory*. New York: Praeger Publishers.

recognition memory, left me with particular perspective that has continued to inform my work on olfactory perception. Along with many other students of olfaction I am in debt to Professor Engen for opening my mind to such an endlessly fascinating topic.

~ ~ ~

Theresa White

I never met Trygg Engen – at least, not in person. Instead, I met him through his work. When I became interested in olfactory memory as a graduate student at Oxford in the late 1980s, there were very few people actively publishing in the area. I stumbled across a book written by Dr. Trygg Engen at Brown University called *The Perception of Odors* (1982), which I found extremely exciting. Here was an *entire book* synthesizing the available knowledge about the way that people interpret and interact with olfactory stimuli! It was quite a valuable find for a graduate student who was trying to quickly come up to speed on the literature. I appreciated its thoroughness and readability, since it seemed targeted to an audience that was well-educated, yet unfamiliar with olfactory research. Yet, a synopsis of the field is only the beginning for a graduate student. I began to read many of the scientific papers cited by the book, many of which were authored by this same prolific Dr. Engen. Two lines from a 1977 paper grabbed my attention: "Several other lines of evidence indicate that memory for odors differs from memory for other stimuli. First, there is *no evidence for a short-term store* ..." (Lawless & Engen, 1977). These two sentences literally changed my life.

The concept of a sensory system without a short-term memory challenged everything that I knew about memory at that time. I couldn't stop thinking about it. How could memory for odors be so different from the memory for pictures or words? Because I couldn't reconcile this question, I felt that Engen must be somehow wrong in his conclusions, and it made me want to find out the truth of how olfactory memory worked. Engen suggested that there was no short-term memory based partially on the work of Lawless and Cain (1975), who reported the results of an odor recognition experiment that did not show an effect of serial position, and partially on his own work that showed very little effect of time interval on forgetting in olfactory recognition experiments (Engen, Kuisma, & Eimas, 1973). Each of these pieces of research reported findings that were systematically different from the prevailing view of verbal memory, and each expressed the sentiment that olfactory memory was somehow unique.

In the course of evaluating these well-written papers, I noticed that Lawless, Cain, Kuisma, and Eimas had all been at Brown University. This immediately made me think that Engen had single-handedly trained or influenced everyone doing research on human olfaction! This location commonality also made me wonder if Engen's laboratory might be subtly influencing the growth of olfactory memory theory, not just by the very solid results that were generated there, but also by affecting the questions that were asked. In the interest of maintaining the ability for critical evaluation, I resolved to be one of the (seemingly) few people in the field of olfactory cognition trained separately from Engen. I felt that perhaps I would try to work with him after my own ideas about olfactory memory were firmly fixed, but to my regret, I never did. It has since occurred to me that despite my desire at the time to be trained separately, Engen has still been an incredible influence on my work. After all, it was his research that drove me into the olfactory memory lab, and a large part of what inspired me to stay there.

References

Engen, T. (1982). *The perception of odors*. New York: Academic Press.

Engen, T., Kuisma, J. E., & Eimas, P. D. (1973). Short-term memory of odors. *Journal of Experimental Psychology, 99*, 222–225.

Lawless, H. T., & Cain, W. S. (1975). Recognition memory for odors. *Chemical Senses and Flavor, 1*, 331–337.

Lawless, H., & Engen, T. (1977). Associations to odors: interference, mnemonics, and verbal labeling. *Journal of Experimental Psychology: Human Learning and Memory, 3*, 52–59.

Gesualdo M. Zucco

I was deeply sad when I heard that during the night in between Wednesday 25 and Thursday 26, 2009, Trygg departed. This was a grievous loss for our community, since Trygg was one the most eminent and leading authorities in the field of the chemosensory sciences, and author of numerous papers on the human sense of smell and of remarkable books as "The Perception of odors" and "Odor sensation and memory".

But, beyond these academic notes he was for me and for some other colleagues of my generation, a mentor, an intellectual guide and a close friend. In such a way I would like to remember him. Flashbacks come to my mind: my first letter to him (when I was a Ph.D. student), sent after days spent at the library, looking for people involved on this neglected field called olfaction; his availability, helpfulness and generosity in the first answer and in the following interactions with me; the impatience I had in waiting for a letter from him (when, luckily, neither internet nor e-mails nor "the special effects" of power-point still existed!).

Later the invitation to Providence (1987); the warm hospitality at his own home; my first learning experiences with the odorants in his lab; our nice, productive and endless scientific discussions on the perception of odours and odour memory, mixed with the stories of his life and family in Norway: the Second World War, the imprisonment of his father (member of the resistance to Nazi); his lively childhood, his love for nature, skiing, soccer and sailing; his deep and emotional remembrances of the fragrances of the Norwegian woods.

In 1988 Trygg came to Italy with his wife Elizabeth. I invited him for a series of lectures on the human sense of smell. This was also a beautiful time spent with him. I remember that I took him to Venice airport with a very old and ramshackle orange-colour Ford car. He was not disturbed at all, rather he gave me suggestions about how to fix it. How different he was compared to the serious and formal Italian professors to whom I was accustomed! Trygg enjoyed very much his visit to Padua, and I was proud when he wrote to me after saying: "our times together were special both here in the States and in Italy (…)".

Following his visit we started to write letters to each other; the rate was slow because Trygg didn't much like new technologies, such as e-mail. One letter I especially recall was when, already retired, he wrote to me that a paper I had dedicated to him "literally brought tears to my eyes (…)". It was a tremendous loss that the heart failure he suffered the night of 25 of November, was sudden and definitive.

I will keep the memory of my *Magister* and friend in my heart forever.

Index

A
Absolute magnitude of pain 44
Acetylcholine (and olfaction) 159–160
Adaptation 10, 18, 19, 31, 73, 75–78, 84, 88, 89, 144, 145, 147, 149, 150, 179, 201, 217, 221, 225 226, 228–234, 250, 257, 259, 261, 263, 266
Affect 33, 50, 59, 65, 71, 85, 101, 121, 245, 253, 284, 289
Affective gain 49, 53
Affective positive-negative asymmetry 278
Affective valence (or value) 101, 245, 253, 272–274, 279
Affectivity (negative) 33, 50, 65–69
Afferent neural activity 46
Aging 52, 160, 199–211
Agnosia
  Olfactory 117, 180, 184
  Visual 117
Air quality 12–16, 23–24, 30, 33
Allergic reaction (to odour) 146, 203
Alzheimer's disease (and olfaction) 83, 84, 156, 169, 199, 204–210
Ambient odorants 281, 282
Amnesia 179, 181, 182
Amniotic fluid 232, 240, 258
Amygdala 101, 102, 111, 113, 158, 159, 192–194, 200, 205, 234, 257, 261, 270, 283
Analytical processing 78
Annoyance 25, 31–33
Anosmia 52, 155, 156
Aphasia 179–181, 185, 195
Apolipoprotein E (ApoE) 200, 207–211
Aromatherapy 24
Arousal 30–31, 140, 246, 255, 256, 286
Associative learning 58, 90, 95, 97, 98, 101, 102, 109, 110,112, 194, 252, 262, 268, 270, 285, 289 291
Attention
  General XII, 138–142, 159, 194, 209, 249
  Selective 10, 25, 33, 141, 144, 146–148, 239, 244
Attitude(s) 31, 113, 270, 274, 275, 279, 284, 285, 288, 289, 291–293
Attitude conditioning 273–276
Audition 11, 13
Autobiographical/episodic memory 99, 103–106
Aversion (food) 240, 259, 270–272, 273, 279, 281, 289
  see also Learned taste aversion

B
Belief(s) 23–25, 27, 29, 31, 32, 34, 35, 36, 52, 53, 149, 270
Betula study 208, 209
Binaral rivalry 144, 146
Body odour 65, 70, 59–70, 109, 120, 130, 240, 280
Bottom-up processing 26–27, 33, 147
Bottom-up saliency filters 139, 140, 145, 148
Bowlby's theory of attachment 63

C
California Verbal Learning Test 203
Category scale 44
CNS (central nervous system) 50
Change blindness 145
Chemosensory function 199
Child 73, 81, 95, 105, 109, 110, 120, 248, 276, 280–281
Cingulate cortex 145, 147
Classical conditioning 101, 235, 246, 252, 262, 268, 271, 272, 290, 292, 293
  see also Pavlovian conditioning
Close relationships (and olfaction) 59, 63, 64, 70, 71
Cognitive load 78
Cognitive performance (and odour associated with frustrating tasks) 281, 282
Colour discrimination 164, 166, 167, 174
Combinatorial code 77
Comfort 50–70
Competitive selection 139, 140, 148
Conditioned reflex(s) 222, 264, 270, 292
Conditioned stimulus 271, 293
Conditioning of affective responses 30–31, 275, 289
Congenital insensitivity to pain 52
Congruent (vs. Incongruent) 104
Consciousness 24, 76, 86–87, 109, 125, 130–131, 140–142, 272, 274, 278, 284, 285–289
Content addressable memory 76, 79, 80, 81, 84
Contingency awareness 272, 278, 290
Continuous olfactory delayed non-match to sample (cDNM) 185–193

Consumer science (effect of odours on) 284, 285
Cross-contamination 41
Cross-modal
 evaluative conditioning 278–283
 experiment 99
 paradigm (or design) 100, 102
Cross modality matching 42, 44, 45, 48
 response 46

D
Demand awareness 273, 278
Developmental studies (on odour perception and memory) 71, 222–223, 237–261, 280, 281, 289
Development (impact on odour perception) 81, 105, 109, 194, 224, 281
Degraded input 77
Discrimination
 general 47, 48
 of colour 4, 20
 of a conditioned response 271
 of pain intensity 48, 53
Discrimination performance 55, 81, 83, 129
Dopamine (and olfaction) 159
Double-cuing method 103–104

E
Ebbinghaus (Hermann) 96
Efferent neural activity 46
Electroencephalogram (EEG) 144
 see also Event-related potential
Emotion (and odour) 10, 11, 31, 61–64, 85, 99–103, 108, 110, 131, 256, 269, 272, 273, 277, 284–285
 Evocativeness 99–103, 108
 Intensity 104, 108
 Vividness 99, 103–105, 108
Endogenous attention 138, 144
Entorhinal cortex 158, 167, 189, 190, 192, 200, 205, 211
Environmental exposure 29–33, 202–211
Esthetics 12, 13
Evaluative olfactory conditioning 269–290
 Ecological studies on 283–285
Event-related potential (olfactory) (OERP) 171, 200, 205
Evolution (and olfaction) 23, 62, 63, 110, 185, 194, 224, 252
Evolutionary psychology 60, 70, 71, 266, 238, 250, 266
Evolution theory 62
Evoked Potentials 46
 see also Event-related potential
Executive function XIII, 47, 159, 186, 202
Exogenous attention 138, 145
Expectation 23–33
Extinction
 of a conditioned response 271–273, 285, 288, 290, 293
 resistence to 272
Eyewitness testimony 107

F
fMRI 47, 98, 101–102, 141, 142, 144, 145, 146
False alarms 40, 41, 146
False negative (miss) 40, 41
False positive 40
Familiarity 24, 28, 32, 35, 78–80, 110, 116, 129, 143, 206, 211, 244, 246, 251, 255, 285
Feeling of knowing ("tip of the nose" phenomenon) 123–128
Figure/ground separation 77–78
Flavour(s) 6–8, 13, 16, 74, 87, 110, 238, 245, 253 272, 273, 277, 278
Flavour-flavour paradigm 277, 278
Fetus 232, 243–248, 252
Free recall (memory) 97

G
G-protein coupled receptor 75
Galvanic skin response 271, 276
Generalization
 of a conditioned response 271, 273, 274
 gradient 271, 274
Generalized Labeled Magnitude Scale (gLMS) 45
Glomeruli (Glomerular layer) XI, 75, 79, 80, 84, 159, 192, 200, 248, 251, 256, 260
Go/no go task 188, 189
Gustatory 34, 40, 49, 77, 86, 97–98, 121, 143, 145, 148, 239, 240, 242, 251, 259, 261, 271, 272, 277, 281
 see also taste

H
Habituation 144, 146, 221–233, 242, 244, 252, 255
Health symptoms 27–34
Health risk (perception of) 23–24, 33
Hedonic psychophysics 39, 48–50, 53
Hedonic valence (of odours) 270, 289
Hedonic reactions (non-verbal) 281
Hippocampal regions 87, 102, 270
Hippocampus 98, 101, 102, 205
HM case (and olfactory perception) 82–83
Homeostatic balance 48
Human conditioned preferences 271, 273
Hunger 39–40, 48, 49, 52
Hypnosis 52
Hyposmia 52, 165, 208

I
Ideal Type 51
Identification (odour) 47
Infant 60, 62, 105, 109, 222–233, 237–261, 280
Innate-acquired debate 237–261

Innateness 109, 250
(definition)
Insula XIII, 145, 200
Intensity (general) 39, 49, 53, 54
Irritation 25

K
Korsakoff's syndrome 182–184

L
Latent inhibition 284
Learned taste aversions 110, 224, 270, 272, 274, 279, 281, 289
Learning and pattern matching 79
Lewy body disease (and olfaction) 157–162, 167
Luncheon technique 274

M
Magnetic resonance imaging *see* fMRI
Magnitude estimation 14, 42, 44–46
Magnitude matching 44, 45
Malingering 43
Matching task 83, 180
 exact match 180, 181, 193
 similarity match 180, 181
McGill Pain Questionnaire (MPQ) 50, 51
Memory
 bump 105–106
 content addressable 76
 long-term 96, 141, 224
 systems (multiple/separate/unitary) 108, 141
 recognition 96, 97, 98, 119, 182, 194, 203, 206, 210, 280
 recollection 103
 selection 103
 semantic 47, 87, 202
 short-term 85, 96, 137, 139, 244
 odour-evoked autobiographical episodes 95, 98, 99–108, 269
Mere exposure 50, 81, 246, 255
Method of Adjustment 41
Method of Constant Stimuli 40, 41
Method of Limits 40
Method of Successive Categories 44
Mild cognitive impairment (MCI) 207, 209, 211
Milk 110, 240–242, 245, 253, 258
Mitral cells 157, 192, 193, 200, 211, 243, 247, 256
Mononaral rivalry 144
Mood (and olfaction) 24, 156, 284, 286, 288
Motion-sickness paradigm 279, 280, 291
Multidimensional scaling 48, 51, 52, 53
Multimodality (multisensoriality) 71, 81, 86, 87, 129
Multiple chemical sensitivity 36, 271, 282, 283, 292–294

N
Naming (odour) 83, 89, 115–123, 125–129, 132, 134, 179, 180, 181, 184, 192, 206, 214, 195
Neonate 236–261, 280
 see also newborn
Neuritic plaques (NP) 160, 200, 204, 205, 207, 211
Neurofibrillary tangles (NFT) 200, 204, 205–208, 211
Newborn infant 106, 109, 222–233, 237–261
 see also neonate
Neutral stimulus 255, 258, 269, 270, 272
Non-parametric 51
Normal aging (and olfaction) 199–203, 213–216
Novelty 79, 100, 147, 239, 244, 246, 255–256, 258, 285

O
Object recognition approach 73, 75, 77, 85
Objective measures 46
Odd ball paradigm 146
Odour cue 111, 269
Odour classification 8
Odour detection (detection threshold) 40, 41, 47, 56, 82, 83, 138, 145, 146, 156, 159, 182, 183, 192, 201, 208, 210, 213, 215
Odour discrimination (and odour quality) 47, 51, 56, 79, 80–84, 85, 86, 89, 90, 91, 113, 118, 119, 120–122, 129, 132–134, 143, 144, 159, 164, 165, 167, 173, 179, 182, 183, 185–193, 195–197, 201, 202, 205, 210–213, 233, 235, 244, 251, 263, 265
Odour experts 98
Odour hedonic(s) 13, 31, 33, 52, 53, 58, 85, 109, 111, 113, 133, 147, 204, 238–241, 245, 247–249, 252–257, 259–261, 263, 267, 269, 275, 279, 280, 281, 287, 289, 291
Odour identification 8, 47, 56, 85, 88, 115–123, 128–132, 134, 138, 149, 165, 173, 175, 176, 179–181, 183, 202, 203, 206–217
Odour/olfactory imagery 97–98
Odour intensity perception 11, 12, 15, 18, 34, 50, 58, 132, 149, 150
Odour memory 36, 83, 85, 89, 91, 95–99, 101, 103, 105, 106–113, 132, 134, 149, 150, 152, 182, 197, 206, 215, 233, 291
Odour naming
 see Odour identification
Odour perception 3–6, 10, 269
 Impact of culture 81
Odour preference(s) 121, 122, 129–131, 269, 270, 280
Odour quality 3–17, 119–121
Odour sensitivity 35, 201, 206
Odour space 7–17
Olfactometer 41
Olfactory agnosia 117, 184
Olfactory bulb 144
Olfactory comfort
 behavior 59, 60
 and attachment 62–65

Olfactory cortex 57, 58, 88, 90, 101, 144, 145, 151, 152, 179, 180, 185, 189, 192, 194, 195, 197, 205, 234
Olfactory ephitelium 141, 192, 204, 207, 230, 232
Olfactory event-related potentials (ERP) 200, 205
Olfactory impairment 90, 164, 167, 174, 183–185, 188, 191, 199–201, 203, 204, 206–211, 215
Olfactory Lateralization 50
Orbitofrontal cortex 77, 87, 98, 101, 141, 144, 145, 146, 147, 181, 197, 200
Orthonasal 87, 140, 145, 201

P
Pain 28, 29, 34, 39–43, 45–54
Paired–associate paradigm 97, 100
Parametric 51
Parosmia 52
Parkinson's disease 155–171
  Olfactory symptoms of 156
  Physiological basis of olfactory dysfunction in 157–160
  Environmental and genetic determinants olfactory dysfunction in 160–167
  Olfactory function in different variants of 167–171
Pavlovian conditioning 36, 270, 272, 291, 292
  see *also* Classical conditioning
Percept (odour) 84, 85, 86, 144, 147
Perceptual coding 142
Perceptual expertise 81
PET 159
Phantosmia 52
Picture Identification Task (PIT) 184
Picture-picture paradigm 272
Pleasantness (odour) 8–17
  see also Odour preference
Pollution (perception of) 23–24, 33
Positive and Negative Affect Schedule (PANAS) 65, 66
Power function 44
Positron emission tomography *see* PET
Post-traumatic stress disorders (PTDS) 281–283
Prefrontal cortex 150, 186, 187, 189, 191, 193–197
Prejudice 276
Primacy 143
Primary odour 13
Primary olfactory cortex 101, 142, 144, 145
Primary taste cortex 145
Proactive interference 97, 106
Proust (Marcel) 95, 101, 107
Psychophysics 39, 40, 43, 44, 47, 48, 50, 53
Pungency 25
Piriform (or Pyriform) (cortex) 76, 77, 83, 84, 87, 98, 142, 144, 145, 147, 181, 189, 190–194

Q
Qualia 69
Quality *see* Odour quality

R
Ratio scaling 43, 44
Recency 143
Reference memory 183, 186, 188, 192, 193
Relationship questionnaire (RQ) 65
Remote odour memory 206
Respiratory epithelium 200
Retroactive interference 97, 106
Retronasal 87, 140, 145, 201, 204

S
Stevens S. S. 43, 226
Salivary conditioned response 271
San Diego Odor Identification test 207
Second order conditioning 271, 272, 276
Semantic
  conditioning 273
  description 7, 13, 16, 45, 51
  differential scale 275, 284
  labeling 96
  memory 47, 54
Sensory hyperactivity 31
Serial position curve 143
Sniffing 76, 140, 144, 145
Signal Detection Theory/ Sensory Detection Theory/ SDT 40, 41, 47, 55, 57
  tasks 182, 188
Similarity judgments 52
Simple discrimination 185–193
Social odour *see* Body odour
Somatic effects 30–31
Somatosensation 86
State-Trait Anxiety Inventory (STAI) 65, 66
Subjective (response) 46, 47
Subliminal perception (of odorants) 133, 273, 285–289, 290–292
Suggestion, suggestibility 19, 25
Suprathreshold (sensation, assessment, scaling) 43, 45, 47

T
t-test 47
Taste 34, 40, 49, 77, 86, 97–98, 121, 143, 145, 148, 239, 240, 242, 251, 259, 261, 271–272, 277, 281
Taste aversion(s) 110, 111, 271, 272, 279, 280, 290
Temporal lobe 83
Thalamus 101, 139, 141, 144, 150, 151, 168, 185–187, 189, 190, 191–193, 197
Threshold (odour detection) 17, 40, 41, 45, 47, 57, 82, 84, 120, 131, 134, 138, 156, 165, 169, 171, 176, 182, 183, 192, 201, 205, 208, 212, 213, 215–217, 222, 224, 228, 229, 234, 252, 288

Threshold sensitivity 43, 53
Top-down processing 27, 30, 85–86, 147
Top-down sensitivity control 139, 140, 144, 148
Toxic environmental compounds 204
Toxicity perception 35
Trigeminal perception 12, 225, 232, 240, 242–243, 253, 254
Tween 20, 277
Two-back task 143

U
Unaware learning (of odours) 273, 278
Unconditioned response 271
Unconditioned stimulus 270
Unconscious cognitive conditioning 275
Unconscious evaluative odour conditioning 285–289
UPSIT (University of Pennsylvania Smell Identification Test) 156–171
Unpleasant event 270, 272
Unpleasantness 8–17, 24, 31, 48–54, 239–260, 270, 273, 274

V
Verbal coding 96, 142
Verbal conditioning 273–276
Verbal Descriptor (scales) 45, 51, 54
Visual Analogue Scales (VAS) 45

W
Well-being 24
Working memory 132, 139, 140, 142–143, 145, 148, 183, 186, 192–195